D1878323

SURVIVAL
BY
SEA WATER

SURVIVAL BY SEA WATER

(VOLUME - 2)

Dr F. A. Ryan

GYAN

© Author

All rights reserved. No part of this work may be reproduced, stored, adapted, or transmitted in any form or by any means, electronic, mechanical, photocopying, micro-filming, recording or otherwise, or translated in any language, without the prior written permission of the copyright owner and the publisher. The book is sold subject to the condition that it shall not, by way of trade or otherwise, be lent, resold, hired out, or otherwise circulated without the prior publisher's written consent in any form of binding or cover other than that in which it is published.

The views and opinions expressed in this book are author(s) own and the facts reported by them have been verified to the extent possible, and the publishers are not in any way liable for the same.

Published, 2021

Published by

Gyan Publishing House
5, Ansari Road
Daryaganj, New Delhi-110002
Phone: 011-47034999, 9811692060
www.gyanbooks.com
E-mail: books@gyanbooks.com

Printed at: G. Print Process, Delhi.

Cataloging in Publication Data—DK
Courtesy: D.K. Agencies (P) Ltd. <docinfo@dkagencies.com>

Ryan, Felix A., author.
Survival by sea water / Dr. F. A. Ryan.
volumes cm
Includes bibliographical references.
ISBN 9788121213615 (HB) set
ISBN 9788121219761 (HB) vol. 2

1. Seawater—Physiological effect. 2. Seawater—Physiological effect—India. 3. Survival at sea. 4. Survival at sea—India. I. Title.

LCC G540.R93 2019 | DDC 910.54 23

Contents

Chapter - 1

A Drought Shrub Which Grows in Saline Soil

Botanical – Calotropis

Sanskrit – Arkand of Arka

Tamil – Erukkam

Hindi – Duddar or ok

Family EUPHORBIACEAI

Description

A common shrub in many tropical countries with bunches of pale white flowers. Abundant and uncared for in the Indian sub-continent. Found all along road sides in urban and rural areas and waste lands. When the broad pale green leaves are plucked or branches broken a milky juice drips. This milk is called uddar in Norths India and Yerukku in South India and it is applied on bools, bruises and cuts for quick healing. Left in the wild it grows as high as 2 metres with branches about an inch in thickness. The shrub is extremely drought resistant and cattle do not eat the leaves.

Uses

1. According to Himal Botany (page 275) the plant is reported to yield sugar (so far not commercially exploited).
2. The leaves are excellent plant manure cum insecticide and they are commonly used by farmers as trashn mulch.
3. The stem yields fiber, which is spun into thick thread (count 1 or 2) and woven into coarse cloth by the tribals and used as rugs and carpets.
4. Ropes made out of this fiber are used in tying thatched roofs as insects don't attack them.

5. Roots, when dried and powdered are excellent as an insecticide.
6. The poisonous oil in the leaves is steam distilled by the village process and the oil is rubbed to cure skin diseases. The oil is applied on fiber ropes and wooden poles to protect them from white ants.
7. Tribals mix the oil with many other edible oils from nuts and seeds found in the forests and use the mixed oil to make flame torches and carry the burning fire when they cross jungles for light and warmth and also to scare away wild animals and snakes.
8. They also pour the oil in termite hills to kill dangerous ants.
9. Branches of the fully grown shrubs can be used to make frames to make oven-sack tiles. Frames are put into used oven sack cement bags.
10. Dried branches are made into briquettes for domestic cooking.
11. The roots of Calotropis, Vasaka, Deatura and Vinca Yosea, dried, powdered and mixed in equal proportion has the effect of high quality insecticide.

The manufacturing of this root insecticide is an excellent rural industry and can provide employment for thousands of families in hundreds of villages in Third World Countries.

The leaves of Calotropis, dried and Ground to granules also have properties of insecticides but powdered roots are much more effective and ideally suited for orchids and home gardens. As they are purely organic they can be used inside the house, in boxes and storing places to control moths and mosquitoes, pests and parasites.

If Planted During the Rainy Season the Climate Takes Care of It.

Chapter - 2

Zachun Oil Tree

Botanical – Balanites aegyptiaca

Grows in very arid conditions, deserts and on sea shores. More drought resistant than Acacia. Has thick and shiny leaves with a thick coat of wax, which protects against hot dry air. The plant has a twin root system and goes into the ground vertically deep and horizontically broad, spread to several metres. When the leaves fall, photosynthesis continues through the branches and spines. The leaves make fodder. The wood which is not attached by white ants is used for making low quality furniture and agricultural implements. The spines are used as pins.

The fruits which are hard and not very sweet or fleshy, inferior to dates in quality. They yield a sugary syrup when boiled in water. The syrup is used in making sweet meats, and for washing silk.

The leaves and bark of Zachun tree are said to be medicinal for treating jaundice and rheumatism.

The kernel contains 50 percent oil by weight and more protein than groundnut, sunflower or cotton.

The tree must be introduced on an intensive scale in droughtridden and semidesert areas and on the coastal belt and along the sea shore of warm and dry countries.

> The genes and chemicals contained in the world's store or biological diversity offer an abundant and virtually untapped supply of raw materials with potentially path-breaking and profit-making applications in medicine, agriculture and other fields.
>
> The ancient knowledge of seeds and sustainable agriculture is being cast aside as folklore and the multinational manipulation of genes is worshipped as scientific invention and rewarded with a patent.

Chapter - 3

Parched Lands

The plants, trees and weeds listed here give the "12 Fs", namely, food, feed, fodder, flowers, fruits, fragrance, fiber, fence, furniture, foliage, fertility of soil and fine chemicals.

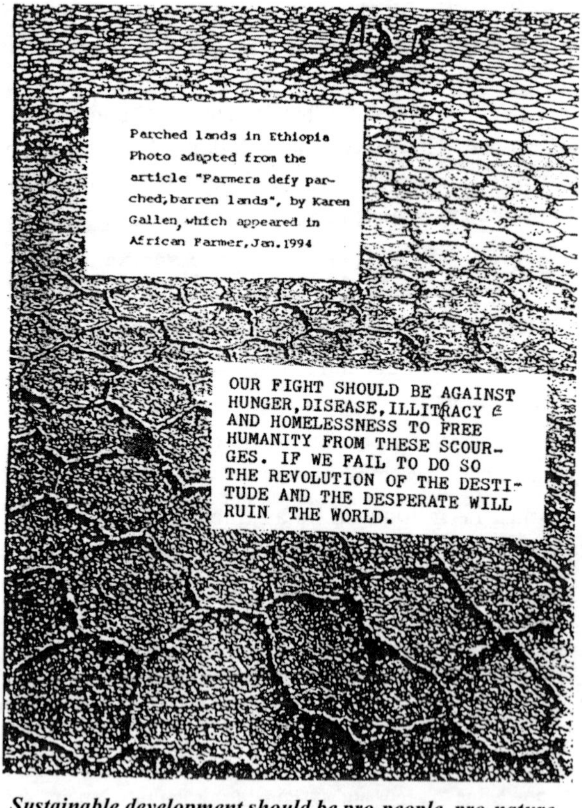

Sustainable development should be pro-people. pro-nature, pro-joba and pro-women.

Ryan foundation is basically interested in the foliage that they give in abundance out of which briquettes are made to boil sea water on a domestic scale to convert it into potable water four by using the four mud pot still.

Also, to use pots gigantic stainless steel stills to hold 10,000 gallons each for desalinating water on a commercial scale using solar, wind, wave, briquette and waste oil energy, all combines as described in the publications of the foundation citing.

Chapter - 4

Micro Hydro Power

People of Nepal have made a mark in the world for deriving motive power for turning various machines to meet domestic needs. Most of the micro hydro power units of Nepal are owned and managed by individual villagers in the neighbourhood. People use water powered mill to hull rice, press oil seeds, grind flour and even operate a band saw in some places. The journal Hydronet from Germany dated March 1989. From which the

A typical water mill in the hills of Nepal.
Photo. FAXT HYDRONET-3/89.

photo reproduced here is taken with permission, reported that atleast 600 such hydro powered water wheels were in existence in Nepal in 1989 and indicated that they were becoming more and more popular.

Water to turn the turbine is diverted from a streamlet or rivulet by piling a row of rocks or poles to make a barrier. The diverted water is led to a pipe to flow on the turbine. A simple hut or shed is build around the turbine to call it a mill. The water filling the pit below the picture turbine inside the hut is led to a lower level through a gutter, as shown in the picture and out into the open to join other rivulets or to form new streamlets, take its course and flow away.

The cost of a micro hydro power unit depends on the length of the lead-pipe, the size of the turbine, the number and size of machines attached and the installation of additional attachments such as generators, bandsaws or grinding wheels.

Water wheels/generators can be erected on slopes of mole hills, slopes of lake or pond bunds or catchment areas. A bottom level of the receiving slope or pit is adequate to make the system work efficiently. If depth is not adequate the length or the lead may be increased so that the water flowing through it gains force. Alternatively water may be lifted into over-head tanks and allowed to fall on turbines. On the spot investigation by Dr. Felie Ryan (Photo by Fakt).

The Best of All Governments is That Which Teaches us to Govern Ourselves. People Need Low Tech Solution to High Tech Problems.

Chapter - 5

Sea Water Energy for Grinding

From village to village sea-water canals and channels may be dug atleast one metre below mean sea level and about 5 metres broad. Bunds on either side must have a slope of 45 degrees or more to prevent slope sliding. (The main sea waterways may be three metres deep and 10 metres broad to be navigable by country boats and steam launches.)

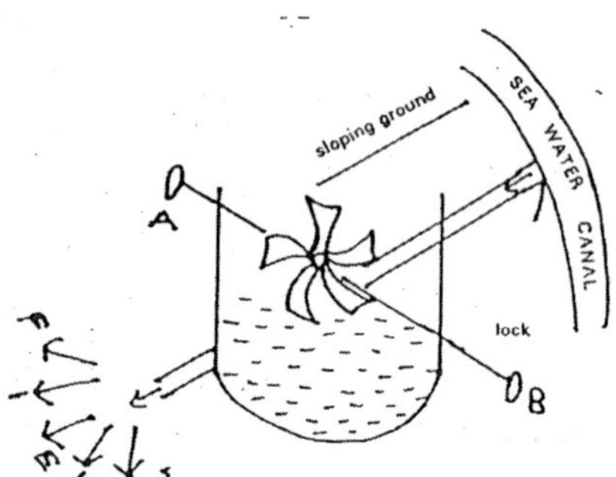

In villages where land or catchment areas lie more than two metres below sea level, the canal and channels may be dug in high ground and through cement, earthernware or bamboo pipes water may be taken down the slope into a pit of about three metres in diameter the length of the pipes should be atleast 10 metres so that the water flowing inside gains force.

In about half a metre below mean sea level mini turbines may be erected inside the pit as shown and an outlet pipe provided below the level of the turbine for water to flow out of the pit, spread and drain into open fields and get evaporated.

Where the pipe connects the canal there may be a concrete gate with iron or concrete slab control lock arrangement to allow or stop the flow of water.

The DTCD News Quarterly of the UN reports in the second issue of 1991, "The water table in parts of South Moroco has dropped nearly 100 metres in the last 20 to 30 years. As a result of lowered water tables due to ever exploitation, pumping costs have soared, springs, streams and wells have dried up, waste lands have disappeared".

On either end of the axle of the turbine attached pullies may be connected to grinding wheels or flour mills with fiber rope belts with knots, as done in Nepal.

The flow gate must be opened, closed and regulated by a named person in the village for a remuneration and he should keep the grinding wheels/ mills working for 2 hours in the evening.

With small cycle dynamos some light may also be provided in the area.

This system is possible and easy to construct where the land or catchment area is below mean sea level. If the land is not low enough, the catchment area may be deepend or the length of the flow pipe increased or water dropped through a pipe from an overhead tank holding sea water such mini generators, water wheels and grinding mills powered by water from streams and rivulets are efficient and successful and can be seen in several villages in Nepal. There is no reason why they cannot be promoted using water dropping from streams and rivulets of sea water from man made canals and tanks.

Chapter - 6

Fish in Sea and Sewage Water

Two-thirds sewage and one-third sea water mix in ponds or lagoons or swimming pools like water holders will be good and convenient to raise Thilapia fish.

Thilapia are one of the fastest multiplying kinds of fish and they are not only edible but also serve as a raw material for several cottage and farm industries. They are sturdy and need little or no care and are very popular in Bangladesh as edible fish. They yield fish oil for village lamps shed for black anti-corrosive paint. Fish gelatin for domestic and industrial use. Fish feed, chicken and pig feed and fish fertilizer for paddy and vegetable fields. In course of time, the sea water mix may be increased.

Milk fish, another fat multiplying fish may be promoted similarly. This being a sea water fish, the sea water in the pond must be two-thirds and sewage water must be one-third.

After studying the progress in a given environment the mixture of water and sewage may be altered. That is the sea water may be increased for the Milk fish and the sewage water increased for Thilapia. As one is a fresh water fish and the other a sea fish, they will have to be released in different ponds. In a third pond the two may be tried together to see if they co-exist.

Things must be done differently today if we want a different and more progressive tomorrow. At the farm, field and grass root level we must try and improve the methods that are to be followed.

Promoting fish farms, as explained, can harness sea water and promote different types of edible fish for several farm and industrial uses. Villagers, farmers and fisher-folk must experiment, learn, produce and progress.

More about Thilapia and the Milk fish is given in the following pages.

There is yet another variety of fish called the Tarpon Fish (Megalops Cyprinoides) (In Tamil, South India language) it is called Morankudai.

TARPON and THILAPIA (REOCHROI IS MOSSAMBICUS) breed well in the same pond but THILAPIA matures faster and must be harvested early. Details of TARPON are given in a chapter.

The very sturdy KAKI CAMBEL ducks may be reared in fish ponds and their droppings serve as fish feed. Duck rearing is also explained in a following page.

> The fish called Sea Cucumber which is quite common in many coastal areas, is used to treat arthritis and gout.

Chapter - 7

Milk Fish

In coastal villages sea water should be taken about 30? away from the shore by underground bamboo pipes (rigid PVC pipes or flexible garden hose pipes) into 2 to 3 deep man-made ponds and chanos and Calcarifer should be introduced in the ponds. People can eat fresh fish, smoke and preserve, make chicken feed pig meal, fertilizers gum (glue) Halophytes may be grown around the ponds and thousands of jobs created based on

them. The water may be taken from one pond to another in open cut canals. (This project is part of the Survival By Sea Water Movement launched and monitored by the Ryan Foundation, India) Chanos is known in English as "milk fish". In Tamil it is called "Poal Meen" and in Malayalam "Poomeen". It is ideal for rearing in backyard farms in rural homes in sea or saline water. It lives on decaying organic matter breed in the high seas and can be collected from back water and estuaries. When tidal waves are brought into land by man-made sea water canals, spawn and fry of milk fish come with the waves and form excellent nuseris and hatcheris. In Tamil, it is also called Pal Kandai.

One hundred square meter ponds in depth are excellent to rear milk fish. The water temperature should be between 25 and 35 degrees Celsius and pH 6.0 to 9.0 degrees Celsius for Phyto and Zooplanktons form a major food item for milk fish. Part of the old water must be replaced by new water every third day.

In a well managed pond milk fish grow to a weight of 750 grams in about a year and the yield will be about 500 kilograms of fish in a one hectare farm per year.

Usually later calcarifer, commonly called "Betki fish", is introduced along with Chanos chanos as a measure of biological control of smaller weed fish. Betki is a voracious predator, multiplies fast and can be harvested along with milk fish. Betki is edible. Like Tilapia, milk fish and betki can also be smoked and preserved. Oil can be extracted and chicken and fish feed can be made. RYFO fish glue can also be made.

Ryan Foundation recommends promotion of milk fish in man-made sea water canals and ponds in coastal villages.

Mullet (Madaval in Tamil) can be introduced along with Chanos. Both are surface breeders and both thrive in inland brackish water as well (not sea water).

Adequate Availability of Water, Food and Energy is Critical For Global Security.

Chapter - 8

Problems After the Catch

It is already dusk (6 p.m.) She must sell the fish within an hour or two. If she does not, the fish will become stale and stink and nobody will buy them. She has no storage facilities. Transporting the fish is also a problem and costs money. Rich middlemen who own cold store plants take advantage of her helplessness and cut down the price late in the evening.

The solution to the problem is to encourage fish cultivation in sea water lagoons, lakes and canals, fish in them and leave the catch inside the nets as shown in the picture below, until the market opens demands the next morning. There is no better place to store fish than the water in which they live.

Fish Trap Store

When many fishes are seen around the heap of stones, the two ropes are pulled together from the canal bank the net closes and the fishes are

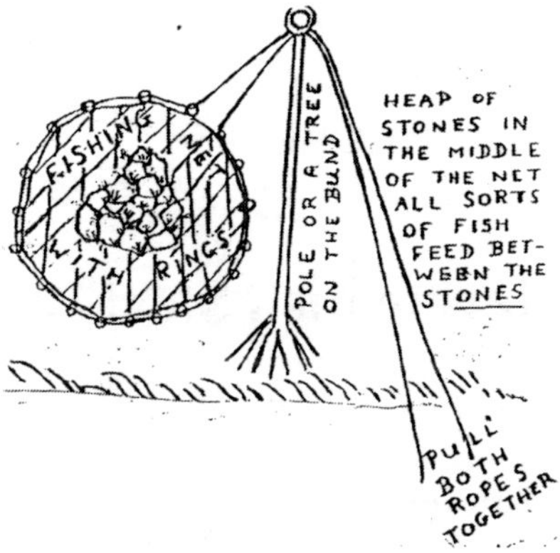

trapped. As the market opens the fishes are removed from the net (trap) and sold. Thus, the net serves both as a trap and as a store and the consumers get fresh fish. This method works very well in sea water canals.

The poor will remain poor if they depend on the technology of the rich.

India has a Coastline of 7,500 Kilometres and 100 Million People Depend on the Sea For Their Livelihood.

Chapter - 9

Declining Marine Fish Stock Worldwide

Reducing Fisheries Subsidies

A group of international marine scientists have appealed to World Trade Organisation (WTO) Director General Pascal Lamy, asking him to push for deep cuts to the fisheries subsidies, which they blame for declining marine fish stocks worldwide.

Signed by a group of 125 fisheries experts from 27 countries, the declaration was delivered to the WTO head on 24 May in Geneva. It calls on Mr. Lamy to use his "skill and leadership" to encourage a successful outcome to the fisheries subsidies talks. As part of the Doha Round, member governments are negotiating new disciplines on fisheries spending.

Not only is fish a heavily traded commodity, it also plays a vital role in sustaining the welfare and livelihoods of hundreds of millions of people worldwide, mainly in developing countries. Yet, scientists have reported that fish stocks are dramatically decreasing in significant measure due to government subsidies that provide incentives for too many boats to catch too many fish. It is estimated that if overharvesting is not halted soon, the world's fish stocks face irrevocable collapse within 50 years.

Signatories to the letter addressed to Mr Lamy assert that the WTO has the opportunity to reverse this cycle through Implementing stronger fisheries disciplines globally. Their letter notes that 90 percent of all the "big fish" large bodied sharks, tuna, marlin and swordfish have disappeared as the result of industrialised fishing; the decline of many ocean species is increasingly impairing the ocean's capacity to provide food, maintain water quality, and recover from perturbations; and despite massive technological advances and increased effort, global catches of food fish have been declining for more than a decade.

230 Million Go Hungry Daily in India – Un Report Dated 18.05.2013.

Chapter - 10

Freedom Without Food

Nowhere in the world people want freedom without food or food without freedom.

Four important towns of Somalia are Mogadishu (the capital), Jubba, Valley, Mera and Beled Weyne. All of them are thirsting for water and starving for food. The whole of Somalia is a semi-desert. The only way to help the people of this country is to take sea water cannals to the towns in the country, to begin with. Somalia has a long coastline of about 2000 kilometer and the coastal land is only about 4 to 5 feet sea above sea level. It is easy therefore to take the sea in and give the people sea food distilled and desalinated water, and help them to grow halophytes plants and trees that grow in sea water. But the ruling and waring tribes of Somalia don't care author worked in Somalia.

The Least Developed Countries (LDCs) numbered 24 about three decades ago. By 1991 the number increased to 42 and in December 1991 five other countries joined the ranks of LDCs which are Cambodia, Madagascar (now called Malagaski) the Solomon Islands, Zaire and Zambia. The addition of Angola, Eritrea in September 1994 brings the number to 49, of which 34 are in Africa.

Hundreds and millions of men and women in these and other LDCs whose daily survival on this planet is a daily challenge have little time to wait.

There is no excuse for famine or poverty. They are not just Bad Luck but are totally preventable by cutting sea water canals and taking the sea water into the land by man-made canals to promote sea food, navigation, boat markets and kitchen gardens in boasts, as in Japan, and to distil or desalinate sea water in different sectors of the countries. The distillation may be done on a community scale, village scale or family scale. For village scale distillation of sea water, (see chapter 188) which explains how large scale distillation is done in Saudi Arabia, which is a Ryfo concept.

It is clear that an improvement in living conditions of a large part of the population of the globe could undoubtedly be achieved by means of more efficient and more equitable be achieved by means of more efficient and more equitable utilisation of the resources and wealth already available in the poor countries and what is considered to be useless salty sea water.

The front line technologies (easily transportable and inexpensive) and a better use of collective know - how should constitute the spearheads of the fight against poverty in both cities and the countryside.

As The Year 2014 Unfolds California is Realing under the Worst Drought in Modern Times.

Chapter - 11

Ducks in Fish Ponds

Ducks are happy and healthy in fish ponds and provide manure as feed for fish. Fish also eat the food wasted by ducks. About 20 ducks are recommended for a pond of 100 mts × 100 mts. There are Campbell must be preferred as they are sturdy and give plenty of eggs.

A duck pen on fish pond is shown in the picture. About five ducks may be kept in a floor area of 1 metre by 1 metre. The floor of the pen should be made of poles to let the dropping fall into the water as feed for the fish. The pen must be fenced to confine the ducks to one portion of the pond.

Design by the Author

It is economical to purchase 2 or 3 days old ducklings and raise them for about 5 months when they begin to lay eggs and get the eggs hatched.

Duck feed is prepared by mixing cheap broken rice and rice bran in equal quantities. One to 5 months old ducks may be taken to paddy fields to give them natural feed. Each khaki Campbell 1 to 5 months old will eat about 150 grams of feed every day. Ducks drink plenty of water during the night and clean water must be supplied everyday.

Usually about the half the number of 4 to 6 months old ducks will lay eggs and they will lay well for one full year and thereafter irregularly for about 6 months. So ducks must be sold for meat before they are 2 years old. To have a continuous supply of eggs every month half a dozen new ducklings must be introduced into the pond. Eggs may be set in RYAN Foundation Incubator designed to hatch quail eggs.

The ducks in the pen must be counted every day, observed if they are healthy and the door of the pen locked to prevent theft.

The World Over 40 Percent of Water used in Toilets and For Washing Farm and Pet Animals is Rain, Ground or Treated Water. Why Not use Sea Water, as in Hongkong and By Some Hotels and Lodges in the Kutch Region of India? and Why Not Have Sea Water Swimming Pools as in Malagaski and Some Gulf Countries. Milk Fish or Pearl Spot in Sea Water Swimming Pools Where Water Flows Back into the Pools from the Sea Through Filters and Flows Back Into the Sea Can Provide Abundant Sea Food for the Rich and the Poor and Water for Distilling into Safe Drinking Water.

Chapter - 12

Motes for Sustainable Development

All over the world most of the ancient forts and fortresses have a deep mote around them meant to stop warning enemies from gaining easy entrance into them. Besides this basic purpose of defence, motes had other purposes and advantages, such as production of fish, fruits, flowers and vegetable gardens inside the forts. However, they serve little purpose under modern warfare and certainly they don't serve to keep away enemies from gaining easy entrance into the forts and there isn't enough fresh water to fill these motes.

Under the order of the day small scale sea water motes have to be dug around large institutions, homes, villages churches, play grounds, mosques and temples. Some ancient mosques and temples with motes exist already in India but not with sea water. These motes, which may be 10 feet deep and 10 feet broad around villages and big institutions and smaller around small institutions and residential settlements can.

1. Replace a compound wall.
2. Offer security to the inhabitants from terrorists and intruders.

3. Keep the atmosphere cool and breezy.
4. Provide sea food.
5. Encourage saline agriculture and xerophytes (salt loving plants and trees).
6. Serve as a swimming pool if kept clean.
7. Help fire fighting in case of fire.
8. Help domestic desalination in the surrounding areas.
9. Prevent breeding of mosquitoes.
10. Entertain children in boat rides and so on.

The Palace of Sheiki (Sheik's sister) in Dubai, near the Dubai clock tower has a sea water mote.

The picture in this handout shows human settlements surrounded by small motes. Ryan Foundation wants as many sea water motes as possible, to begin with around villages, and thereafter around large institutions – wherever possible in lieu of compound walls. These motes will bring about the much talked about sea water resolution, which had a good backing by the Earth Summit held at Rio. Will any coastal country come forward to dig a sea water mote around a dry village with a broad bridge to enter the village and another to exit? That is a million dollar question, a challenge for sustainable development.

Chapter - 13

Why Tilapia is Recommended

Tilapia is an exotic freshwater fish. It normally spawns when it is 3 months old in water temperature between 20 and 30 degrees Celsius and salinity less than 3 percent. The fish spawns to ten times in a year with an intervel of about 20 days between two spawnings. The sex of a Tilapia can be determined when it attains a weight of 50 gms. The male has a dark body and usually it is larger than the body of female of the same age. The fins of the male are deep red.

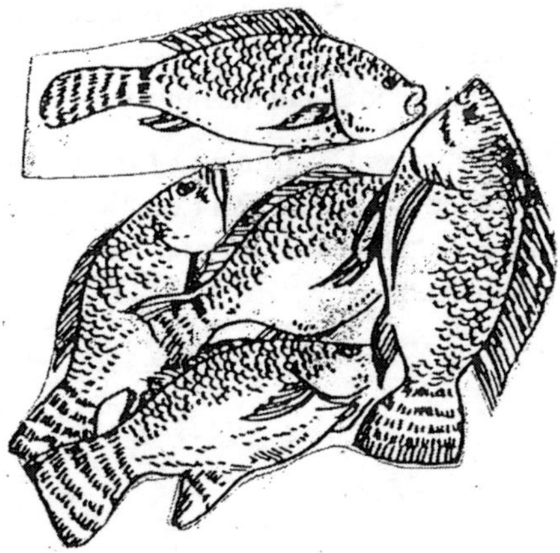

The female has two orifices in the genitalia, as against one in the male. Before spawning the parent fish make a saucer shape depression in the stomach 30 to 90 centimetres long and 6 cm deep. The fish use the mouth to make this depression in the stomach of the female.

At the time of spawning, a single female lays over 1000 eggs which are fertilised by the male with it is sperm. The fertilised eggs are yellow in colour with a diameter of 2.5 millimetres. As soon as the eggs are fertilised the female collects and hatched into larve in 60 hours at about 30 degrees Celsius. A week thereafter the larve swim out of the mouth of the mother. It takes only 5 months for a 2 to 3 centimeter fingerling to grow into a fish of 200 grams.

Tilapia and tarpon are an excellent combination. The yield is always more when the fishes are mixed in a pond fish and allowed to cross breed. Both the fishes eat almost any type of food, including small plants and insects and table waste.

Both are tough fish and can be stored well and require less storing space than most other fishes. As Tilapia usually breeds in the pond, the fry can be collected when the pond is drained.

Tilapia lends itself well for smoking and preserving. Smoked and crushed granules make good chicken and fish feed. They serve well as fertilizer for vegetable gardens. The fish is very suitable also to make gun (glue) as explained in chapter 245.

There are several species in Tilapia and the Nile Tilapia is very common in Africa and has been popularised in several countries. The family called Oreacheromis Niloticus continue to be the focus of development in tropical countries. The Food and Agricultural Organisation of the United Nations is doing it's best to promote and popularise this variety in many countries.

Surprisingly, the Fisheries Department of the Government of Tamil Nadu is doing it's best to eradicate this fish from the state of Tamil Nadu as it multiplies too fast and reduces the scope of promoting tastier varieties of fish such as the common carp.

Chapter - 14

Replenishing Forests

If sea water canals are cut through our forests halophytes (salt loving trees) and plants and mangrove 3 will grow there and multiply and we will benefit by all the bounties which only forests can give us. Even if people cut from the trees ruthlessly, the sea water will replenish the trees. Moreover, there will be water available for tribals and forest dwellers for distillation and fish for good. Reforestation is possible only if make an effort to make them survive and grow –

Dr Felix UN – declared Global Environmentalist.

Chapter - 15

What to Do When There is Mass Death of Fish

(See Next Page for Picture)

Put them into a large oil of the type you see in petrol bunks to a height of the feet and pour on the fish water to height of on foot. Close the drum with a rubber sheet (a heavy vehicle sheet may be cut to size and used). The sheet must be tied closing the opening of the drum.

Then put the drum on building bricks or hollow blocks (campers stove) and fire from below. As the water boils, the contraption functions like a pressure cooker. When the rubber sheet bulges high, remove it and let out the steam.

Now, inside the drum you will find plenty of fish oil for which there are many uses. After tilting the drum and removing the oil you will see a thick white substance looking like frozen ghee. Scoop out this substance, put it in an open pan and heat it on a low fire and keep turning the substance with a ladle and you will get high quality fish glue. Put the glue in an air tight container like any other glue and use it for domestic or industrial purposes. The left cover residue can be used as fish meal, chicken feed, pig meal, fertiliser and so on as it is dried or granulated, packeted and marketed. Several drums may be used in a row for quick and large scale processing.

The Aral Sea, once the fourth largest lake in the world, has shrunk in size by half, and its salinity has tripled since 1960s, mostly because the incoming water from feeder rivers was diverted to irrigate cotton and rice crops of dubious benefit.

We are not only outstripping the precariously limited freshwater stock of our planet but also poisoning the scarce and fast depleting water resources that sustains all life.

Chapter - 16

Dead Tilapia

Yields plenty of oil, glue and fertiliser

When you boil all unwanted parts of the fish and the whole of the non-edible fish in a large pressure drum cooker (shown in this book,) fish oil floats on top, below the oil there is fish gelatine glue and below the glue there is fish residue to be used as a fertiliser or as chicken, duck and fish feed. The oil is used to lubricate the dies that make briquettes and to burn cooking stoves to steam distill sea water as shown in this book. The glue is used for bingind foliage briquettes (also shown in this book chapter 247).

Chapter - 17

Tarpoh Fish

(Megalopa Cyprinoides)

This is a carnivorous marine fish having wide tolerance in varying salinity. It attains maturity when it gains a weight of about 400 gms and spawns twice a year. Seeds or fingerlings could be collected when it migrates to rivers and estuaries between April and October.

Tarpon fry (like those of Tilapia) feed mainly on Zooplankton, such as daphnids, Cyclops and fotifers and therefore the nurseries have to be fertilised with cow dung for the production of these organisms. Chemical suppliments need not be used.

Fingerlings of tarpon are stocked at the rate of 3000 per hectare tender weeds of Napler grass may be thrown every day as extra feed. Rice bran, ground nut oil cake and tapicca flour are also given to tarpon in equal proportion and Tilapia will also eat these feeds. The total amount of this feed given every day may be about 2 percent of the body weight of each fish. Thus fed, aoh fish will attain a weight of about 500 gm in 5 months.

Tarpon and Tilapia, especially, breed well in the same pond but Tilapia mature faster and they must be harvested early. Peculiar characteristics of tarpon and Tilapia are as follows: (1) They have a wide tolerance to water temperature (8 to 42 degrees Celsius) and to salinity upto 7.5 percent pH (5 – 11) and Oxygen level as low as 0.1 ppm, (2) both the fished produce a large number of fingerlings throughout the year, (3) both feed on a very wide variety of plants, water weeds, insets and planktons, (4) both are tasty and nutritious and (5) both take about the same time to get smoked. In short, tarpon, which is strictly speaking a marine fish, will comfortably adjust itself and cohabit with Tilapia in fresh water ponds and in saline water and with bad soil and changing weather. When these two types of fish are mixed together for rearing freshwater and sea water may also be mixed together if there is shortage of fresh water.

If sea, water canals, streams and streamlets are made available to the rural poor, as recommended by the Ryan Foundation in it's Global movement Survival By Sea Water the rural poor can be given plenty of Tarpon to eat.

Sustainable development is development that meets the needs of the present without compromising the ability of the future generations to meet their own needs.

Chapter - 18

Deep Water Fishing

Small scale fishing by individuals and families is traditional unproductive methods in third world countries. Frequently one can see anglers sitting on banks of rivers, ponds and lakes for hours together waiting to hook fish. Anglers sitting on banks can catch small fish and only after long intervels. For more and bigger fish they must go at least Two meters deep into the water and this is difficult without a boat and the poor cannot afford a boat.

The diagram gives a simple idea for productivity in bank fishing but it is recommended only for those who know to swim. Two ladders made of bamboo or causarina of almost equal size, say, 3 meters long may be tied together at one end as shown. At the other end blocks of wood or about a meter long tree branches may be tied horizontally to prevent the bottom of the ladders from sinking into the soil.

The top ends are tied in such a way that the inverted 'V' can be folded to get the bottom ends of the ladders together and spread out when required. The angler may drag the ladders and swim and mount them in a convenient spot a depth of about 2 to 3 mts., climb and sit comfortably on top as shown and angle from there. Baits and begs to collect the angled fish must be tied to the upper end of the ladder.

What is seen in the diagram is one pair of fishing ladders. If three pairs are planted, as in the three corners of a triangle the three top ends being of equal distance from one another, and if the three top ends are connected and tied with caseearina poles or bamboos, the three pairs will stand strong and steady as one unit and can be left to stay in the water. Three anglers may swim to the structure and sit on top of the ladders as shown and angle from up there and are sure to get a profitable catch.

If the anglers are good swimmers, they may follow the same system to catch from the sea but should not go beyond three meters deep. As a safety measure one end of a strong long tree or a firm peg on the shore. Catching by nets or baskets may also be tried from this position.

> The main engine of development is and always has been self help and hard work and not aid.

Chapter - 19

Lake Victoria Plagued

by the water hyacinth

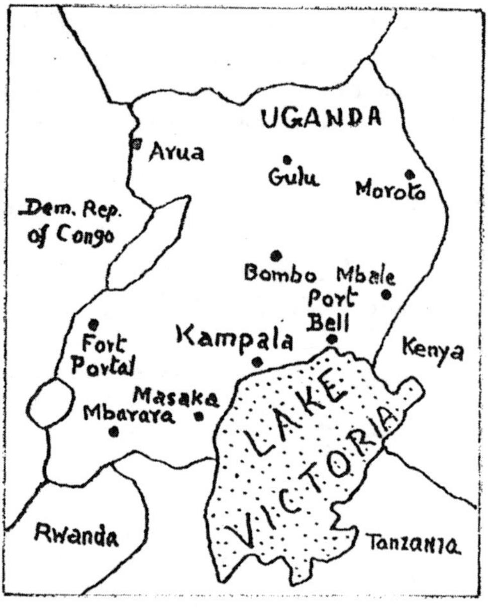

There in an over abundance of water hyacinth in Lake Victoria, causing several problems to the many countries around the lake. Efforts are on by many agencies to clear the weed. Uganda, Kenya and Tanzania are almost equally affected by the over growth of water hyacinth in the lake. Out of the water hyacinth and tilapia fish in the lake briquettes can be made to meet cooking fuel requirements of the surrounding countries. Cottage and village industries must be organised for briquette making as co-operative societies and federated like the handloom industries of South India, Briquettes must be used to steam distil water from the same lake into safe drinking water.

Chapter - 20

Sea Water Fishes Recommended for the Indian Ocean Countries

Mullets

1. Mugil cephalus (grey mullet)
2. Liza trade (trade Grey mullet)
3. Liza macrolepis (Bornieo mullet)

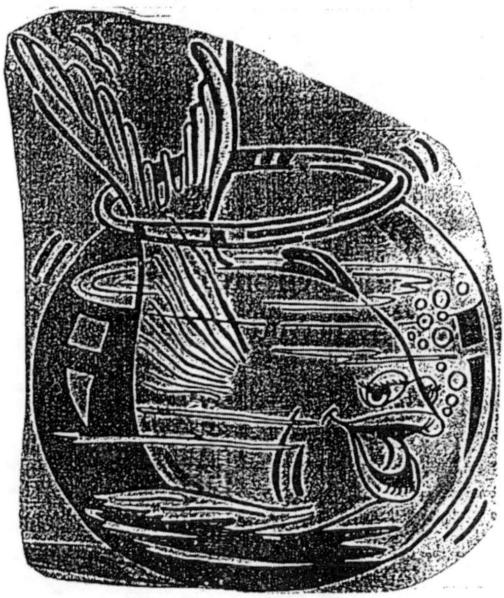

4. Liza passia (gold-spot mullet)
5. Rhinomugil corsula (corsula mullet)

Others

6. Chanos chanos (milk fish)
7. Hilsa Ilisha (hilsa)
8. Lates calcarifer (thetki)
9. Etroplus suratensis (pearl spot)

Prawn Varieties (sea water)

1. Penseis Indicus (white prawn)
2. Pemoeus monodon (tiger prawn)
3. Penaeus semisulcatus (green tiger prawn)
4. Metapenaeus monoceros (speckled prawn)
5. Metapenaeus dobsoni (flower tail prawn)
6. Metapenaues brevicornis (yellow prawn)
7. Metapenaeus affinis (jinga prawn)

Chapter - 21

Oil By Crab Fermentation

Oil can be extracted from nuts, kernels, barks, leaves, roots and seeds by several methods. Merely by boiling some oil can be extracted from certain items, crushing with a little water is another method. Steam distillation is a third and solvent extraction is a fourth. All these methods, using different devices are explained in several RYFO handouts, except solvent extraction for which large investment is required.

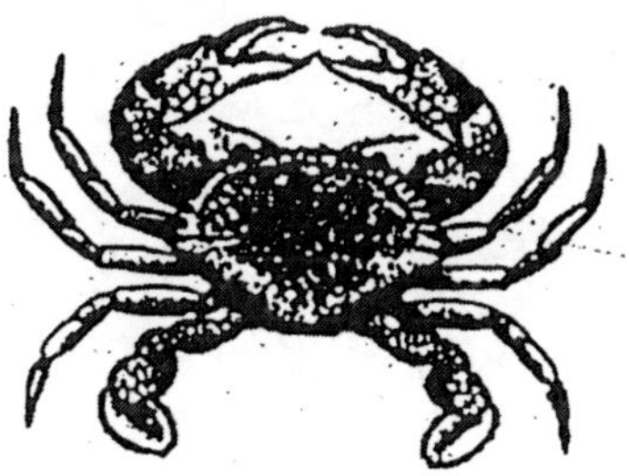

This handout explain an easy method of extracting oil from coconut, by crab shell fermentation.

1. Grate ripe coconut.
2. Remove the top shell of a river crab and grind it well. You may use a village grinding stone.
3. Mix the shell paste with the grated coconut in a vessel and leave it overnight (two crab shells for one big coconut).
4. Expose the mixture to sun light the next day for about 4 hours and turn it over while drying.
5. In the evening squeeze the mixture and the oil will come out.
6. Use 10 crab shell to extract the oil from 5 coconut of large size.

7. The mixture is the culture or seed, before sun drying. A part of it may be taken and mixed. With fresh grated coconut and the fermentation will spread. In other words, if only four or five crab shells are available the fermented first mixture may be used for fermenting subsequent lots of grated coconuts.
8. One quarter of the fermented mixture (unexposed to the sun) mixed with three quarter of fresh grated coconuts give good results.
9. The carried over culture is good enough for five cycles of extraction. Thereafter, fresh crab shell must be ground again.
10. The oil when filtered and bottled is as good as oil extracted in an expeller, keeps long and tastes good a small onion may be peeled and put into it for longer preservation.
11. The coconut residue may be fried and eaten or mixed with curry or cutlets.

NOTE:- In Indonesia one can commonly see people in homes and villages extracting coconut oil using the crab shell technology. Down south in Tuticorin also some people follow this method and save money.

Chapter - 22

The Cage is Made of Aluminium or Tin

(Avoid Plastics)

Cage Culture in community wells

Fish culture in community wells is better manageable than in rain water ponds. Hardy fish such as milk fish, catla, silver carp, tilapia and chinese. White amur are suitable for cage culture. It is advisable to start with SG (stunted growth) fingerlings if available. If tilapia is cultured, other fish should not be mixed.

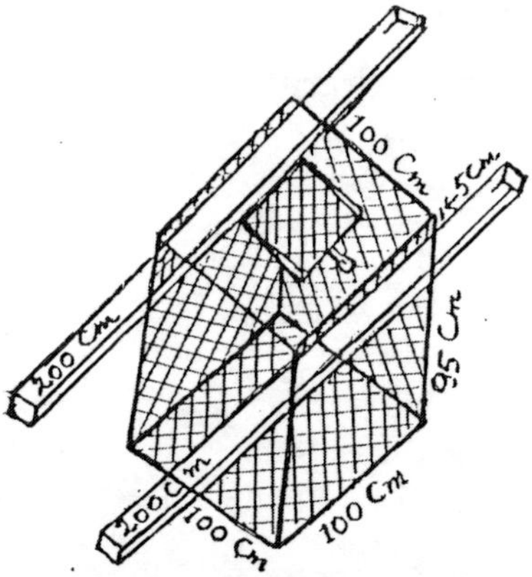

The bottom soil of many wells contains acid sulphate unsuited for fish. To remove this during summer, as much water as possible is removed from the well and lime stones thrown in. Harrowing can quicken oxidation of the acids present. Tilling, liming and flushing may be repeated thrice with an interval of 5 days between each operation. Finally, lime may be spread as a layer and left to stay. Oxidation must be done every year.

A wire mesh cage (0.4 cm. mesh) of the size 100 cm. × 100 cm. × 100 cm. with a lid on the top side fixed on two wooden arms (floats) as shown in the diagram may be used as the cage. To the four ends of the 2 arms nylon ropes may be tied (hooks may be used) and the cage powered into the well to float. The other ends of the ropes may be joined and tied to a post or hook in the wall of the well.

Feed mix may be prepared out of farmyard manure, soaked oilcake, rice bran, boiled corn or maize, ground soya bean, coconut flesh after the extraction of the milk, manioe leaves, pressed peanuts, soaked bread and algae and daphnia may be grown in shallow ponds separately and mixed. About a kg. of the mixture may be filled into perforated transparent plastic bags (like milk bags) heat sealed and placed into the cage. Three or four bags may be placed at a time and the fish will draw the feed through the perforations.

The fish named above reach harvestable size in about three months and the big ones must be removed first by lifting the cage out of the well. After culturing for three months a family can harvest enough fish every week for the family.

Design By the Author

Release Gambustia Fish into Wells and Large Pools of Water; They Eat Larvae of Mosquitoes.

Chapter - 23

How To Net The Fish

There are several ways to catch fish but netting is the most popular method. Women and children who cannot go deep into ponds and lakes to catch fish must be given a safe and easy method. A net bag like the one shown in the diagram can be easily made by village women. Sisal (cactus)

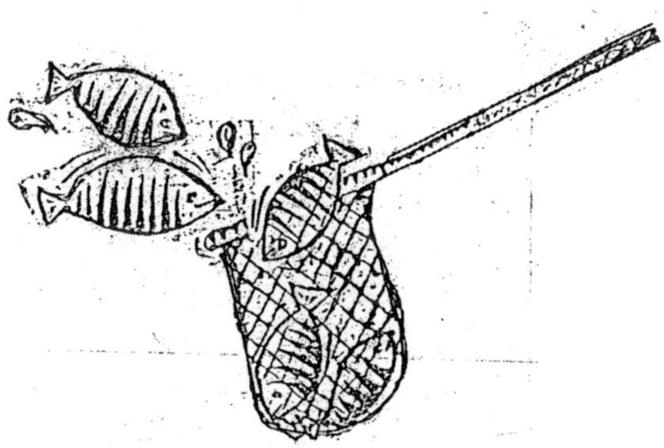

fiber is recommended as the fiber is long and strong and gains strength when soaked in water.

Sisal sheaths can be soaked in water and retted (beaten) with a wodden hamer to separate the flesh from the fiber. The fiber may be twisted by hand and tied to make nets and attached to bamboo or any light pole as shown in the diagram. With smoked fish or bread inside as bait, if the net is lowered into water, several fishes rush to catch the bait. The net bag is then pulled up with a sudden jerk and the fishes get trapped.

If a rectangular frame made of sticks of the size 45 cm × 30 cm. is tied to the opening of the net, it can be lowered into wells with a long rope and fishes from wells can be netted too.

Tilapia, the sturdy fish which is fast multiplying and usually over populated in a well, easily taken out by this method.

This fish, which is good enough to eat, is the favourite fish of the poor in West Bengal and Bangladesh.

The Net is Made of Sisal Fiber Which the Fish Cannot Easily Bite Open and Escape.

Chapter - 24

Fish Glue as Binder for Making Water Hyacinth Briquettes

Take all unwanted parts of edible fish and the whole of non-edible fish. If you include salted or dried fish wash and remove the salt first. Keep the collected parts soaked in water at least for a day and wash them at least four times with drinking water. Then chop and mince the contents and wash again in potable water twice or thrice.

Transfer the contents to a wooden board, chop and mince, including the bones, and put the contents into a suitable vessel and don't put a lid on the vessel so that the unpleasant smell escapes while boiling. For boiling place the vessel with the contents inside a simple double wall drum (take a big drum, place a brick in the middle of the bottom of it, put a smaller drum on the brick and fill a vessel in the middle of the small drum) close the big drum with a lid and put a low fire under it and the steam will circulate and cook the contents. Once or twice remove the lid and stir the contents with a ladle and cook until the contents become jelly like. Allow the cooked matter to settle down and then skim and remove the top layer of oil. Filter and separate the liquid in the middle that is between the oily top skimmed and remove the cooked matter below. Filter this semi liquid through a plastic or wire mesh or thin cloth. Bandage cloth or mosquito curtain cloth may be used. As you heat the liquid outside the drum under a low fire it will thicken and give you a fine glue continue heating to get the desired consistency.

The solid portion left after the first filtering may be boiled again inside the drum for a second filtering. The left over residue makes an excellent fertiliser both in wet and in dry form and it is odourless.

The liquid glue thus prepared will remain in liquid form always but must be kept in a closed container. It has a permanent usage life. On wood this glue penetrates deep and dries slowly and if made into thick paste, it is very strong but not water resistant. It has a pale yellow colour and no odour. As it does not solidify unexposed to air it can be packed, stored and marketed in cheap plastic sachet (like the milk sachet).

The Ryan Foundation Survival by Sea Water Movement want women in coastal fishing family settlements to prepare common purpose glue and market it for consumer and industrial use. The glue is excellent for binding foliage briquettes for domestic cooking and for steam distillation of sea water into safe drinking water.

> The Significant Problems We Face Cannot Be Solved at The Same Level of Thinking We Were at When We Created Them.
>
> Do Not Fear To Be Eccentric in Opinion or Concept, For Almost Every Opinion Or Concept Now Accepted Was Once Considered Eccentric.

Chapter - 25

Ryan Foundation Still (Table Model)

Instructions For Construction

1. Take a large flat basin or tray like vessel 60 centimeters. It may be made of mud, metal or plastic but nothing like having it in aluminum, which is rust proof. If made of mud or plastic, use small mirror or plain glass sheets at the bottom of the basin and some pieces to line the side wall. Fix a small outlet pipe or tube at the bottom of the basin as shown.

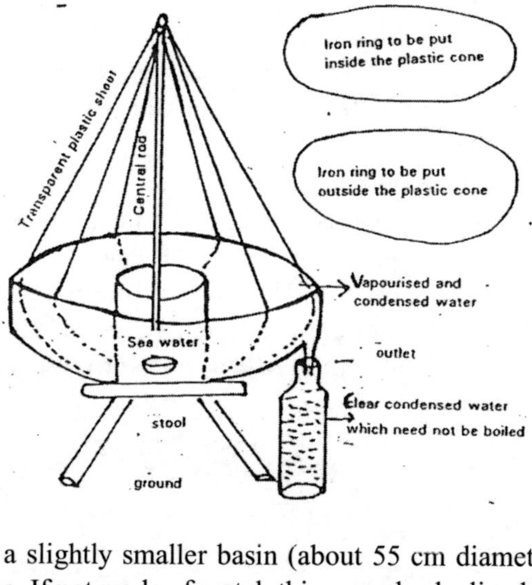

2. Place a slightly smaller basin (about 55 cm diameter inside the big one. If not made of metal, this may also be lined with suitable pieces of glass or mirrors.
3. Place a wooden, non corrosive or painted metal stand fixed to a base plate in the middle of the smaller basin as shown, The length/height of stand may be 60 cm and one cm. thick off (4) Take a thin

circular transparent polythene sheet 2 mts. in diameter and put it on top of the stand which will have a nob or ring on top and allow the sheet to fall like a cone into the bigger basin touching the side wall uniformly off. (5) Place on ring inside the cone and tuck the sheet (bottom of the cone) under the ring to close all gaps and avoid air getting into the cone. (6) Before setting the still thus, fill the middle container with sea or saline water. Place the still on a stool or elevated place in the open sun and connect a rubber or plastic tube to lead the condensed water into a bottle as shown. To prevent air from entering place the other ring (59 cm in diameter) outside the cone to sit on top of the ring inside the cone. The two rings should be of the same size. If some mirrors, glass sheets or shining metal sheets are available, set them on a higher level and reflect the rays of the sun into the small basin.

4. Water from the small basin will evaporate, hit the plastic cone, get condensed and drop into the big basin and flow into the receiver (you may also stick silver paper or aluminium foil on a cardboard to reflect the sun light.)
5. One still is enough to give pure water for a small family and the water need not be boiled. Two buckets of sea water will give nearly one bucket of pure water. If sea water is used salt will stand as a residue inside the small basin which can be collected and used in the kitchen.
6. The diameter of the big basin may be 60 cms and that of the small one may be 55 cms. The height of the big basin may be 15 cms and that of the small one may be 20 cms The diameter of the outlet pipe may be 3 cms. This may be a metal pipe welded to the basin or a plastic tube drawn through a hole made in the basin The (invention by Felix Ryan.) Alleviate Poverty Without Disturbing the Environment.

Chapter - 26

Air-Filled Solar Boiler

Designed by Ryan Foundation

(fabrication and working description-next page)

(Designed by Ryan Foundation)

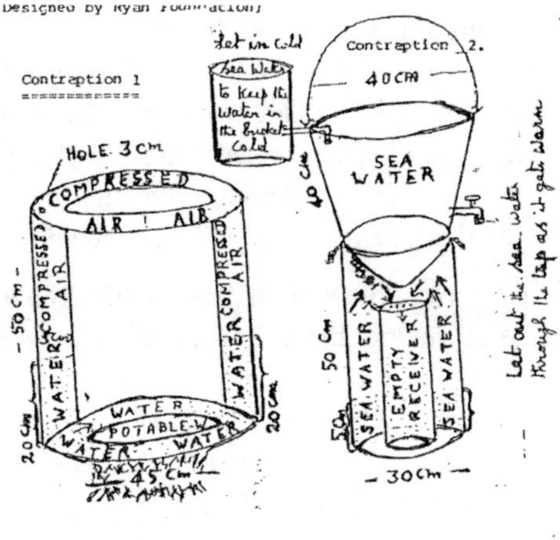

THE COOLING SYSTEM

When a hot dish is kept on the dining table with a lid, a good amount of water drips from the lid when it is lifted. The science behind this is simple to understand.

When hot vapor comes into contact with atmospheric temperature the vapour condenses and drips down as beads of water. Colder the temperature that the hot vapour meets, quicker the condensation.

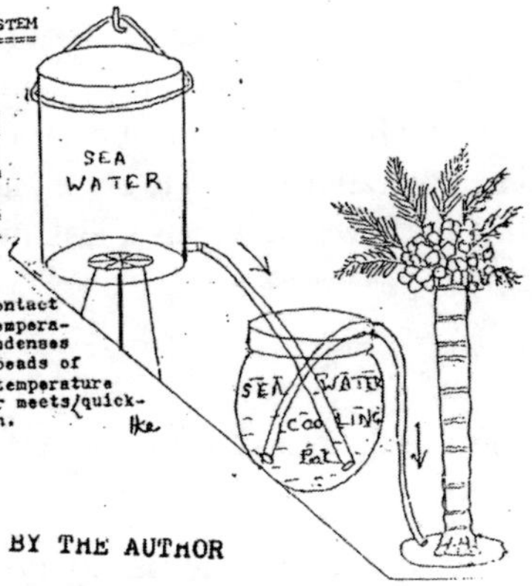

DESIGN BY THE AUTHOR

Chapter - 27

Fabrication and Working Description

Contraption 1 is a double wall drum (a small drum inside a big drum is welded at the bottom), with a 3 cm. gap between the floor of the big drum and the floor of the small drum, with a gap of 5 cm. between the side walls of the two drums. The gap at the opening of the drums (at the mouth) is closed with a circular plate and welded. A 3 cm. diameter hole is provided on the top plate that closes the drums to make it possible to pump air into it. A metal valve removed from a discarded car tube is welded blocking the hole to close it air tight. Then air is pumped in through the valve and the hole in the valve is welded and permanently closed (sealed).

Contraption 2 is again a double walled drum, where the drum inside is about one-third the size of the big drum. This contraption is placed inside the air filled boiler. The sea water in Contraption 2 gets heated by the boiler, vapourizes, hits the bottom of the metal bucket on top, which holds the cold sea water, gets condensed and drips into the receiving container. Contraption 2 is also painted inside and outside. The condensed water that falls into the receiving container is clean and potable.

Note that the bucket meant to hold sea water in contraption 2 has a V or U shaped bottom to make the condensed water fall right into the receiver which has a saucer lid with 4 holes. Cut a 20 cm. diameter hole in the middle of a towel and spread the towel to close the big drum in contraption 2 in such a way that the towel serves as a leak proof washer between the bucket and the drum.

How it Work

The boiler works on the principle of solar energy heating water under air pressure.

Heat, under the ordinary sun, can rise to three or four times the heat outside the drum. The entire contraption is painted black inside and outside to absorb and retain heat. The air under pressure to be pumped in and store will depend on the quality and strength of the drum. Too much air can explode the drum even as a car tyre bursts when there is too much air inside. It is advisable to have the drum physically tested for its bearing capacity and the amount of air (pressure) that can be put into it.

Vacuum or air pressure of 0.25 kilogram on the cubic centimeter of water will make the water boil even in diffused sunlight or very low temperature but it is not easy to achieve high vacuum pressure without the help of machines and controlled conditions.

Therefore, air pressure is recommended as in a pressure cooker. The lesser the water, is the more is the pressure and more the heat. After about an hour in the sun, the entire water between the walls and the floors of the two drums boils, converts into steam and circulates between the two drums and heat is generated by the circulating steam.

Stage II

Condensation takes place the whole night and the next morning the drum has plenty of potable water. Under bright sun there will be 5 litres of potable water obtained out of 6 litres of sea water. The metal or mud saucer on top of the receiver has three or four holes in the middle to allow the condensing water to drip into the receiver and to prevent the potable water from evaporating again and getting out of the receiver. The advantage of this method is that desalination takes place even when the heat of the sun is mild. Economically, a big gain for the poor and lower for the middle and upper classes of any nation.

In case the solar heat is inadequate, a small metal can or a small bottle filled with waste oil, or non edible oil, with a wick, may be lighted to heat the water in the boiler because of the air pressure resting above the water inside.

Every morning the bucket may be lifted the water in the receiver removed, and the drum set for the next day with fresh sea water. Every week the salt deposit is then removed from the drum, and the drum is cleaned and protected from rust by painting it black once a month.

> One of the most attractive tourist spots of Kerala has the lake Vembanad. The lake is suffering from chemical run-off from paddy fields, waste from households, over fishing, sand mining and lack of drinking water. Isn't there a way to save this and similar beautiful back water lakes?

5. Cost

If fabricated out of 18 gauge galvanised steel sheet both the contraptions will cost about ₹ 2,000 (or about US $70) on a commercial scale: and thrice as much if fabricated out of stainless steel of the same gauge. Both the contraptions may be fabricated in aluminium too for cost reduction. The life of an aluminium gadget will be shorter.

6. Make it tasty and healthy

Potable water must contain small amounts of hydrated lime, carbon dioxide and hypochlorite. When ordinary water containing these is boiled and distilled, these properties are lost. Then, a little of sea water may be filtered, boiled and added to restore the lost properties.

One liter of sea water to 10 liters of distilled water is recommended. Pure distilled water is tasteless and insipid and takes away from the body, needed minerals and vitamins. Pure distilled water alone is lethal without the vitamins, minerals and nutrients. But when some clear sea water is added, it gets a fine taste and the lost nutrients, a quarter teaspoon of salt, sugar or jeera (cumin) may be added to improve the quality.

Also, add some lime stone, charcoal or ash of burnt rice husk (whatever is easily available or all) to the desalinated sea water to make it safe and healthy.

> Professor Amartya Sen, who won the Nobel Prize in 1998 for his work on welfare economics, said that elimination of poverty depends very much on public policy. "I see poverty in terms of wide spread illiteracy, a poor health care system, incomplete land reforms, gender differences, deprivation of women and neglect of children than even, perhaps, sub-Saharan African, he lamented.

> So many years after the world becoming Modern, water scarcity has become an acute global problem. Hundreds of countries of the world are drought ridden or water stressed. Humanity does not and cannot put back even half the water it draws from the ground. Almost the whole of Africa is practically dry and states like California in USA are suffering from severe drought for the past several years. Brazil is craving for water. Our earth is in a vacuum sans water and can suck the earth at many time as it happened in Gujarat in the first decade of the present century. Small earthquakes will expedite the sucking and calamities after calamities will follow.

Chapter - 28

Ryfo Family Model Still

This distillation is done by the ordinary water steam process and therefore the retort or the main body of the still should have about 2/3 space for holding sea water. The steam will collect on top of the water and then pass into the bent tubes.

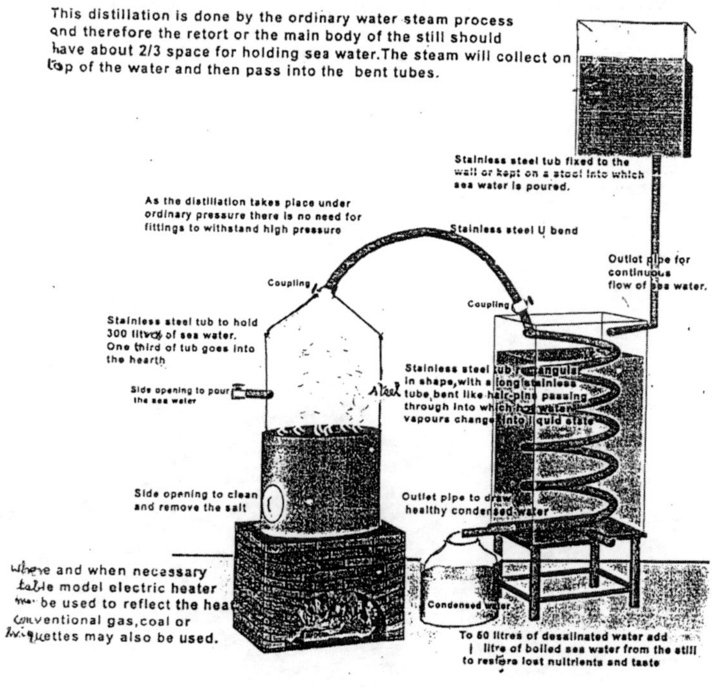

The RYFO Family Model Still is basically meant for the urban rich and not for the rural poor. It will cost as much as a big size colour TV. A middle or even a lower middle class family can well afford it.

The principle on which the still works is shown in the diagram but the dimensions are not given because it may be made into any size, 10 gallons per charge to 100 per charge capacity 20 gallons per charge will be ideal for any average size home of about 5 members.

The still may be mounted on the wall if small and the wall model containers should be rectangular in shape. If it is to be located under the open sky all the containers should be round and may be mounted on concrete pillar. The condensing pipe may be a coil in a round tank and bellows type in a rectangular tank.

An open air unit is recommended if heating is to be done by a hearth and folliage briquettes are used for firing.

Somewhere in the middle of the boiler there should be a small tap to draw the boiled sea water to restore taste and lost nutrients. This tap is not shown in the diagram for lack of space.

The boon of this system is that the outlet pipe gives absolutely clear and safe water devoid of heavy metals destroying evil bacteria.

Design by the Author

Chapter - 29

Using Solar Energy to Convert Saline Water into Drinking Water

The author here suggests the design of a solar still which uses solar energy in combination with heat from foliage briquettes to extract condensed water from sea water. This may become a more economical way of getting drinking water from sea in comparison to other available techniques so far.

The working mechanism is explained as below (see sketch):

Sea or saline water tank is on the terrace. A stainless steel motor (½ H.P. ₹ 15,000) lifts the water from the sump into the tank. B.C.L is a 2" high density plastic pipe taking the tank water the usual way. At M a ball valve may stop the flow of water into the flats. At B, a T-joint leads the seawater to D, entry point into solar plate collector. J and K are stainless steel solar reflectors to reflect. The solar heat on the glass of the plate collector. These two reflectors nearly double the heat inside the flat collector. Without the reflectors the heat may reach only about 60°C when the atmospheric temperature is about 50°C. The solar plate reflector which is about 2 m long and 1 m broad is kept 1 m above ground level and a stove fired by rice or wheat or any husk for foliage briquettes is placed under the heat collector to take the temperature to nearly 250°C. The steam produced by the plate collector flows out at E and enters a metal cooling pipe at F. The pipe is about 30 m long, bent like a coil spring inside a stainless steel drum of 500 litres capacity. The pipe is of ¾ inches diameter.

The water from the main water tank which flows through C fills the drum, cools the coiled pipe in it carrying hot steam and flows out through a rubber or plastic hose pipe into the toilet/bath through the window, at G the distilled (condensed) water flows out through another hose to the kitchens in the block (8 or 12 flats) as safe and pure drinking water.

It is absolutely healthy to drink pure distilled water. However, for good taste a quarter teaspoon of salt and quarter teaspoon of sugar may be added to one litre of the distilled water. Some tulasi leaf (Holy Basil) or cumin seeds may be added for pleasant aroma. The flow, steaming, cooling evaporation is a continuous process and on from about 10 A.M. in the morning to 4 or 5 P.M. This process is maintenance free with no renewable fuel.

If a quicker yield of a larger volume of water is needed, additional solar plate collectors may be inter-connected with an increased number of solar reflectors. Alternatively, 3 or 4 independent units may be set on the terrace. Existing sump and overhead water tank may be used to begin with. Thereafter a separate sea water sump and separate sea water overhead tank must be used.

In Hong Kong, sea water is being utilised for supplying to the people.

Chapter - 30

Terrace Still – Cost Estimate

1.	Solar heat collector made out of 18 gauge stainless steel (alloy with 9 percent chromium) with heat resistant Tuffe glass sheet and inside fittings of pipes and metal FEATHERS	Rs. 10,000
2.	The heat exchanger (barrel with coil pipe inside, also made of stainless steel and painted black outside.	Rs. 10,000
3.	Poly – propylene (Plastic) pipes 1000 metres long to lead the condensed water one side of the four four building (8 kitchenes and to take the sea water that cools the coils pipe inside the drum to the toilets on the other side of the building.	Rs. 10,000
4.	Four stainless steel 'T' joints stoppers and taps four	Rs. 1,000
5.	Two stainless steel solar reflectors of metre diameter	Rs. 2,000
6.	Plumbing work	Rs. 3,000
7.	Underground concrete sump of 15,000 litre capacity	Rs. 50,000
8.	One Horse Power stainless steel motor sea water resistant	Rs. 20,000
9.	Cast iron rice husk stove or foliage briquettes stove	Rs. 1,000
10.	One day requirement of sea water to be purchased and brought by a tanker to fill the sump (two loads)	Rs. 1,000
11.	Sea water tank on the terrace	Rs. 40,000
	Total	Rs. 1,75,000

To give atleast 50 litre of pure and safe drinking water without any dissolved salts, heavy metals or toxic of human waste every day for eight flats (families).

Only the purchase of sea water every day will be the recurring expenditure till the government lays sea water pipes and cement sea water tanks in every colony as in Hong Kong, and interconnects them.

President John F. Kennedy noted that if humanity could find an inexpensive way to obtain fresh water from the sea, the achievement would really dwarf any other scientific accomplishment.

Chapter - 31

Ryfo Domestic Still

The sea water fed through Inlet C may be pre-heated by a solar heat collect dr. and heated further inside the boiler by reflectors, such as opened-out umbrellas and parabolic magnifying glasses. Briquettes or pellets stove may be used when there is no sun.

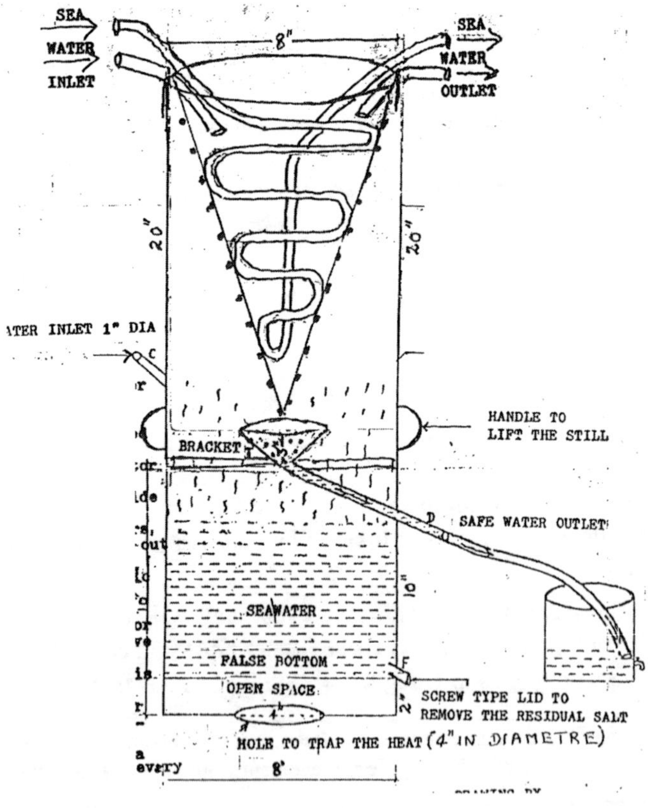

The boiler must be flushed with filtered sea water after every process.

The inverted cone and the ½ pipe inside is made of high grade copper and the copper pipe should make as many circles as possible inside the cone as shown in the computer diagram attached.

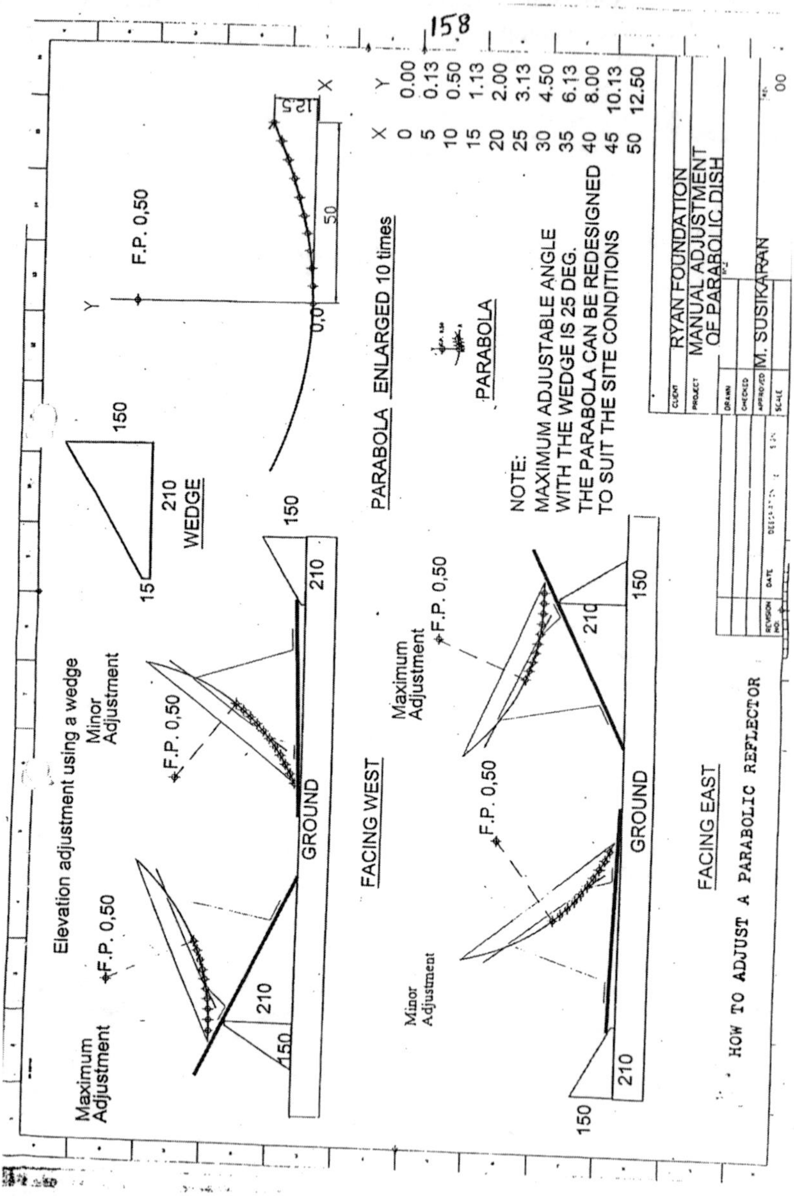
158
Elevation adjustment using a wedge
Maximum Adjustment
Minor Adjustment
F.P. 0,50
210
150
GROUND
FACING WEST
FACING EAST
WEDGE
15
Y
X
0,0
50
12.5
PARABOLA ENLARGED 10 times
PARABOLA
X Y
0 0.00
5 0.13
10 0.50
15 1.13
20 2.00
25 3.13
30 4.50
35 6.13
40 8.00
45 10.13
50 12.50
NOTE:
MAXIMUM ADJUSTABLE ANGLE
WITH THE WEDGE IS 25 DEG.
THE PARABOLA CAN BE REDESIGNED
TO SUIT THE SITE CONDITIONS
CLIENT RYAN FOUNDATION
PROJECT MANUAL ADJUSTMENT OF PARABOLIC DISH
DRAWN
CHECKED
APPROVED M. SUSIKARAN
SCALE
00
HOW TO ADJUST A PARABOLIC REFLECTOR

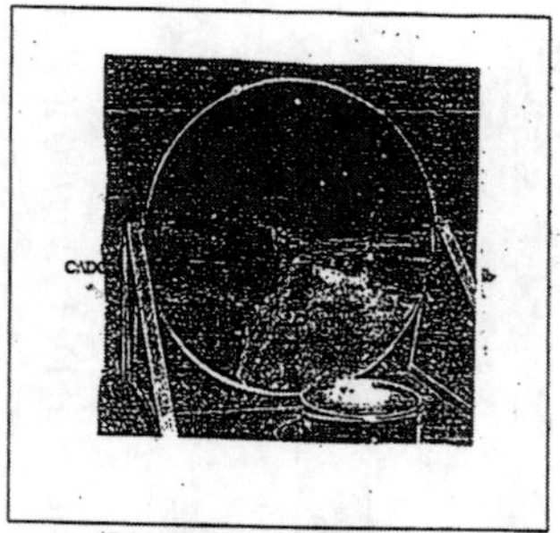

PARABOLIC COOKER
heats to over 500 degrees C

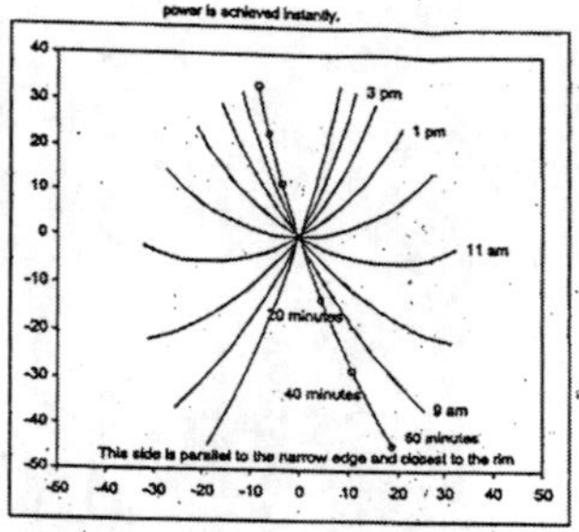

An alignment chart for 10°N latitude at summer solstice June 21

Support of Ref

held → The reflector is ... a rigid frame, which is attached to a flat base.
The frame consists of the following:
A Circular ring on which holds the Reflector A four legged frame attalched to the Ring
The four legs are attalcheed to a flat base (See Photo above)
Principal optical axis of the reflector is inclined with respe ct to the base, at the default elevation of 50. deg
It is adjustable for 25° in either direction Then the range of achievable elevation is between 25° and 75°.
To achieve the desired ±25° range, the wedge should stand on its shorter side. It should also be pushed toward the center of the base. For stability the wedge is only pushed in for about 30% of the base length.
The Details of Wedge are shown separately The wedge width can also be 150 mm.
It can be made of any locally available material.

Alignment Guides
A pinhole is drilled on the reflector, 100 mm above the base. Warning: Do NOT look directly into the sun! When the principal optical axis points directly toward the sun, a bright spot appears on the base. Name and mark this spot M. Alignment can be made ~~better future~~ later without looking at the sun. Simply adjust the azimuth and elevation of the base to bring the bright spot onto this mark. The maximum cooking

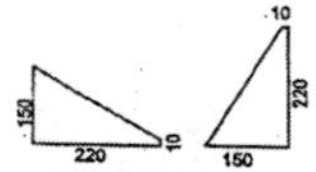

DETAILS OF WEDGE

PARABOLA FOCAL Pt. 36 In COORDINATE		PARABOLA FOCAL Pt. 48 In COORDINATE	
X Inches	Y Inches	X Inches	Y Inches
0	0	0	0
1	0.0069	1	0.0052
2	0.0278	2	0.0208
3	0.0625	3	0.0469
4	0.1111	4	0.0833
5	0.1736	5	0.1302
6	0.2500	6	0.1875
7	0.3403	7	0.2552
8	0.4444	8	0.3333
9	0.5625	9	0.4219
10	0.6944	10	0.5208
11	0.8403	11	0.6302
12	1.0000	12	0.7500
13	1.1736	13	0.8802
14	1.3611	14	1.0208
15	1.5625	15	1.1719
16	1.7778	16	1.3333
17	2.0069	17	1.5052
18	2.2500	18	1.6875
19	2.5069	19	1.8802
20	2.7778	20	2.0833
21	3.0625	21	2.2969
22	3.3611	22	2.5208
23	3.6736	23	2.7552
24	4.0000	24	3.0000

Minor Adjustment | Maximum Adjustment | Minor Adjustment | Maximum Adjustment

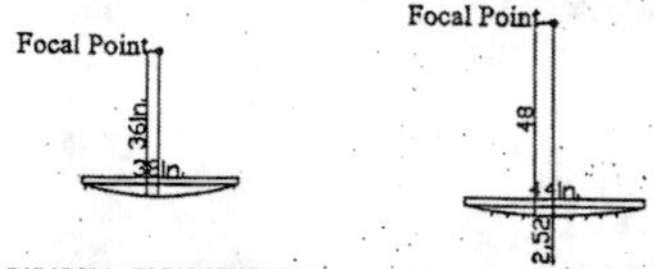

PARABOLA - FOCAL LENGTH 36 in. PARABOLA FOCAL LENGTH 48 in.

REVISION NO:	DATE	DESCRIPTION	SIGN.

CLIENT	COOKING BY SOLAR REFLECTOR	
PROJECT	PARABOLIC REFLECTOR FOCAL LENGTH 36 in. and 48 in	
DRAWN		TITLE RYFO SOLAR REFLECTOR
CHECKED		
APPROVED	M. SUSIKARAN	DRAWING NO: REV 00
SCALE		

Chapter - 32

Mammoth Solar Cooker

It is now working in India; said to be the world's largest solar cooker, which can also boil sea water to attain a temperature of more than 400 degrees Celsius. It is located in Mount Abu in Rajasthan and can cook food for more than 1200 people or provide distilled water every day. It can also be used to generate electricity like a similar mammoth plant in Saudi Arabia, 22 miles from Jaddah.

The system was financed by a contribution of DM 120,000 in German aid to India and administered by the German Development Organisation for Technical Co-operation. It is located at the Brahma Kumaris Education Society and consists of 24 parabolic concentrators covering a total area of 180 square metres.

The system produces 600 kilograms of steam every day and has a capacity of 75 kilowatts. The heat reflected by the mammoth dish is focused on 12 receivers. Water is heated upto 200 degrees Celcius in the receivers. The presurised water generates steam by means of heat exchanger and a solar generator.

A pipeline then leads the steam into the main kitchen. Mr Deepak Chada, an engineer and manager at the Eco Centre in Valsad town in Gujarat fabricated the parabolic dishes under the guidance of the German project co-ordinator, Mr Golo Pilz.

The basic design, solar steam generator and the receivers are being made by a company in Germany called Hochtemperature Technik, located at Herford. The project which this author studied in person in Valsad is excellent for centralised steam distillation of sea water into safe drinking water and should be copied the world over.

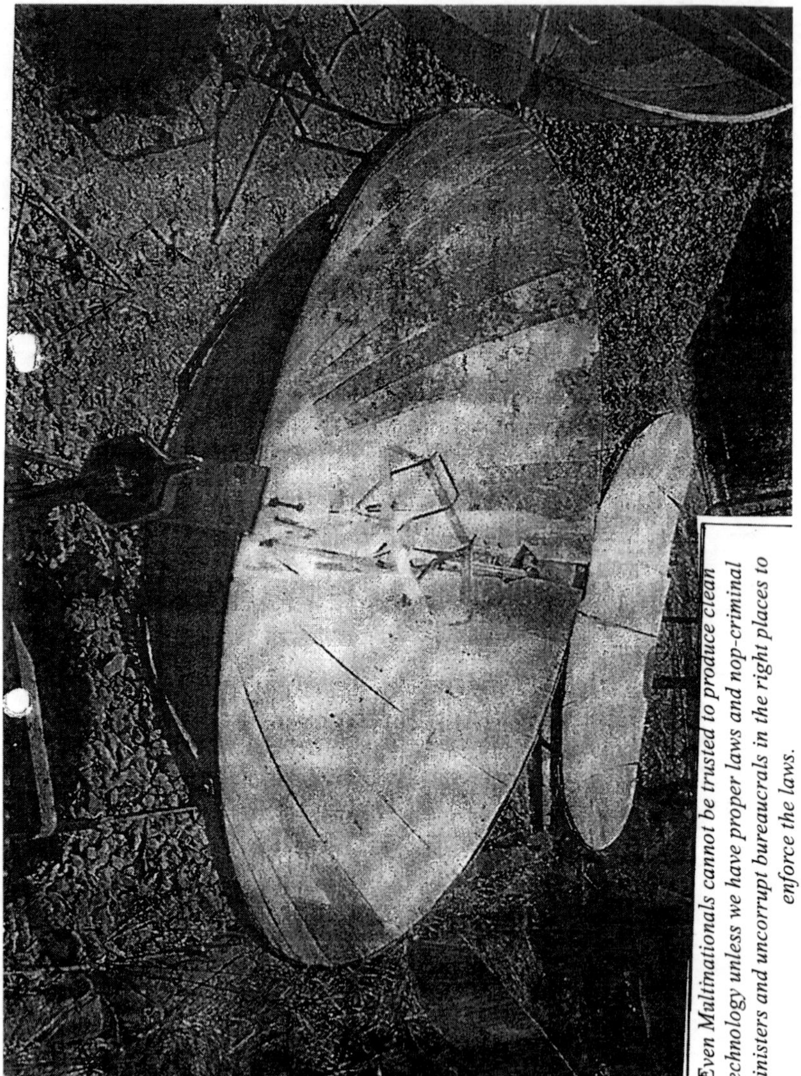

Even Multinationals cannot be trusted to produce clean technology unless we have proper laws and nop-criminal ministers and uncorrupt bureaucrals in the right places to enforce the laws.

Chapter - 33

Lake Out of Sea Sand

For the past several years about 8 lakhs cubic metres of sand is being dredged annually from the mouth of the Ennore estuary near the city of Chennai in South India. The sand that is dredged is dumped here and there without a specific plan or purpose.

Similar dredging operations are done in several countries and only in some places the sand is used thoughtfully. Usually it is used as land fills or sea defence.

According to the Ryan Foundation (Survival By Sea Water Movement) it will be ideal to heal the removed sand as shown in the picture and to build a chain of heaps to make the bund of a tank. Saline shrubs and desert shrubs may be grown on the bund.

The tank may have a diameter of about 1 kilometer atleast. Sea sand or estuary sand may be heaped between the heaps (seen in the picture) to make the bund regular and of uniform height.

Sand from the inside of the lake thus formed should be removed to a depth of three meters and thrown on the saline bund to further increase the height of the bund and for making the soil of the bund good for growing useful trees like the neem, coconut and in particular, acacia holosorcpa of Australia, now introduced in many developing countries. The Auroville township between Madras and Pondy in peninsular India has thousands of this trees. These drought trees are not grazed by cattle. They are nitrogen fixing and produce pods which yield 30 percent edible oil by weight. The wood gives a high calorie of heat.

If rains are adequate, rain water may be stored in these lakes. Alternatively, sea water may be led into them by man made open cut canals. Even parks may be promoted on the bund.

Suitable fishes and sea food may be developed in the lakes. Only sea and saline agriculture can sustain humanity in the 21st century.

Design by the Author

Chapter - 34

Ryfo Kitchen Distiller

The top drum with an inverted cone fixed to the bottom and having grooves or threads at the bottom is screwed into the opening of the bottom container, which has grooves or thread at the top as shown. If it is just

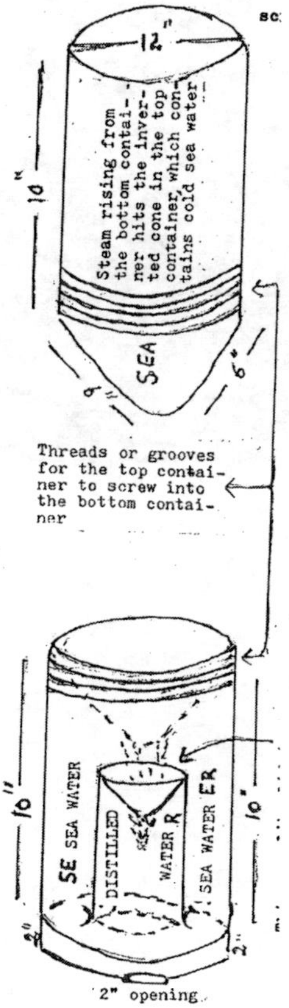

inserted as done in a domestic coffee percolator much of the steam will escape through the crevices and produce less water.

Water condenses around the big cone and drips into the small cone inside the bottom container. The receiver is closed with a cup like lid with three small holes at the bottom of the cup to prevent the already condensed water from missing using again as steam.

There is a 2" false bottom in the big bottom drum with a 2" hole in the middle to trip the heat of the stove. Trapped heat is much more than untrapped heat.

Some common vegetable matter is added to the condensed water and left overnight to restore the lost nutrients in the condensed water.

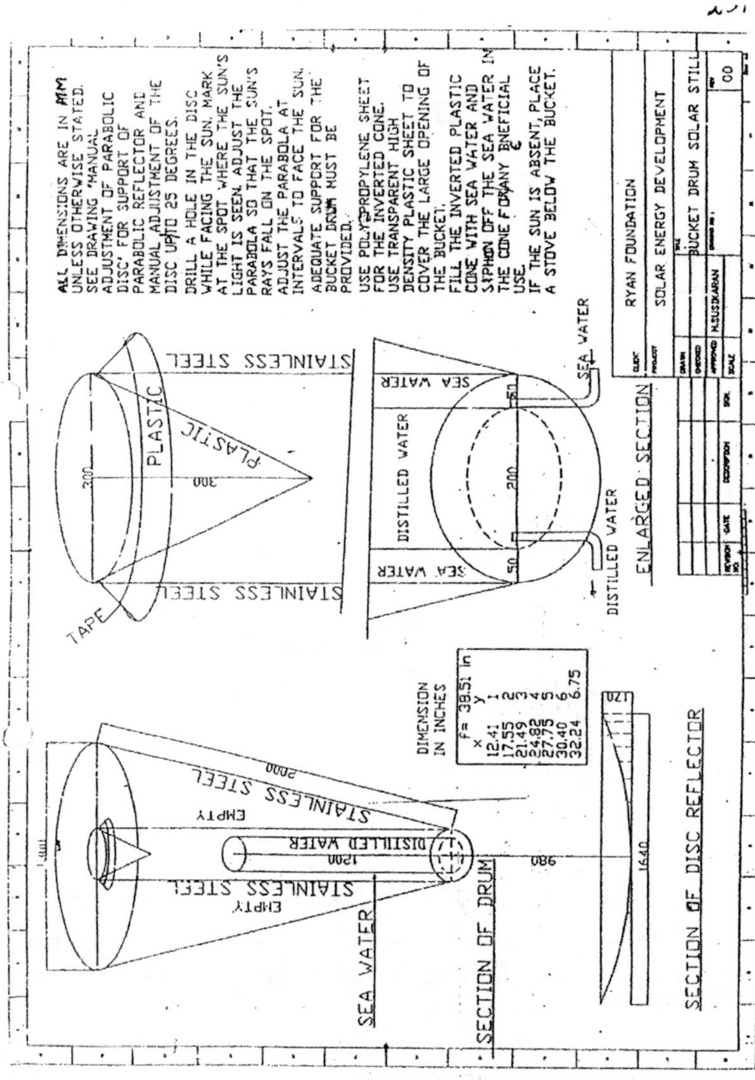
ALL DIMENSIONS ARE IN MM UNLESS OTHERWISE STATED.
SEE DRAWING 'MANUAL ADJUSTMENT OF PARABOLIC DISC' FOR SUPPORT OF PARABOLIC REFLECTOR AND MANUAL ADJUSTMENT OF THE DISC UPTO 25 DEGREES.
DRILL A HOLE IN THE DISC WHILE FACING THE SUN. MARK AT THE SPOT WHERE THE SUN'S LIGHT IS SEEN. ADJUST THE PARABOLA SO THAT THE SUN'S RAYS FALL ON THE SPOT.
ADJUST THE PARABOLA AT INTERVALS TO FACE THE SUN.
ADEQUATE SUPPORT FOR THE BUCKET DRUM MUST BE PROVIDED.
USE POLYPROPYLENE SHEET FOR THE INVERTED CONE.
USE TRANSPARENT HIGH DENSITY PLASTIC SHEET TO COVER THE LARGE OPENING OF THE BUCKET.
FILL THE INVERTED PLASTIC CONE WITH SEA WATER AND SIPHON OFF THE SEA WATER IN THE CONE FOR ANY BENEFICIAL USE.
IF THE SUN IS ABSENT, PLACE A STOVE BELOW THE BUCKET.
STAINLESS STEEL
PLASTIC
TAPE
SEA WATER
DISTILLED WATER
ENLARGED SECTION
SECTION OF DRUM
EMPTY
DIMENSION IN INCHES
SECTION OF DISC REFLECTOR
RYAN FOUNDATION
SOLAR ENERGY DEVELOPMENT
BUCKET DRUM SOLAR STILL
M.SUSIKARAN

Chapter - 35

Ryfo Still

The tub and pot still is very simple and inexpensive. Only five components are required to set it:

1. The floor of the bottom tub is 60 centimetres long, the sides, 45 centimetres high, the angle between the floor and the sides 135 degrees. Fabricated out of 18 gauge mild steel, it may either have

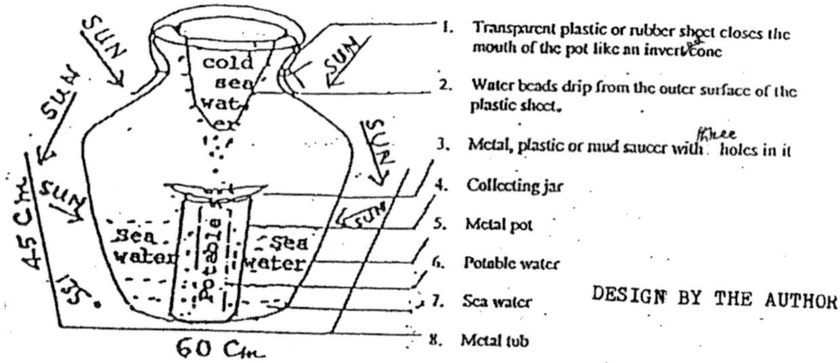

four slanted sides or be a circular tub. A circular tub is more efficient and cheaper. This is the only part that is to be fabricated.

2. The longish collecting jar should hold at least 4 liters of water and may be made of metal or plastic. Such a container is usually available in households.
3. The big metal pot, preferably made of stainless steel is 60 centimetrs high and 45 cm broad. Such a pot is usually available in homes.
4. The thin transparent plastic sheet which covers the pot like an inverted cone, should sag into the pot. It is tied around the neck of

the pot to prevent leakage of the evaporating water. The sagging cup is filled with cold sea water.

5. The saucer shaped lid, which closes the "receiver", made of metal or plastic, has three or four holes in the middle to pass the dripping water into the receiver and to minimise evaporation out of the receiver.
6. The sun falls directly on the two shoulders of the big pot and on the sides of the pot. If heated only by sun pour only two litres of water in the pot. Parabolic metal reflectors may be used.
7. The still is placed in open sun and removed only the next day as condensation takes a longer time than evaporation.

Grass Roots Corruption Has Defeated The Bare Foot Revolution. We Now Need A Self Help Revolution To Defeat Grass Root Corruption.

Chapter - 36

Demonstration at Valsad

Only the top container closed with a saaing plastic cloth (transparent) is the still which distils sea water. There is an empty jug inside the top container and sea water around the jug. When the top container is heated

Livelihood for women

with a solar reflector, the sea water in the container gets heated. At 60 to 70 degrees Celsius the vapour rises and touches the closed surface of the plastic sheet which acts like a funnel and beads of condensed water drips into the receiving jug in the middle. There may be five or six reflectors all around to a ment the heat. The middle and bottom vessels are empty. They are used here for want of a stand or stool to heighten the still to get the focus of the solar rays. This demonstration on a hot afternoon in Valsad, Gujarat, India, gave about 300 millimetre of safe condensed water. Many poor people in the coastal belt of Gujarat have now taken to this method.

Chapter - 37

Pentagon Still

This is the best domestic still innovated so far by the Ryan Foundation. It is not as complicated as it looks in the diagram. All the arrows only indicate the rays of the sun.

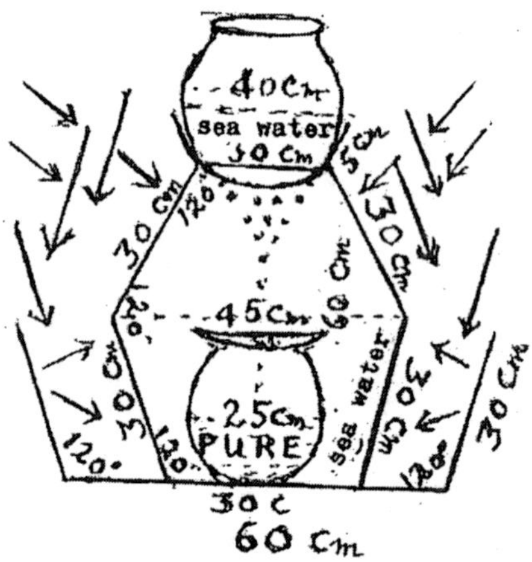

The five sided container is made of 16 gauge stainless steel as per the dimensions in the diagram.

The tub like container is made of 16 gauge mild steel as per the dimensions shown and painted black inside, outside and at the bottom.

The bottom pot, which can hold about 4 litres of water may be of any metal, preferably aluminium. The aluminium lid that covers it should have four nail holes right in the middle.

The pot on top may be of mud or metal and painted white inside and outside and filled with cold sea or saline water. It should be large enough to hold about 10 litres of sea water.

Five litres of sea water is poured into the large five sided container. The pot inside is kept empty and clean. Only the five sided container should be made of stainless steel. The working principle is very simple. The sun's rays directly hit the top two sides of the large container fall on the sides of the tub and reflect on the top, bottom and sides of the big container. Thus, the big five sided container is heated on all sides and the sea water in the big container gets heated, evaporates, hits the bottom of the cold pot on top, condenses and drips into the bottom pot through the holes in the lid. The already condensed water will not evaporate from the bottom pot and go out as vapour because of the saucer. A rubber hose siphons out the warm water from the top pot as cold water is filled into it. A kerchief or thin cloth may be used as a washer between the big container and the pot on top to reduce the steam escaping.

> Banana Skin, Fresh Lime and Drumstick Seeds Act as Natural Flocculator and Coagulator in Water to Make Water Fairly Potable.

The still is set and left in the open sun throughout the day, removed and opened for collection only the next morning to get about 4 litres of healthy drinking water. As this water will be tasteless of the sea or saline water heated to a high degree in the big container may be filtered and mixed with the distilled water to restore the lost nutrients and taste. Thus, one gets 4-4.25 litres of safe water. If more water is needed, set two units.

On days when there is no sun the five-sided container may be taken out of the tub and put on three stones (campers stove) and fired from below. It may also be heated on a regular kitchen stove. RYFO foliage briquettes (explained elsewhere may be used for heating. Small stainless steel reflectors or metal bonds, which construction workers carry on their head, may be used but they must be painted with aluminium paint inside and black outside. A husk stove (vumi aduppu) will be ideal and trouble free.

> The Unep Annual Report 2000 says. "More than 5 million people die every year as a result of poor water quality-ten times the number killed in wars. More than half the victims are children. In short, water is life. Clearly, the unsustainable use of water is the most serious and immediate environmental problem facing the planet at the beginning of the new millennium.

> China Daily reported last May that Beijing's two main reservoirs were diminishing rapidly. Water tables beneath the capital drop 1 to 2 metres per year and a third of its wells have gone dry. India's 835 million people face severe shortages stemming as much from gross mismanagement of land and water resources as from nature's fickleness. By the late 1970s, the government had invested more

than US$ 12 billion in 1,554 large dams, but deforestation has now undercut much of that effort. Denuding of watersheds causes more water to run off in floods and less to percolate into the ground.

Chapter - 38

Reverse Osmosis

It is not right to convert sea water into potable water by the reverse osmosis Ro process, using costly Japanese machines, Gulf oil, and the Haban diaphragm, then filling the water in plastic bottles, each. Costing about ₹ 2 made in large scale factories, using heavy machines, scarce and costly electricity, inorganic chemicals and toxic raw material, which cannot be easily disposed of in the form of bottles or pouches; and on top of it all, spending ₹ 15 or more to transport the bottle with the so called potable water to a distance of 5 km or less in road polluting vehicles and selling the water to the urban rich at ₹ 15 or more per bottle, which includes publicity, promotional expenses, service tax and profit. And RO water does not reach the poor.

RO is an inefficient process 25 percent - 45 percent - of the total water drawn is rejected in the form of concentrated salt solution and several impurities. Moreover, pumping costs are high and do not always yield safe drinking water without metal and chemical impurities within the limits prescribed by the World Health Organisations (WHO).

In February 2002, the Kasimedu RO Plant (Chennai, India) which was installed in January 1999 at a Capital cost of ₹ 55 lakhs to serve a population of 5,000 people, was shut dwon due to high TDS level (total dissolved salts.)

In Oct, 2001, the Kasimedukuppam RO Plant at Chennai, India, installed in Oct, 2001 to serve a population of 7,500 people at a cost of Rs 24.58 lakhs was also shut down due to high TDS (Total Dissolved Salts).

Plants installed at a cost of ₹ 41 lakhs at Nochikuppam and at Velachery, Chennai, at a cost of ₹ 75 lakhs and at Ayodhyakuppam at a cost of ₹ 32.75 lakhs are both working unsatisfactorily and yield little water to long queues of women waiting with plastic pots.

One cannot understand why governments are not converting sea water for farm, factory and domestic use by the RYFO methods using wind, wave and solar energy explained elsewhere by the RYAN FOUNDATION and also make sea water available to people through man made canals and streams especially in remote drought areas for domestic distillation in RYFO do it (-) yourself stills which are assembled in mud or metal pots. From time immemorial villagers have been using crude, unproductive stills for distilling country made illicit alcohol and they certainly know the principles of steam distillation. RYFO stills are very productive, and convenient and work on conventional and non-conventional energy.

Innovating for a sustainable planet is the best job on earth.

A down to earth water revolution is the need of the day and it has to be a sea water revolution of the people, by the people, for the people. Governments cannot usher in a peaceful revolution and have never done it. They can only keep revolving around problems the way they have been doing over the years and keep saying that the water table is falling year after year and the over the years and keep saying that the water table is falling year after year and the water quality getting from bad to worse.

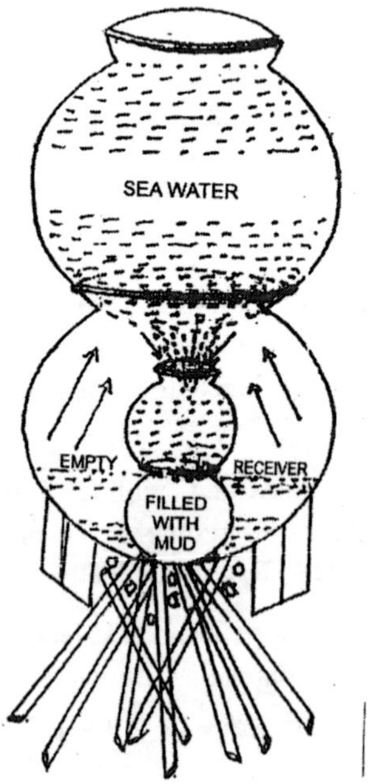

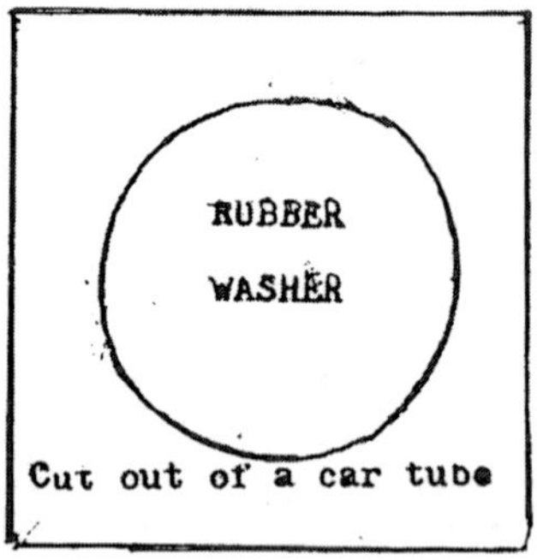

DESIGN BY THE AUTHOR

They will spend a billion dollars to install bore pumps to get saline water but will not spend ₹ 100 to get down to brass tacks. Third World governments like to promote scarcities and problems to stay in power and to make their money. They derive their strength as rulers from poverty, problems and scarcity.

Desalination projects in the United States and around the world have had mixed results.

Santa Barbara city leaders spent $ 34 million to build a desalination plant in 1991 during California's last drought, and then they mothballed it over high cost when the drought ended, Distillation is the answer and not desalination.

It will be impossible to achieve an ecological economy, without a sea and solar resources base.

Instructions

1. The top pot should not be closed. If somebody is around the warm water in it may be siphoned down to plants and trees and fresh cold water replaced before one goes to bed.
2. Car or lorry tube must be cut to size with a big hole cut out in the middle as shown in the diagram in chapter to serve as a washer. Two or three such washers stuck together will be fine to prevent leakage of steam and to yield more condensed water.
3. The three-stone camper's stove must be high (longish) to take in a large quantity of charcoal or foliage briquettes or firewood so that there is enough heat to steam all the water in the big bottom pot during the night.
4. The setting can be left in the garden, on the terrace or in the sun shade or any open area or inside the kitchen or balcony, fired and left there over night. When the fire and heat die down, water will condense in the cool breeze during the night.
5. Fill the small bottom pot with mud to prevent the already condensed water in the small pot from further evaporating.
6. Man, animal or birds cannot survive by drinking distilled water always as it is rid of vitamins, enzymes and other nutrients needed by the body.
7. To replace the lost nutrients, add to the condensed water (after removing it from the setting) some kitchen ingredients mentioned in the next page.

If a glass of milk, a glass of ink and a glass of whisky are mixed with a pot of water the water alone cannot be separated from the mixture. Likewise water mixed with heavy metals, dissolved salts and toxic and human waste cannot be separated by the RO or any mechanical or chemical processes. Only steam distillation can separate water from all impurities. In India, ground water contains plenty of arsenic and fluoride. A arsenic leads to cancer and fluoride weakens the bones and the teeth.

In 2010, The United Nations Recognised Access to Safe, Affordable Water and Sanitation as Human Right. Yet The Reality Remains Stark. More People Today Own Or Use a Mobile Phone Than Have Regular Access to a Toilet. Unclean Water is the Greatest Killer on the Globe.

Chapter - 39

Ryfo Pot and Bucket Still

Instruction For Use

This is a very simple and very easy to assemble and dismantle, still innovated by the Ryan Foundation.

1. Take a big bucket of about 40 centimetres opening diameter.
2. Take a small bucket of about 30 centimetres opening diameter.
3. Take a large pot of mud or metal, of about 20 centimetres.
4. Wrap an old towel around the pot and place it inside the big bucket as shown.
5. Before placing it pour sea water into the big bucket upto the brim of the vessel inside the bucket.
6. Place the setting on your terrace or in the garden in such a way that the rays of the sun fall on it.
7. The sea water in it will evaporate, the steam will hit the bottom of the pot which contains cold sea water, condense and drip into the container kept in the middle of the bucket.
8. You may use a reflector made of aluminium or stainless steel to reflect the rays of the sun. To begin with the metal bond commonly used by construction workers to carry sand or cement may be used to reflect the rays of the sun. if necessary three or four bonds may be used.
9. For more heat set a one inch diameter magnifying glass which will quickly take the water to boiling point.
10. If a magnifying glass is not available, put the whole setting on a three stone campers stove and light the stove.
11. Now and again remove the warm water from the pot and introduce fresh cold water. The colder the water the quicker the condensation.

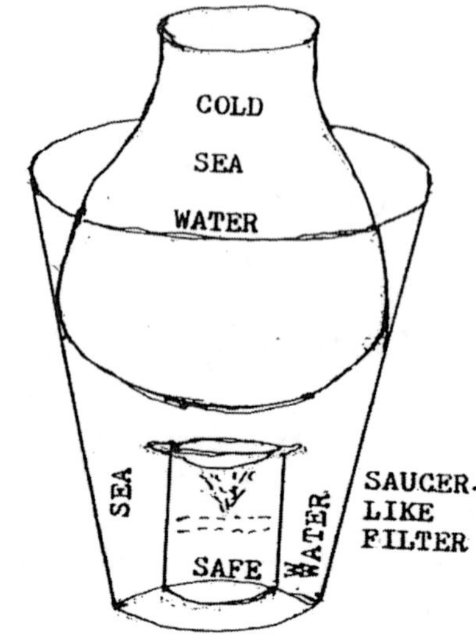

An Old Bath Towel is Conveniently Folded is Put Round The Pot Before The Pot is Put into the Bucket.

12. A saucer, like lid with holes is necessary to prevent the already condensed water in the receiver from evaporating.
13. Pack the cavity between the big bucket and the big pot with an old towel to prevent the steam from leaking out. The packing must be compressed hard all around the pot.

Design by the Author

Chapter - 40

Bucket Hole Still

This still is very easy to assemble with available kitchen ware and yields enough water. A large bucket, a pot to go into it and a suitable receiver are required. All of them may be of galvanised iron sheet but covered with cashew oil paint to prevent sea water corrosion. The bottom of ships and boats is painted with cashew oil paint. The only known alternative is titanium paint or spray but this is very costly. Those who can afford may use it. Distributors of titanium in India are Mishra Dhatu Nigam Ltd. P.O. Kama Chandbagh, Hyderabad – 500258.

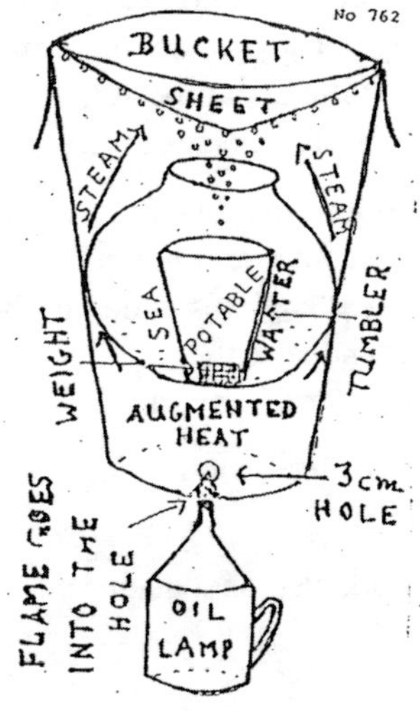

After cashew oil paint has been applied they must all be painted black inside and applied outside. The receiver alone need not be painted black.

Poor rural families may even have all the vessels made of common clay. A big garden flower pot, a rice pot, and a small clay pot will do. A lid made of clay with four nail holes in it may be used to close the small pot (receiver) to prevent the condensed water from evaporating again.

A transparent plastic sheet is used to close the bucket and it is tied around the bucket with a broad ribbon.

A hole of 3 to 6 centimetres diameter is made right in the middle of the bottom plate of the bucket (or pot) and the pot and tumbler (receiver) are put inside the bucket as shown in the diagram.

A poor man's oil lamp (see diagram) with any non-edible oil and a wick made of old cloth is lit, kept under the bucket and the flame directed into the hole as shown. Of course, the bucket is mounted on three stones so that the lamp goes under it. In place of the lamp a beer can or a small bottle may be used.

Heat is trapped between the bottom of the bucket and the bottom of the pot and the trapped heat is very much hotter much higher calories than untrapped heat.

So the sea water boils quickly and the steam enters the crevices on either side of the pot, hits the inverted plastic cone, which holds cold sea water, condenses and drips into the receiver.

If the capacity of the receiver is 2 litres a little more than 2 litres of sea water, is poured into the big pot. The lesser the water, quicker and more is the evaporation. The contraption may be left in the open under bright sunlight sun but even then the flame is necessary (may be a candle flame).

What is set in the morning should be removed only the next morning but the flame will go off in about 2 or 3 hours. The boiled sea water from the big pot is filtered and a little of it is added to the condensed water to make it regain lost nutrients and taste the residual water may be evaporated in open sun light to get common salt.

One may five or six of these contraptions to get safe and healthy drinking water for the family.

This innovation will work also on a giant scale.

Chapter - 41

Sea Water Swimming Pool – A Model

Unfermented palmyrah or coconut tree juice (from the sap), called "pathaneer" contains glucose, calcium, iron, thiamine, riboflavin, ascorbic acid, protein and calories. A recent study has revealed it also acts as an antibolic. HINDU 3rd May, 1997.

Coconut trees flourish when washed by sea water. Some homes and institutions plant and protect them on the shore of the sea. The swimming pool of the hotel in the picture circulates sea water through filters. In these days of acute fresh water scarcity sea water pools are being promoted in some countries. A pool that is drawing world attention is in the Sheikdom of Shajah, near Dubai, Gulf, and it is meant exclusively for ladies. The Survival by Sea Water global movement of the Ryan Foundation encourages the promotion of sea water pools, where fish and galine agriculture may be promoted too to supplement the inadequate food and to relieve hunger in the area to some extent. In India, we say, ten coconut trees can sustain a family.

Experiments carried out in the Coconut Research Station (TNAU), Veppankulam – 614906, India showed that the east coast Tall exhibit higher drought tolerance by recording more epicuticular wax (1043 uc. cm^2 and lower chlorophyll stability index (CSI) (28.9%).

Chapter - 42

Coconut Sugar

The juice or liquid found inside a tender coconut is a fine substitute for safe drinking water and it is also an organic health tonic.

There are not 85 countries in the world where coconuts grow naturally. Other countries, like the Gulf countries, are introducing coconuts as they grow well on the sea shore and saline soil. Production of sugar from the sap of the tree is a profitable business but most of the countries do not make sugar. However, in Indonesia, Thailand and certain parts of Tamil Nadu and Kerala, sugar is produced in the villages using open pans.

The experience in countries like Thailand, Indonesia and Sri Lanka is that earmarking trees for tapping sap for sugar is more profitable than leaving them for producing nuts and copra. In Indonesia, farmers leave about 30 percent of the trees for sugar production.

Trees may be used for both production of nuts and production of sugar. This is done by tapping only alternative "spathes" or by restricting tapping to about 6 months of the year.

In the Philippines, experiments demonstrated that the same spathe could be used for production of sap and nuts by tapping only half the length of the sheath and leaving the other half for production of nuts.

Research results in many countries showed that it is more profitable to leave about 25 percent of the trees only for the production of sugar.

In the provinces of Samut Selcngkhram, Samut Prakhan and Samut Sakhon of Thailand, coconut gardens are maintained only for the production of sugar and when the trees are 5 years old, the sap is tapped for sugar in these gardens. For about 10 months the tapping is done and then the trees are left to recoup for the next cycle of production. Thus, they are tapped every year for 10 months only.

In about 30 provinces of Indonesia the government encourages production of sugar and sugar making is a popular cottage or rural industry in these provinces.

Local organisations help these sugar units and organise the sale of the sugar for the domestic and export markets.

It is reported that the annual production of coconut palm sugar in Indonesia is more than a million tonnes. In 1985 the industrial sector of the country absorbed 36% of the coconut sugar produced locally. The major consumer was the soya sauce industry.

In 1991, the average monthly demand of coconut sugar from West Asia, New Zealand, The USA, Australia and Europe was over 1,000 tonnes. More and more people in the advanced countries prefer to consume no chemical, natural sugar in preference to chemically refined cane sugar.

As more people demand coconut sugar, the production will increase and the cost of sugar will come down. Cultivation of coconut palm on coastal areas will pick up and provide livelihood for people directly and indirectly.

As coconut grows luxuriously when washed by sea waves coastal countries must promote the best variety of short duration coconuts all along the coastal belt. Each family living on the coast must be given ownership of 10 coconut trees on the strict condition that they will not tap the trees to get KALLU or TODDY, which is illicit alcohol ruinous to health and family well being.

Where TODDY is tapped the tree will yield very few coconuts and will have very little water or juice inside. God given natural wealth must be used and not misused for destruction.

Chapter - 43

Tapping Palms for Cheap Sugar

A report produced in 1981 said that out of about 60 million palmyrah trees in India the State of Tamil Nadu in peninsular India alone had 40 millions. Out of these only 6 million trees were tapped for neera (juice) and 5 million were utilised for making products from the fiber and leaves. Nine million were reported to be unfit for use due to negligence and the remaining 20 million still remained unutilised.

A recent study reveals that a man can climb at least 50 trees everyday and get about 3 litres of juice per tree.

Women heat the juice for several hours before pouring them into moulds for getting them into blocks, (Palmgur). The maximum that a family can get out of 100 trees is about 15 kilograms gur or sugar, on an average only 100 litres of neera is tapped during season. But a male tree yields only about 50 litres.

The difficulty faced by the tappers is that most of them do not own the trees that they tap but share the profits with the owners and the sharing is usually on 50 – 50 basis. Merchants purchase palm gur for ₹ 10 (1994 price) per kilogram and sell it for double the price. Thus, the middlemen take a big profit and also put up the market price. Another difficulty of the trade is that it is seasonal- January to April. A third difficulty is the tax which the government levies on palm sugar instead of treating palm sugar as an agricultural commodity.

Ryan Foundation study reveals that making products like baskets, brushes and beams is not as profitable as making sugar. Marketing sugar is easier than marketing other products made from the tree, as sugar is always in demand and gets or can get a good profit. If more and more palm tree tappers are encouraged to make palm sugar, the price of crystal white open market cane sugar will also fall. People know that palm sugar is (or can be) cheap and healthy. Their demand has to be met to the advantage of both the producer and the consumer. Where there is no demand it has to be created. More palmyrah trees must also be promoted as they are highly drought resistant, tolerate high salinity and can be grown on sea shores and on the banks of sea water canals, lagoons lakes and ponds. Palmyrah are drought trees but grow healthy and well in a sea water borne atmosphere, yield and more juice, fruits out of which jams and pickles are made and broad leaves for thatching huts, village toilets, cattle sheds and so on.

The tree trunk cut lengthwise into four can be used as beams and rafts if painted with molten tar (bitumen) or soaked for a week in a cemented tank containing copper sulphate solution.

For over half a century the All India Khadi and Village Industires Commission has been promoting palmyrah products but it does not make briquettes out of the waste of the tree which will give cooking fuel for millions of poor people. Briquettes can be made by the Ryfo no-investment methods.

Chapter - 44

Palm Trees for Sugar and Livelihood

There are trillions of palm trees in India and several trillions in the other 85 palm growing countries of the world, palmyrah and coconuts grow very well on sea shores washed by sea water. In the palm family there are three well known species which yield plenty of sugar.

No Watering if Planted During the Rainy Season.

1. Palmyrah botanical name Borassus laballifer Linn in Hindi, Bengali and Marathi it is called tal or tar.
2. Wild date palm: Botanical Phoenix Aylvestris Roxb.
3. Coconut coco ucifera Linn.

Several products are made cut of palm as can be seen in the picture. The floral bunches of the three varieties mentioned above possess plenty of sea which is sweet and tasty and drip readily when the boat-like bract is slit with a knife at the end of it. A mud pot tied under the dripping slit and the juice collected overnight becomes intoxicating "toddy" by the action of the yeast floating in the air. But when a handful of mineral lime (*chanam*) is applied inside the pot the sap does not ferment but remains as a sweet juice called NEERA which is a healthy and delicious drink. Sugar or jaggery is obtained by boiling the unfermented juice (neera) into a syrup semi-liquid and then allowing it to cool and harden.

Jaggery, brownish in colour, is formed into lumps or caste into moulds and sold. It melts in warm weather and to have it bleached and crystalised. it costs much more than cane sugar. Producing if only the governments of the countries which have palm promote and encourage the use of liquid brown sugar of palm packed in sachets as they now market intoxicating drink, milk, fruit juices and ghee they will be lowering the price of cane sugar and creating millions of jobs for the people. Instead, governments are now promoting capital intensive and centralised cane factories and care cultivation. Is it because there is money for politicians, bureaucrats and middlemen in centrelised factories for making sugar and fertilisers for cane cultivation and in their packing, transportation distributiours and licensing, creating scarcities in rural areas and developing black markets? It is a tragedy that many palm growing countries import white sugar, topple in balance of payment and turn to the IMF or the EEC for loans and remain as LDCs ternally. The USA is trying to market protein rich synthetic THAUMATIN, which is 1,000 times sweeter than sugar, to wipe out cane cultivation. Whither development?

Palm sugar is good for health. The economy and rural progress.

Chapter - 45

Briquettes from Coconut Pith

When coir fiber extracted from coconut pith is allowed to fly and float around it contributes to dust and pollution. The pith is Hygroscopic and can be compressed into hard briquettes. The pith contains a high percentage of Lignin and a small percentage of Nitrogen. Polyphenol is found in its raw form. The cellulose content and high level of Lignin does not make the pith bio-de-gradable.

Foliage – dried and ground dry leaf, paddy straw, cob corns, wood shavings, charcoal dust, groundnut shells and so on may be granuled and added to the coconut pith. More of coconut pith the better are the briquettes.

Male and female dies are made as big as a building block (one going into the other) the dies are fixed to a fly press, ball press or turn press. The briquette mixture is filled into the female die and the press is turned by hand. When the handle or wheel is turned and held tight for a minute briquettes are formed.

Cassava, yam, rice or tamarind seed starch or fish glue (gelatine) may be mixed to make the briquettes bind harder and burn longer.

In a coconut grove it will cost less than 5 paise to produce a briquette and two briquettes can cook a days meal or heat sea water to get about 5 litre of safe drinking water.

Palmyrah and many other trees in the palm family are drought trees but grow and flourish well in a sea water borne atmosphere, yielding and more juice, fruits yielding (used for making jams and pickles) and broad leaves for thatching huts, village toilets, cattle sheds.

The tree trunk cut lengthwise into four can be used as beams and rafts if painted with bitumen (molten tar) or soaked for a week in a cement tank containing copper sulphate solution. All waste from the palmyrah tree can be used as cooking fuel or for making briquettes or pellets, thus saving on cooking gas or kerosene.

For over half a century the All India Khadi and Village Industries Commission of the government of India has been promoting palmyrah products but does not promote the making of pellets and briquettes as an alternative source of energy.

Nearly 70 percent of the world's population lives within 80 kilometres of a sea coast. The World Bank and the Environment Fiscal 1992.

> When common people don't make decisions nothing changes and there is no progress without change.

Chapter - 46

Sea Water Through Rocky Soil

Some people say that if sea water is brought inland in man-made canals the soil will get more and more saline and salinity will spread to other areas. It should be remembered that the land is not friendly and fertile every where. In many sectors it is just dry and dead, cracked and parched. Some areas are rocky and full of gravel, clay and pebbles. For generations they have not been used and given up as not of use.

Sea water canals may be dug across such wasted and parched lands to let moisture cover the area and allow halophytes and cactil, sea weeds and salt bushes to grow there.

Sea fish will automatically come in there providing food and livelihood for the people. Crabs and turtles will also find a place there.

The water may be used for washing cattle and farm equipments and, most important of all, for saline agriculture, which is now a fast developing science. The book haloph by J. Aronson published by the office of the

Arid Land Studies of the University of Arizona in 1981 has listed 1,560 plants and trees that grow in saline soil.

If there is a dry river or rivulet in the area the canal may be connected to it. Alternatively, after winding through all the patches of parched lands it may be connected back to the sea just as the Buckingham canal is taken out from the Bay of Bengal from Beddagangam in East Godavari District in Andhra Pradesh, India, and connected back to the same Bay of Bengal near Marakhanam a distance of about 420 kilometres. It is 165 kilometres long in Tamil Nadu.

Chapter - 47

Substitute Organic Nutrients

Safe drinking water must contain a certain amount of protein, carbohydrate, calcium, iron, vitamins, enzymes, minerals, sodium, potassium and so on within the limits prescribed by the World Health Organisation.

What to Substitute

A pod of onion, a pod of garlic, small pieces of ginger (4 grams) two cloves, fresh banana skin cut to bits (about 8 grams) fresh orange peels cut to bits (about 4 grams) neem leaves, ten tulasi leaves (holy basil), 50 grains of rice, 50 grains of dhal, 1 gram of salt, 10 murunga pods (drumstick seeds) or leaves, cut lime skin two bits (2 grams each).

Most of these culinary items are readily available in the family kitchen. Take about six to seven of them that are available and make a thin cloth bundle as small as a table tennis ball and place the bundle inside the distilled water collecting pot to restore the lost nutrients in the steam-distilled water, which is lifeless. The old bundle must be replaced by a fresh one about once a week. Pure distilled water is lethal. Human beings, animals, birds and plants cannot survive by drinking pure distilled water for days together.

Tie the cloth bundle to a string and holding the other end of the string lower the bundle into the receiver which collects the drinking water. The loose end may be tied to a weight such as a metal curtain ring and left outside for lifting the bundle conveniently. The residue in the used up bundle contains insecticidal properties and so may be thrown into flower pots or croton plants as insecticide.

> Water must contain some protein, carbohydrate, calcium, iron, vitamins, enzimes, minerals, sodium, potassium and so on to iodise water add about 30 milligrams of potassium odate to litre of distilled water.

Chapter - 48

Ryfo Community Water Supply System

Note

On a miniature scale only one stainless steel boiler is used. If a I inch pipe connects this boiler to a sea water supply tank, there will be a continuous flow of sea water into the small boiler. In this case the boiler will yield 2 litres of water every minute that is 120 litres per hour or 1440 litres in 12 hours. Taking time into consideration from 6 a.m. to 6 p.m. the yield will be about 1,500 litres. If three boilers of the same size are clustered together the yield will be 4,320 litres in 12 litres or atleast 4,000 litres.

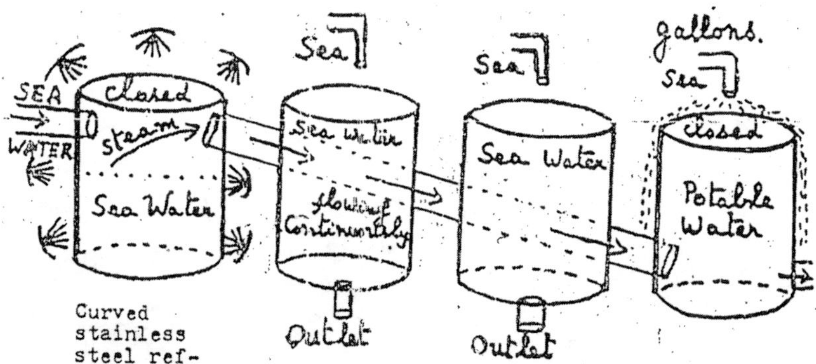

The cost of 1 litre will be approximately 10 pais a that is 10 litres will cost a rupee, which the present day poor can afford, Innovation by Dr Felix Ryan.

Curved stainless steel reflectors and parabolic magnifying glasses around this tank only.

The cold sea water in the two middle tanks condense the steam and renders freshwater into the last tank.

Only one stainless steel tank is shown in this diagram. Three such tanks each of 10,000 gallons capacity, should be clustered together for commercial-scale production of potable water. The long tube which passes through all the tanks is also made of stainless steel.

Chapter - 49

Ryfo Renewable Energy Burner

When oil is not used pack the burner with shavings, charcoal rice or wheat bran corn cobs, groundnut shells, coconut pith or broken foliage briquettes and light it.

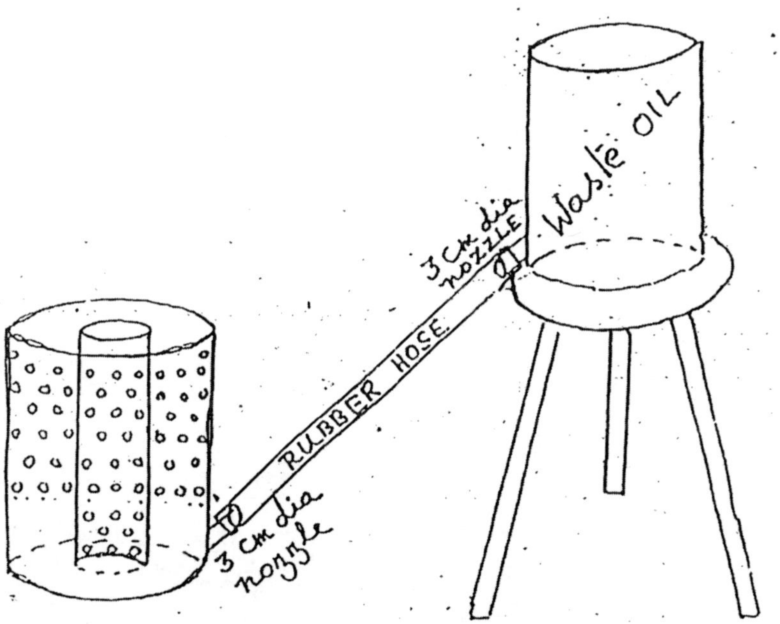

Dimensions

1. Height of stool 90 cm.
2. Tin on top 9 cm × 9 cm.
3. Length of nozzle 3 cm × 3 cm.
4. Length of rubber hose 100 cm. Diameter 3.5 cm.

5. Size of burner (tin) under the distiller 9 cm × 9 cm.
6. Size of each hole 0.5 cm.
7. Size of middle cylinder 7 cm × 6 cm diameter.
8. No holes in the outer cylinder to a height of 6 cm.
9. Burner cylinders both packed with tailors waste or rags cut to pieces.
10. Some oil is poured on top of the burner to light it.
11. The most suitable and cheap oil is the monoxide castor oil, which is easily available in large quantities, Castor is a drought plant and animals don't brow them. May be cultivated in the kitchen garden.
12. Next in preference is the Pongamia and Jatropa also available easily in large quantities at a low price. Pongamia may be cultivated as avenue tress and Jathopha as a commercial crop. Castor Grows Wild in Open Lands.

Chapter - 50

Renewable Energy Burner With or Without Oil

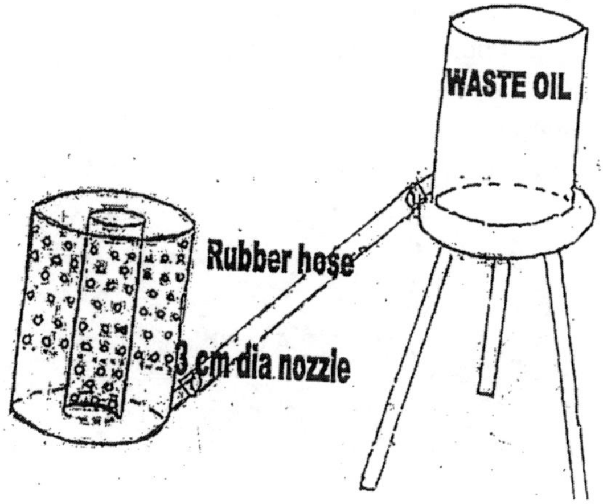

Chapter - 51

Ryfo Double Stove Foliage Burner

Cylinders AB and EF are welded together side by side. Slightly smaller cylinders CD and GH are inserted into AB and EF to serve as pistons. Q, R, S and T are the stands. In circle P holes are bigger and closer front portions of the two cylinders AC and EG are ferforated (one cm dia holes). The front portions AE and EV should also be perforated. Working instructions are attached.

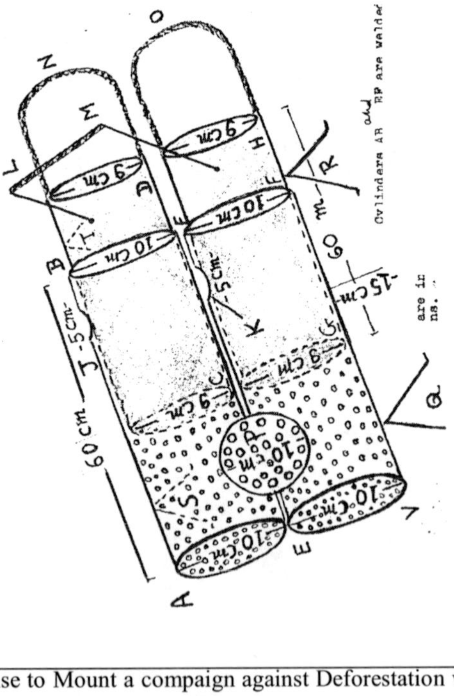

It makes No sense to Mount a compaign against Deforestation without making available alternative fuels and Resources.

Chapter - 52

For Joint Family Cooking With Foliage Briquettes

Where three or four families in a village can join together and cook food to save time and energy, it is convenient and economical to use this cooking platform to steam distill sea water.

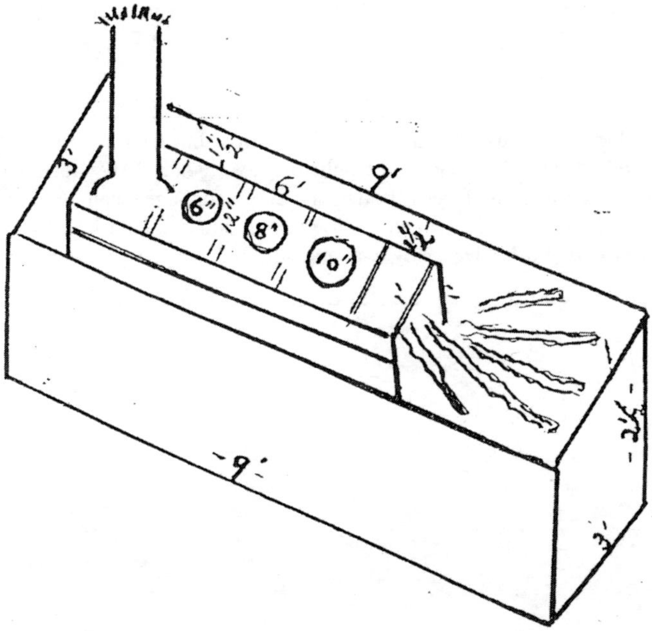

In rural areas of the Third World cooking is almost always carried on inside the small, low hut, which has scanty accommodation and ventilation and the whole family lives in the same smoky room where the cooking is done. Rural communities should be advised to cook, as far as possible, out in the open before the sun sets or after sunset. They should also get into the habit of cooking in standing position instead of squatting near a low fire.

The open air cooking platform may be built out of bricks, mud cement or clay building blocks. To form the top of the platform, quarter inch thick mild steel rounds or squares may be laid across the two long parallel walls and the bricks or slabs may be arranged on them and plastered. Alternatively, a reinforced cement slab may be pre-cast and put on top. A small second platform with circular openings to hold pots as shown in the diagram may also be hollow inside with side opening and firewood may be stored in it.

Families that can afford to put up this construction in front of their hut may do so. Three or four families may do well to join together and put up a common platform for their common use.

Aid agencies would do well to build open air community kitchens for the collective use of selected families. Different times may be assigned to each family to use the platform to prepare their meals. Such open air kitchens will not only make life inside huts liveable but also save time, labour and fuel for the rural poor.

Living in degraded lands mean an extraordinary work burden for women. Over worked mothers need assistance to help them out with their daily chores which results in the education of the girl child being sacrificed. So the main issue ahead is to get rid of this constraint, which means regeneration of India's lands, making water. Firewood and fodder easily available and reducing the work burden of women.

Chapter - 53

New Gasifier Produces Cooking Gas from Kitchen Waste

(Excellent Renewable Source of Energy to Steam Distill Sea Water)

An inexpensive gasifier has been designed to produce cooking gas from the kitchen waste of a household.

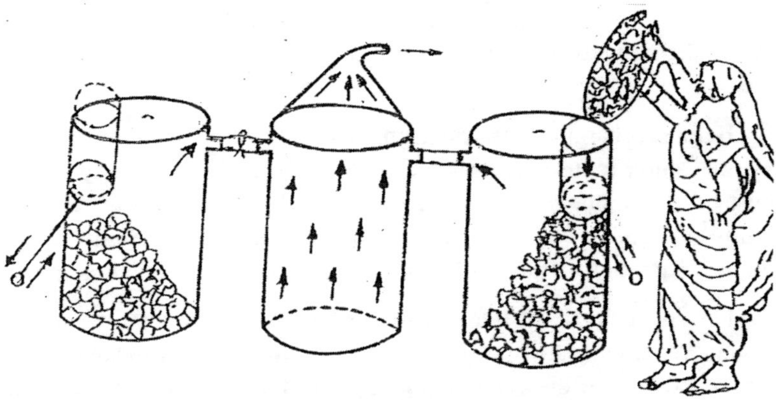

(Fig.1. Cooking Gas From Kitchen Waste)
Farm Waste, Paper Waste, Foliage, Halophyte and Garrage Etc.

The gasifier is easy to fabricate. Take three drums, each of about 100 litres capacity, and two drums of 5 litres capacity each. Weld the small drums inside the big ones and provide doors (shutters) at the bottom of two big drums (see sketch). Make the lid and bottom of the small drums revolve on a point axle. Now weld 6 gas outlet pipes near the upper rim of the big drums and connect them with 30 long bicycle tube bits. Attach to the revolving bottom of the small drum a thin rod, like an umbrella rib, and take the rod through a very small hole made in the side wall of the big drum. Fix tight rubber / leather washers on either side of the hole (one inside and one outside) to prevent leakage of gas. Make sure that the lids of the big drums also fit tightly, finally, weld a conical lid with an outlet pipe on the middle drum.

Fill the first and the third feeder drums with raw kitchen and cooked table waste. Rotten eggs and fruits, spoil milk, curd, banana skins, rejected root crops, corn cobs, paddy straw, food grains farm yard and chicken house manure, piggery and rabbitry sweepings, green leaves, cow dung, grass, weeds, and so on may all be added for more gas. If cheap country – brewed liquor is available, add a pint of it to the "charge" once a month. Cheap vinegar or yeast may also be added to quicken gas formation. The volume of gas in the drum can be judged from the bulge and feel of the cycle tubes connecting the drums.

Connect the main outlet in the middle drum to a domestic cook stove. Tie a ribbon around one tube, arrest the flow of gas from it into the middle drum and use the gas only from the other drum. When this gas is used up, shift the knot to the other tube. Thus, gas from the two drums may be drawn alternatively and both the drums will accumulate much gas when idle.

Before recharging a gasless drum, open its bottom door to remove the digested slurry for use as manure. Charging, recharging and slurry removal must be a continuous process. If the gas obtained is inadequate for the family, add one or two more feeder drums or set up another unit of three drums. Empty asphalt (tar) drums may be conveniently used for this purpose.

For convenience of transporting the second drum could be made 3 centimetres smaller than the first and the middle drum 3 cm smaller than the second so that they can be put one into the other, if the connecting rubber tubes are made longer and the middle drum is raised by about metre, the leakage in the system will decrease.

> World Wide 80 Million People Enter the Workforce Every Year. They Can Survive Only By Exploiting the Sea, the Tree, Foliage and Garbage.

Chapter - 54

Renewable Energy for Ryfo Domestic Still

1. The contraption (drum) can be placed on top of any cookstove – electric, gas, oil, kerosene, and so on but all these are non-renewable energy and may be used only when non-renewable energy is not available.
2. Two open containers are provided along with the still and both are provided with a 1 inch diameter nipple. One of them has a perforated cylinder fixed in the middle. There are two rows of small holes on the upper portion of this container for air to blow in.
3. Set the container without any holes on a stool or in an elevated position – about 3 feet higher than the other container which serves as the stove.
4. Connect the two nipples with a rubber hose or an old cycle tube (do not let the tube sag).
5. Fill tailors, waste and cut pieces of rags into the container with holes.
6. Pour any unwanted oil into both the containers – the oil may be waste engine oil, castor oil, or any unedible oil obtained from crushing nuts.
7. When unwanted (unedible) fish and fish intestines and waste are put in a drum and heated from below, plenty of fish oil is obtained which floats as a top layer and can easily be separated from the water. Oil thus obtained is free and ideal for heating in the burner provided.
8. In coastal areas, ponds (lagoons) may be dug on the coastal belt and connected to the sea either by open cut gutter or connected by

garden hose or rubber or bamboo pipes. Into these ponds and lagoons the fast-growing milk fish (chanos chanos), which multiply and grow very fast in sea water, may be introduced and exploited. These fish are edible and tasty and also yield plenty of oil for burners.

9. Alternatively, fill the burner (only outside) with paddy husk or saw dust, granuled cob corns, groundnut shells or any burnables and fill the cylinder in the middle with waste paper or dry paddy straw and set fire to it. The burnables in the container around the cylinder will catch fire and burn for a long time. Air (oxygen) will enter the container from outside and also through the middle cylinder.
10. Another alternative is to use dried foliage compressed into lumps or briquettes.
11. For binding dried foliage either fish glue (explained in the page attached) or tamarind or cassava starch may be used (also explained).
12. If sea water is on the boil in the open, the sun's rays may be reflected on the container using either mirrors, tin or aluminium sheets.
13. About four reflectors made out of very thin stainless steel, like an open umbrella, reflects the sun's rays efficiently.
14. While setting, the open side of such reflectors must face east and west and not south and north.
15. Now and again, during the day, the reflectors must be adjusted to focus the sun right on the container holding the sea water.
16. Large size magnifying glass as if available, may also be used to focus concentrated heat of the sun.
17. Under normal summer conditions, where the sun is in the range of 100 degrees Celsius, solar heat alone will yield only one-third the capacity of the receiver. Therefore, a combination of heating methods is recommended.
18. Burners fired with waste fish oil are the most economical and productive and next in preference is castor oil (paper attached).
19. In no case should edible oil be used.
20. Children should be kept away from the fire.

Drumstick tree grows well in saline soil and tolerates sea water mixed with sewage water. Leaves of Moringa (drumstick tree) are rich in vitamin A, B, C, and also in Calcium, Iron, Magnesium, Potassium, Sulphur, Phosphorus and many other micro nutrients. From time immemorial in India, Moringa leaves have been used to treat headaches, anaemia, skin infection, dizziness, insomnia, joint pain and constipation.

Chapter - 55

Garbage in Sweden

(an inspiring case study)

There is no garbage left out in Sweden. It is surprising that Sweden has been importing garbage from neighbouring Norway to feed its waste to energy projects that provide heat and electricity by incineration to thousands of homes in the country with a population of more than 9 million.

Garbage in not only burnt in this country to get power but also recycle, recover and reuse. The country leaves only 4 percent of the total garbage to end up in land fills.

In 2005, Sweden made disposal of garbage by farms and factories, homes and communities by sending them to land fills illegal and ordered organic garbage to be treated biologically to make compost, fertiliser and bio gas, the bulk of the collected garbage is used to produce motive power by the environment friendly and sustainable incineration process.

By various laws, incentives and educational programmes the country has made it possible to recover and reuse 90 percent of the garbage out of the 118 million tonnes of waste turned out in the year 2010 and thereafter over the years, starting from the early 1990s, the country has reformed and improved the system of collecting the garbage and using it well and wisely. Are not other countries of the world inspired by Sweden? This is a million dollar question to be answered by global governments.

The European Industrial Revolution was an energy technology revolution. The steam engine made it possible to replace human and animal labour with machines to bring about mass production and to facilitate the concentration of capital. Steam engines were the motor of industrialisation. They gave rise to a concentrated demand for fossil fuel. Which is why the first, industrial centres developed where the coal mines were, that is in England, Northern France, in Silesisa, in Germany's RUHR district and in the American Mid-West.

Economic access to food can be improved only by creating multiple livelihood opportunities based on micro enterprises supported by micro credit.

Chapter - 56

Successful Bio-Gas Plant to Boil Sea Water

The Mamallapuram town panchayat has installed a bio-gas plant on a small stretch of the East Coast road down South Chennai, India. Installed at a cost of ₹ 20 lakhs the capacity of the gas chamber is to take in 500 kilograms of biodegradable waste, mostly consisting of kitchen refuse, dining table waste and dead fish from the nearby sea to produce 40 cubic metres of gas every 12 hrs every day. The gas produced is used to run a generator to produce electricity, which powers about 30 street lights. Each light is a fluorescent bulb. This unit has been functioning from 6 pm to 6 am without a break and it has been working successfully for the past 6 months and attracting schools, NGOs and Rotary and Lions Club members.

About 500 kilogram of garbage is generated daily on an average in the areas under the jurisdiction of the panchayat (local self government) and the waste is collected regularly.

It is reported that before this plant was commissioned the panchayat used to pay ₹ 6,600 every month to the State Electricity department for keeping the 30 street lights burning and by this gas lighting method the cost of lighting the stretch or road has come down to ₹ 600 only.

The panchayat officials state that the bio-degradable waste put out by several establishments, homes and restaurants would be more than sufficient to commission three more biogas projects in the same area and the gas and electricity produced by them will be adequate to power nearly 100 more street lights.

Tamil Nadu is known for regular power cuts on and off and all other towns in India and other countries where adequate garbage can be gathered must install such plants wherever possible. NGOs should take up the mission which not only gives power but also cleans the environment on a regular basis. The method will also produce for the local unemployed, jobs in the collecting and supplying system. This is a clear call to youth who are interested in development work and development is the most important work in our planet.

Similar biogas plants can very well steam distill sea water and provide safe drinking water to people in cities, towns and villages.

Many cities are drowning in their own sewage. There are leaks and blocks in the sewage pipes and the drinking and cooking water frequently get mixed with the overflowing and leaking sewage, leading to disease and death.

> Chennai City generates 4,600 tonnes of solid waste every day. The collection and disposal of the waste cannot and should not be centralised. Only colony, road side, slum and poorly managed large countries like India with indifferent and undernourished labour.

Chapter - 57

Power From Lagoon

(for inspiration and immitation)

In January 2014, the Gaardian Newspaper of England announced that the world's largest unit to generate electric power from lagoons and atools using tidal waves has been launched in Southwales. The project will spend during the first stage about 12 million British pounds.

The promoters are planning four other projects also which will cost about 850 million British pounds, around the SWANSEA Bay with a capacity of 7,300 megawatt enough to meet 10 percent of the electric power needs of Britain.

The promoters are also planning to put up shore wind mills to generate electricity.

Mr. Shorrock, the man behind the venture is very optimistic and says that the full project encompassing 4.5 square miles, will have an installed capacity of 420 gigawats and a design life of 120 years.

The incoming and outgoing tides will generate the power and an enormous amount of sea water will pass through and turn the turbines.

Many countries are waiting for the results of this project and if proved a success will replicate it in their own countries. The world needs power for progress and to distil sea water into safe drinking water.

Now the source and method is known to produce abundant cheap power without polluting the atmosphere.

> Some are inclined to write off Africa as hopeless. Others see Africa as a special case to which the principles of economics do not apply. I reject such views. It is our approach that has failed, Africa has not UN Secretary General Boutros Ghali, chairing the first meeting of the Panel of High Level Personalities on African Development on 28.12.1992.

Development is more than growth; it is an improvement of human well being.

Our planners and politicians seem to go by the KEYN's philosophy, that we will all be dead in the long run and therefore problems like water and fuel for the coming generations should not worry us.

Chapter - 58

Magan Chulha Distils Sea Water When Food is Cooked

The Magan Chulha is about the best cook stove for the rural poor. Metal parts are not really necessary to make a Magan Chulha and it can be shaped or moulded with good clay or Ryfo mortar mix. Ryfo foliage briquettes and waste oil burners perform well in Magan Chulha.

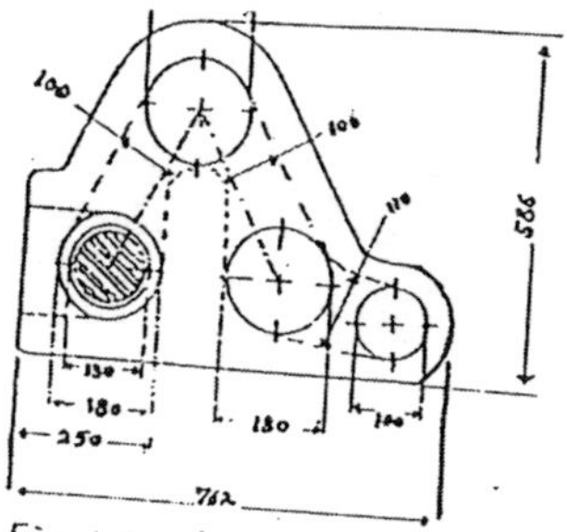

Figure 1

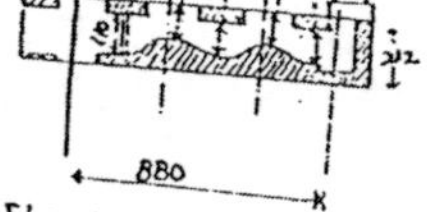

Figure 2

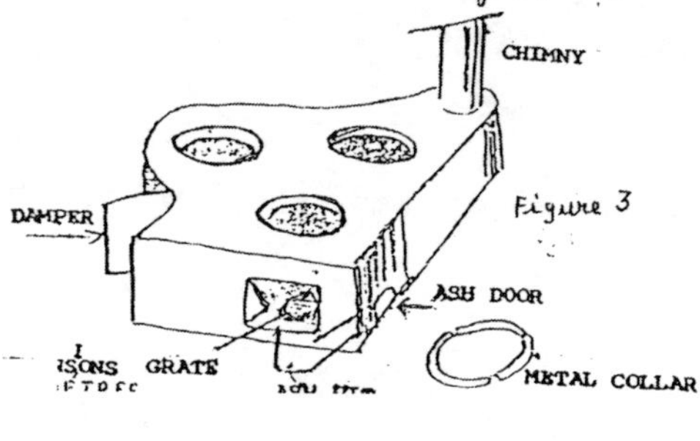

Figure 3

Desalination on the Magan Chulha in the Ryfo pot still yields excellent results. Every rural home should have a Magan Chulha to desalinate sea water using Ryfo briquettes.

Kill two Birds with one stone.

Chapter - 59

Magan Chulha

Description, Magan Chulha is a smokeless wood fuel oven which relieves women of eye diseases and headaches caused due to heavy smoke in the kitchen. It saves 30 percent to 40 percent of the fuel used in a conventional three stove chulha.

Materials: Clay, Cow dung, horse dung, straw cuttings, ash or sand, fire clay, chimney, pipes, grating and dampers (or Ryfo mortar mix).

Construction: The Magan Chulha is basically a mud stove fixed with a metallic (or clay) grate, a damper and optional metal collars on potcholes. It accommodates three pots of 22 centimetres, diameters has a fireplace for long wood and grate for charcoal, a chimney and a fuel gas damper. A proper admixture of clay, sand, horse dung and so on is made to stabilize the soil mud and after due moistening and maturing, a block of 762 × 586 × 212 cubic millimeter is made and allowed to achieve a semi-dry state. The pot holes are marked and tunnel scooped out (fig. 1 and 2).

Process: Once the chulha is lighted, the hot air goes up the chimney, which is replaced by incoming cool air, thus creating a continuous draught to keep the chulha burning. A damper is used between the first and the second pots holes to prevent the flame from rushing into the second pot chamber without heating the first pot properly. The heat flow is controlled by sliding the damper back and forth. (The damper may be made out of clay, like a flat tile.)

The fuel channel is a long zig-zag tunnel to retain the heat for a long period. There is a mud metal grating below the first pot to create a proper draught and allow the ash to fall down in the shallow pit below the grate. An ash door (figure 3) is built on the side of the chulha to clear up the ash from time to time as well as provides a draught from below the grate. The ash door may be a flat tile. Discarded tin containers, if available, may be flattened and used.

Magan chulha was innovated by the magan sangrahalaya samiti, Wardha – 442001 (Maharashtra), India, RYFO chulha is basically a Magan Chulha but it eliminates all metal parts and replaces them with parts made entirely out of clay or Ryfo mortar mix.

World Bank Fiscal year 1994, (Making Development Sustainable) records, subsidising water, while justified in specific situations, promotes waste, almost always benefits the rich more than the poor, and reduces the ability of cash strapped public utilities to extend service to the poorest peri - urban neighbourhood.

The best way of disposing of garbage is to produce gas on large scale away from the city on the coastal belt. The gas may be used to boil, evaporate and distil sea water for human consumption in devices .designed by the Ryan Foundation International, which when tried on a miniature scale produced plenty of gas. The concept and design of a township model are explained below. Perhaps some more R and D work is required.

Garbage Gas Producer on a Large Scale – a Concept.

(Explanation – Next Page)

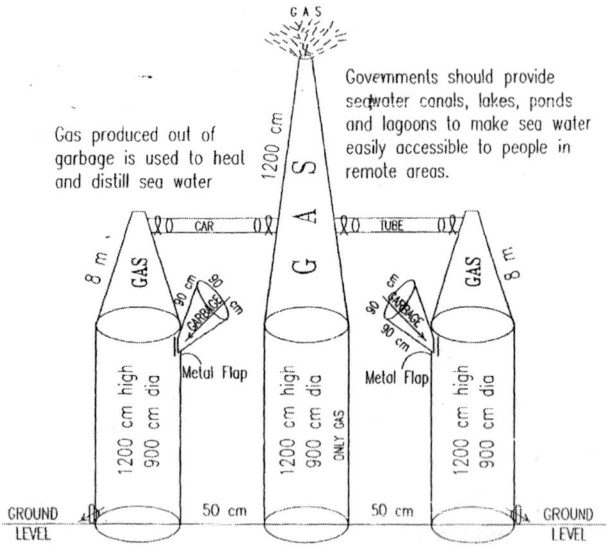

A study in the proceedings of the National Academy of Sciences documents the direct conversion of bio-mass to bio-fuels without pre-treatment using the engineered bacterium.

Chapter - 60

Garbage Gas Producer

1. Three concrete tanks are built side by side in a row with a distance of 150 centimetres between them.
2. All the three tanks are 1200 centimetres high and with a diameter of 900 cm broad.
3. On the open top of each of the tank concrete cones are built. The height of the cones on top of the double sided tanks is 8 metres the height of the cone of the middle tank is 12 metres.
4. The pointed end of the side cones are a little flat on top while the top of the middle cone is open and the diameter of the opening is 5 centimetres.
5. Funnels 90 centimetres long with a feeding side diameter of 90 centimetre fixed for the side tanks. The narrow end opening of the funnel is 30 centimetres.
6. After removing the big and heavy solids, the garbage is loaded into the double side tanks through the funnels. For the first 2 weeks one tank is loaded and sealed tightly to prevent gas leak.
7. During every loading a little gas is bound to leak out. Metal flaps hanging at the end of the funnel inside the tanks prevent the lead to a considerable extent.
8. However, as gas always rises up, most of the gas will be inside the cone on top of the tank.
9. Since, the side tanks are connected to the middle tank with rubber tubes the gas from the double sides cones will enter the middle cone, which is much larger in size.
10. In effect, the gas produced from the garbage will occupy the 3 cones, reducing changes of leakage during the feeding process.

11. Along with the city or road side garbage, hotel, hostel and hospital waste and fresh water hyacinth leaves and garden sweepings should also be fed into the double sided tanks.

> Gasification technology is particularly suited for remote areas which cannot be provided with power through grid transmission lines and fossil fuels cannot reach these places owing to transportation problems.

12. Principles of Biogas Production: Biogas or methane is produced when organic matter is decomposed by bacteria in an anaerobic environment. The bacteria survive and multiply only in an airless condition. In a Bio Gas plant complex molecules are broken down step by step into methane and carbon dioxide through the simultaneous action of different kinds of bacteria. The decomposition process can be divided into two broad categories-acid formation and methane formation.

During the acid formation stage, short chain fatty acids such as acetic acid and formic acid are generated. In the second stage these acids are converted into methane and carbon dioxide.

The bacterial processes in the reactor are mainly influenced by the pH factor temperature. A temperature around 35 degrees Celsius is considered to be the optimum; below 10 degrees Celsius bacteria cease to function. Other factors which influence the methane production are carbon to nitrogen (C/N) ratio, retention time, loading rate, total solids concentration and mixing.

Yeast has the ability to break sugar into alcohol and carbon dioxide. When a raw potato is chopped or mashed and soaked in a cup or mug of coconut water (water found inside either a tender or a ripe coconut) for 4 days or longer, the matter turns into yeast and this yeast can be put into the biogas chamber (fed along with other matter). If a potato is not readily available, one or two slices of bread may be soaked in coconut water for about 4 days to get yeast. After every loading of garbage a bucket of coconut / potato yeast must be poured in.

If about 4 kilograms of mineral calcium (slaked lime or builder's china) is filled in a thin cloth bag and suspended on top or the refuse in the tanks, the calcium will absorb a good amount of carbon dioxide and more flammable gas will come out.

For the world at large progress has come to mean only one thing-rapid industrialisation. Thus, countries which coming out are clustered with factories, smoke coming out through small chimneys and thick blobs of smoke are considered to be developed and progressing. What a tragedy? *Author*

If instead of slaked lime, "quicklime (sea and river shells burnt and powdered) is used, the absorption of carbon dioxide will be more and much more flammable gas will come out.

13. The prime purpose of this gas is to boil sea water to make it evaporate and condense into safe drinking water, as explained elsewhere. So the plant must be located near the coastal belt or near a sea water tank. The amount of gas available for use can be seen from the bulge of the tubes, the more the gas the more is the bulge. In other words, the tubes act as a pressure gauge.
14. After the gas is drawn, the outlet hole on top or the middle cone must be closed tight.
15. The excess gas must be bottled or directly connected to domestic kitchens for cooking purposes.
16. After every cycle of gas loading / drawing, the residue must be taken out through the large screw opening provided at the bottom and sea water flushed through the funnel to clean the double sided tanks.
17. The residue may be collected in a pond adjoining the system and can be used as a fertiliser. It can also be compressed in a moulding machine and made into briquettes. The briquettes may be sold for a small price to slums and nearby villages.
18. Read an extract from "Indoor Air Pollution: Energy and Health for the Poor", World Bank ESMAP Newsletter, Issue No. 1, September 2000 and Issue No. 3, February 2001.

In Germany 65% of municipal waste landfill sites recover energy from landfill gas. In the UK there are over 40 such schemes, which recover the thermal energy equivalent of over 3,50,000 tonnes of coal per year. On a global scale, it has been reported that there are about 600 facilities that uses landfill gas for energy applications.

According to the Rio Women + 20 position statement 2.4 million people in developing countries still depend almost entirely on traditional biomass fuels, namely, wood, charcoal, dung and agricultural residues. It is mostly women who are tasked with collecting and managing these fuels.

Chapter - 61

Solar Reflector

(To be tilted manually)

A parabolic magnifying glass of about one foot in diameter when placed between the reflector and the boiler enhances the heat at least 20 times. The UN reports in its journal Insight (article in chapter 23) that in Saudi Arabia it enhances the heat by 40 times.

The hail on the 'L' bracket should be adjusted now and again so that the shade of the hail coincides with the hall.

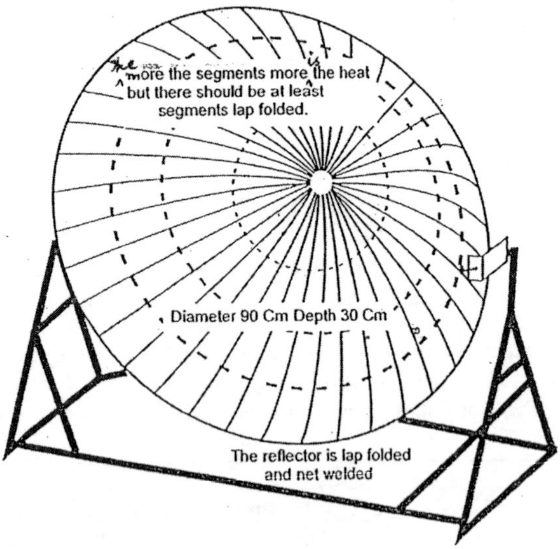

This parabolic reflector is very suited to focus the sun on a sea water boiler for steam distillation. Engineering drawing with complete specifications and adjustable mechanism may be had from the RYAN FOUNDATION (15. West Mada St. Srinagar Colony) for a payment of ₹ 1000 or US$ 100.

Chapter - 62

Home Made Reflector

To cover 15 focal areas from a distance of 2 m materials and dimensions for the construction of solar reflectors.

1. Size of the reflector is one metre diameter (parabolic) and it is of thin (24 gauge) stainless steel.
2. For tracking the sun the RYFO method is followed (diagrams enclosed-chapter 189).
3. Amount of water (output) is 400 litres per day—a continuous process it flows in as steam and flows out as water.

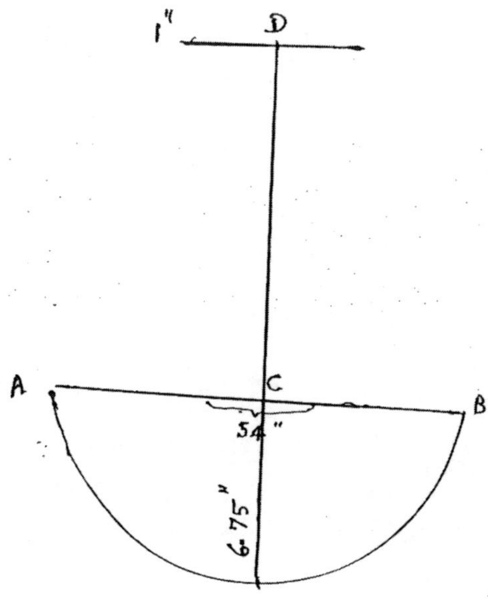

4. Size of the condensing drum – for 300 liters continuous process – flows in as steam and flows out as water.
5. Ambient temperature of 100 degrees Celsius gives 200 degrees Celsius heat (nearly double or more).
6. Distance between A and B one metre.
7. Distance between the middle of the parabolic curvature and the middle of line A-B is 6.75 centimetres.
8. Distance between E and D, where the rays converge, is 38.51 inches. This is the focal point.

Chapter - 63

The Large System at a Glance

(Design by Author)

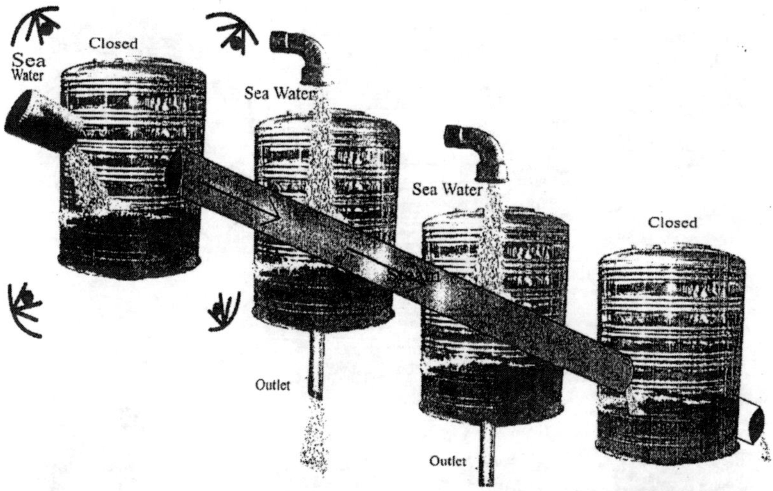

The annual cost of treating drinking water for pesticides in the UK is approximately 120 million British pounds...Each year there are about 250 million cases of water related diseases with roughly 5-10 million deaths. Source, Technology Partnership initiative (TPI NEWS), July 2008, Issue No. 58.

Chapter - 64

Setting the Reflectors in Focus

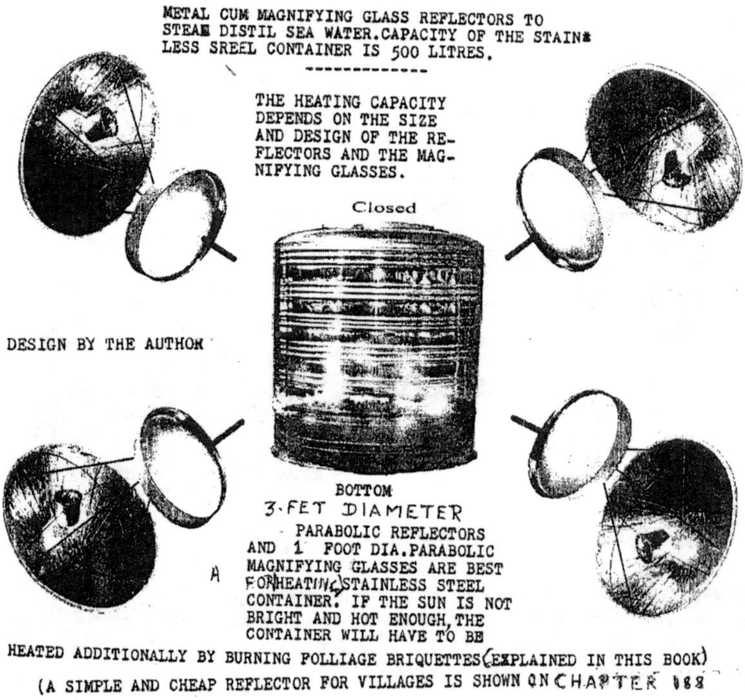

Chapter - 65

Automatically Moving Reflector to Track the Sun

Design by the Author

Parabolic solar reflectors are of different sizes and shapes. The reflectors recommended here is diagrams one and two are 90 centimetres diameter and 45 centimetres deep.

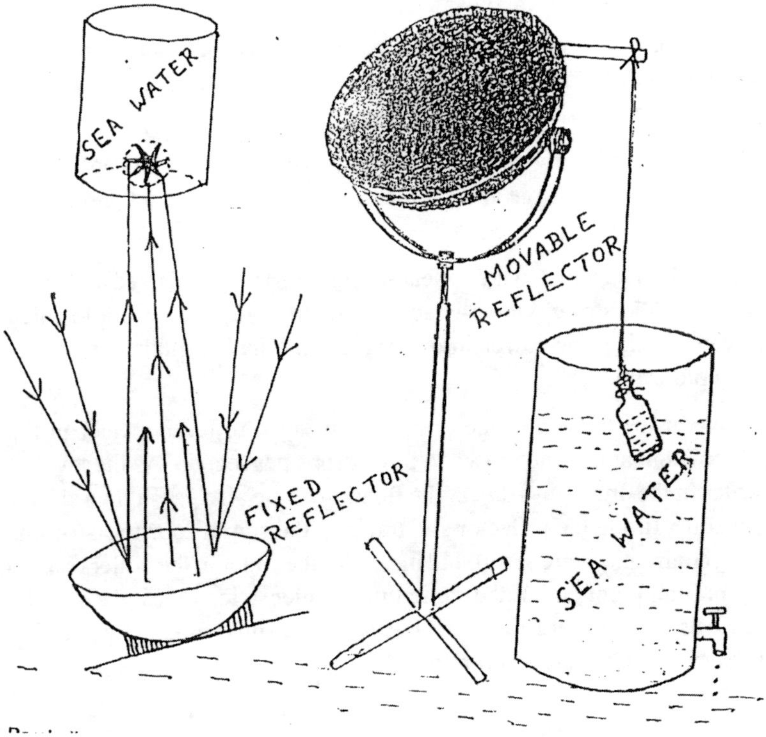

Diagram one illustrates a fixed reflector 90 centimetres away from the water tank focused to reflect the rays on the water tank.

Diagram 2 is a movable reflector fixed in a stand as shown. An arm or a hook is fixed to it and a string like a shoelace is tied to it. On the other end of the string a beer bottle is tied with just enough water to make it float in a drum of water as shown. The drum has a tap which is kept partially open to drop the water at the rate of one drop per second. As the water level in the drum goes down, the bottle also goes down and slowly and steadily tilts the reflector to track the sun to reflect the rays of the sun directly and centrally on the water tank.

In between the reflectors and the tank parabolic magnifying glasses are fixed to magnify and concentrate the heat on the water tank.

The available solar heat is more than doubled by this system.

Tracking the sun – explained

The sun rises in the east and sets in the west. As the sun's rays keep moving, the reflector has to be adjusted once in 5 or 10 minutes to keep the reflection directly on the stainless steel tank containing sea water.

There are several complicated and expensive ways of tilting the reflector slowly but the contraption shown in the diagram based on appropriate technology is easy to set and costs little.

The reflector is only about 15 centimetres deep (not very deep) so that the reflected rays converge about a metre or more away from the reflector and fall on the tank.

In the limb attached to the reflector (diagram) a string is tied and at the other end of the string, which is about 1 metre long, a beer bottle filled with water is tied and lower into a big drum which contains sea water almost upto brim.

A tap at the bottom of the drum is kept open to let the water in the drum come out at the rate of two to three drops per second. As the level of the water in the drum goes down, the floating bottle also goes down slowly and steadily, tilting the reflector and tracking the sun. If four to six of this tracking contraptions are set around the boiler the water in the boiler reaches nearly boiling point and at the same time condensation takes place in the tank safe drinking water flows out. Only when the sun is mild or the weather is cloudy is the light diesel oil burner USED.

The reflector may be made of stainless steel or aluminium plates. Each reflector should be made of at least eight segments and lap folded to get them into shape. A finished reflector will be like a large sized opened umbrella without the handle and the ribs. Very thin guage plates are to be used. Those made out of stainless steel are better than those made out of aluminium. The outside of the reflectors is painted black to augment the heat it receives but the inside must be clean and shining.

The stand may be made of mild steel or wood. For the drum a ready-made tar (asphalt) drum or oil drum may be used.

The amount of water to be put inside the beer bottle may be determined by trial and error, observations based on the floating bottle and the rate at which it goes in.

Large size parabolic magnifying glasses, 1 metre in diameter may be placed between the reflector and the water tank to enhance the heat of the sun. The position of the glass must be adjusted by trial and error during the first setting.

Chapter - 66

The Ideal Soil Binder

Vetiver (Vetiveria zianioides) is an ideal soil binder. It is also good for moisture conservation. Over the years many trees, plants, shrubs, herbs and grass have been tried as soil binders but vetiver has proved to be the best for the following reasons.

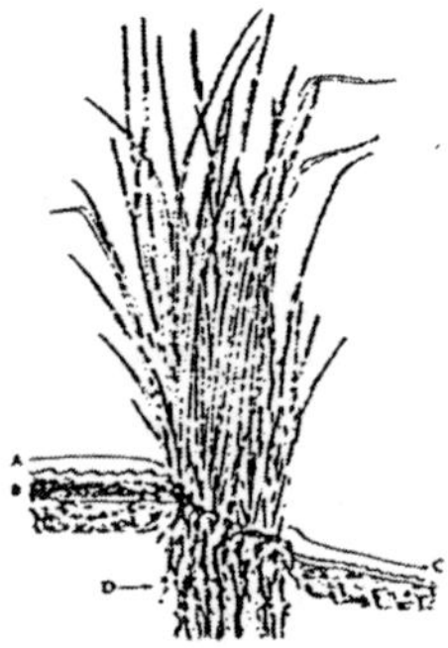

Figure 1. Cross-section of a Vetiver Hedge.

1. It is perennial and requires very little care and maintenance.
2. Its sharp aromatic leaves and roots repel rodents, pests and snakes.
3. It forms a permanent and dense hedge soon after planting.
4. As it's grown below the ground it is protected from overgrazing and fire.
5. Its strong, fibrous roots penetrate the soil to a depth of about 3 metre.
6. It is practically sterile and does not produce rhizomes and stolons and therefore will not become a weed.

7. The roots and leaves of Vetiver are resistant to most diseases.
8. It is said to have associated nitrogen fixing mycorrhiza, which explains its green growth round the year.
9. It is easy to remove when not wanted.
10. It can withstand long periods of drought or waterlogging as it is both a xerophyte and a hydrophyte.
11. It has no negative effect on the yield of neighbouring food crops.
12. It grows in areas with average annual rainfall between 200 and 600 milimetres.
13. It grows in all types of soils, regardless of pH, fertility or salinity and grows well on the seashore.
14. It grows also in soils which have shales, gravel and aluminium toxicity.

Vetriver is an ideal soil binder and water purifier. It also gives a sweet smell to the water in which it is soaked. As it grows wild and spreads fast without care or watering, it should be planted all around sea water canals, tanks, channels, lakes and lagoons. It does not remove the salt from water but improves the quality considerably. It is an excellent soil binder.

Chapter - 67

Scheffler Solar Reflector Technical Specifications

The reflector is of about 3 metres diameter, mounted on an iron frame. It has an arrangement for tracking the sun which is more or less like the time clock mechanism. As the weight pulls down, the reflectors tilts about 60 spaces every minute.

The reflector can be used either for cooking food or for boiling and steaming water. When the day temperature is round about 100 degrees Celsius the reflector throws heat of 400 degrees Celsius. If more heat is required a large parabolic magnifying glass is placed between the reflector and the boiler.

Specifications

1.	Number of facets (Small mirrors or reflectors)	160
2.	Material for construction of mirrors	Polyester film
3.	Size of focal area	40 cm
4.	Focal length	Approximately (2700 mm.)
5.	Dimension	2.4 m x 3.8 m
6.	Area of reflector	7 sq. meter
7.	Tracking the sun	Automatic
8.	No. of people it can cook for	100 at a time
9.	Mode of cooking indoors or outdoors	
10.	Size of cooking pot	50-75 litres (double for boiling water)
11.	Price in India	Rs 1,500 ex-(works including installation)
12.	Transport cost	Actuals
13.	Taxes	As applicable

If ordered from the manufacture in Valsad, Gujarat, it can be installed for ₹ 100,000 in Chennai or anywhere in South India.

Environment News from the Netherlands (March 2000) reports that 2 billion people rely on the world's major rivers for the water they need for irrigation, industry and drinking. The yellow river in China which flows through the country's most important agricultural areas, dried up for 226 days in 1998 and brought about an environmental disaster. Though floods do occur, all over the world, rivers and lakes are shrinking and drying up. So no country can be optimistic about connecting and exploiting its rivers.

On an ordinary sunny day water in a large container reached boiling point, within 30 minutes as witnessed by a group of people. Smaller units can be designed and fabricated.

Chapter - 68

Scheffler Community Solar Cooker

Eco – centre ICNEER takes the pleasure in introducing the Scheffler Community Solar Cooker which can contribute significantly, to firewood and other cooking fuel saving.

This cooker has been designed by Mr Wolfgang Scheffler of Australia. It is being manufactured by M/s Industrial Manufacturing and Engineer Co. of Valsad and St. Xavier Technical School, Anleshwar. Already about 40 Community Solar Cookers based on this technology are in use in India

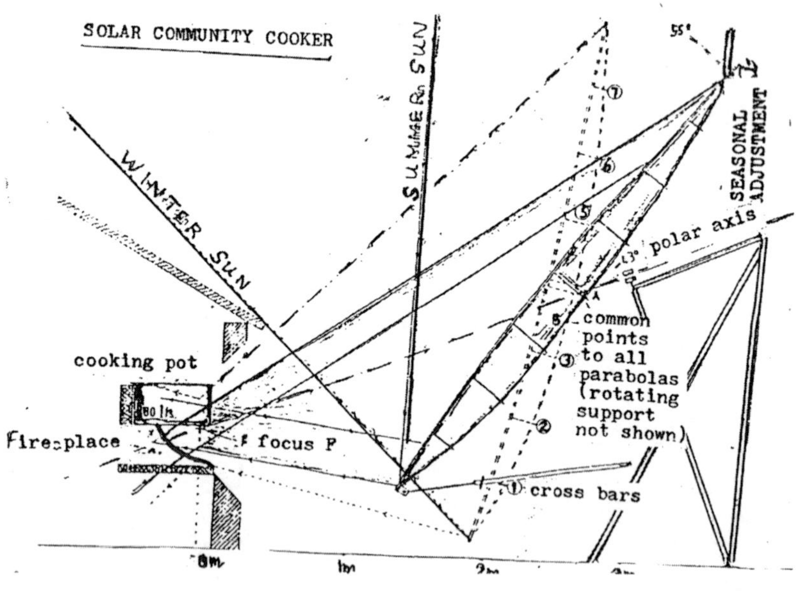

Complete optical design for a location 21.6 degrees North latitude, showing primary mirrors and solar rays. The secondary mirrors reflector for cooking is located inside the kitchen. A fireplace can be used as a back up heat sources.

and have proved its utility. Another 26 cookers would be installed by the end of 1996 at Brahma Kumaris spiritual center at Mount Abu. The cookers are manufacture indigenously, with practically no imported parts except for polyester films which is used as a reflector surface (mirrors).

Sardar Patel Renewable Energy Research Institute (SPRERI) of Vallabh Vidyanagar has recently tested such a cooker on behalf of Gujarat Energy Development Agency (GEDA) and recommended that such Community cookers are ideal for Mid Day Meal Programmes too.

Principle of working

The unique feature of the Scheffler Cooker is that it is possible to cook with Solar Energy but within in house itself. The 7 square meter large reflector standing outside reflects the solar rays into the kitchen through an opening in the wall while a secondary reflector further concentrates the rays on to the bottom of the pot or frying pan, which has been painted black. It is possible to attain a very high temperature of more than 400 degree Celsius.

Advantages of Scheffler Community Solar Cookers

1. *Possibility of cooking for large number of people:* Cooking for about 100 persons is possible with one of these cookers. Same is not the case with other solar cookers.
2. *Indoor Cookers:* Since the solar rays are directed into the kitchen, this enables cooking indoors. The cook does not have to outside in the sun to put the pots on the cooker.
3. *Fast Cooking:* Due to high temperature and power at focus the cooking rate is significantly higher compared with other solar cookers.
4. *Traditional food can be cooked:* Due to the high temperature it is possible to cook almost all traditional dishes including chappatis, poories, dosa as well as doing vaghar / tadka before adding the vegetables, dal or meat to the masala mixture. With box type cookers most of these traditional dishes are not possible.
5. *Automatic tracking of the sun:* There is a clockwork to track the sun automatically. The cook has to set the cooker in focus only once a day, then the rest of the time the clockwork keeps on rotating the reflectors automatically.
6. *Possibility of Seasonal adjustment:* By shifting the two arms of the reflector it is possible to change the curvature of the parabolic mirror for seasonal adjustment.

7. *Easy to maintain:* Since there are no glass parts and no insulation, it is easy to maintain.
8. *Multiple use:* During the period when the cooker is not in use for cooking, it can be used for hot water generation.

Technical Specifications

Number of facets (Small mirrors as reflectors)	: 160
Material for construction of mirrors	: Polyester film
Size of focal area	: 40 cm
Focal length	: Approximately 2700 mm
Dimension	: 2.4 m x 3.8 m
Area of Reflector	: 7 sq.m.
Tracking of the sun	: Automatic
Number of persons it can cook for	: 100 persons
Mode of cooking	: Indoors
Size of cooking pot	: 50 – 75 L
Price	: Rs 51,000 ex-works, (including installation)
Transportation cost	: As Actual
Taxes	: As applicable

There are three prices:

1. For NGOs getting no grant, ₹ 43,000
2. For NGOs getting grant, ₹ 51,000
3. For foreign funded NGOs and Industry, ₹ 62,000

> Sea water can be desalinated and distilled for human consumption and used for flushing toilets and cleaning roads, for saline agriculture, cultivating halophytes (Salt loving trees) for promoting sea fish, sea weeds and reeds and also house boats and floating markets. Further, to make dry areas humid, to make trees and plants flourish, birds to sing and bees to buzz.

Chapter - 69

Rope and Washer Pump

The rope and washer pump can be turned by hand, by animal power (cow, camel or donkey) or by the Ryan Foundation wicket gate arrangement by which for every turn of the gate one litre of water goes into the overhead tank.

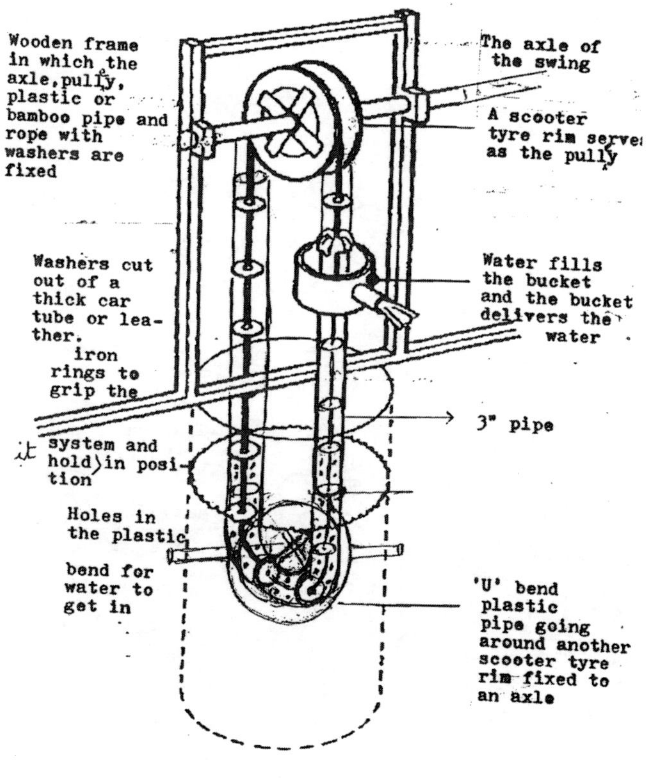

In a fine fiber rope of 5 centimetre thickness knots are made leaving 30 centimetres between two knots and two car tube washers are fixed on either side of the knots. It attached to a swing (diagram attached), when the swing goes backward half the tube will empty into the bucket and when it moves forward the other half will fill the bucket.

Chapter - 70

Sea Water in Open Wells

A billion cubic feet of sea water pumped into this or any well will join the ground water stream and the water in the well will not rise above sea level.

Invited by the Government of India and the Government of Gujarat, the author is seen with a team of officials making arrangements to connect an 80 feet deep dry well in Valsad district near the coastline to the sea, with concreate pipes to get the sea water into the well to raise the ground water table. He explained to the water managers that sea water being heavier than ground water, it will go deep down the earth and push the ground water up and therefore there is no question of salinating the soil or the existing ground water under normal conditions.

Chapter - 71

Gujarat: A Case Study

The Government of Gujarat is making optimistic announcements about the crop situation and that it would not be as bad as it was last year in view of the availability of water in most of the 10,000 or more check dams build as part of the drought relief work last year. Will the shallow check dams raise the water table which has gone down by nearly 2000 feet and still keeps going down ?

Experts are totally convinced that the check dams build last year about which much publicity is being given now, would at best provide some drinking water to people living near the check dams for a month or two as after the monsoon the dams will dry out.

This year, in spite of the much talked about check dams, water is still being supplied by tankers and people are queuing as for a pitcher of water. The situation underlines that the check dams are a short duration surface solution to limited community in a limited area.

A report in India Today (September 11, 2000) gives a rosy picture, hopes and promises of the water situation in Gujarat. The report says that the land that was reeling under a sever drought just two months back is beginning to see a Green Revolution due to more than 1,000 check dams built by the people.

That in a span of just five months the people of more than 2,000 villages in the six districts that comprise Saurashtra have built over 10,000 check dams is difficult to understand and one may doubt and question these statistics. If true, the dams will be merely name sake check dams meant to provide pro-government statistics, to block the seasonal streamlets and rivulets and slow down their downward speed and supply.

What is again questionable is the report that the water blocked by such small check dams, even though the rainfall was less than 15 centimetres percolated into the ground and lifted the water table, which had fallen very low in most parts of the state.

Basically, Gujarat has partly a dry soil and partly a non-absorbing rocky or dome like soil without water holding capacity. Usually the rain water of the first two or three showers is absorbed by the top soil and if the next day is sunny, gets evaporated. Generally speaking 25 percent of surface water evaporates and yet another 25 percent seeps into the soil. What then is left to recharge the deeply fallen ground water table?

Chapter - 72

Soak Pits Store Water

Ryan Foundation recommends two easy ways of tapping and storing run away rain water. If you have an open well which usually dries out in summer, lead the rain water from the roof top into the well. Cement or PVC pipes may be used or a simple open cut channel may be dug for the collecting water to fall into the well.

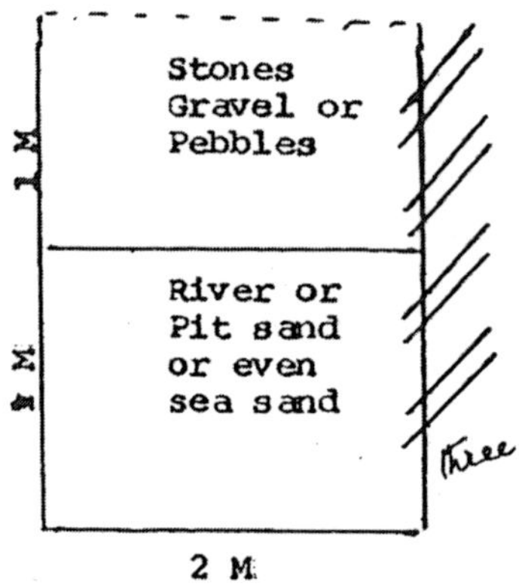

If you do not have an open well, dig a soak pit 2 m x 2 m x 2 m. Fill the lower half of the pit with river sand, pit sand or sea sand or all three mixed. On top of it fill small stones gravel or pebbles or a mixture of the three ram the top to ground level.

If the place you select for the scooters must run over it, then the top must be rammed well with big stones. But if it is secluded, it need not be rammed hard.

If you have enough open space around your house have as many pits as possible; you even have them dug in your car shed, chicken shed or porch, allowing the collecting rain water to enter the pits through pipes provided on the sides of the pit at an angle of 45 degrees. The broad angular pipes may be filled and closed with mud as water will still enter them and go into the pit. In rural areas, where people cannot afford cement or plastic pipes, bamboo pipes may be used. Alternatively, about six or seven asuerina or eucalyptus poles level 1 metre long and about six centimetres in diametres it may be bundled together and buried in the same position as one will place the pipes. Water enters the gaps between the poles and falls into the pit. The pole bundle arrangement is as effective as a pipe and this method has been as effective as a pipe and this method has been successfully introduced in several African countries by the Ryan Foundation.

If a dozen of these soak pits are made all around a well it is most likely that the well will receive and yield water all through the year, even during drought.

Ensure that all the ground around your well that belongs to you is made to slope towards the well so that when it rains the running water automatically flows towards and into the well. Keep your well always closed to prevent evaporation but provide small holes on two sides for air circulation. In warm countries where the temperature averages around 35 degrees celsius in summer about 25 percent of water evaporates from open wells. By keeping the well closed evaporation can be minimised to about 5 percent.

Don't pave the ground around the well and don't grow big trees near the well as they will absorb some water. Specifically, don't have eucalypts tree near the well. Good water management is integral part of family life.

Where there is no Good and Adequate Ground Water, Sea Water must be Led into these Soak Pits.

Chapter - 73

Water Purifying System

Filter the water you drink through a cloth and then boil it. If you don't, you will drink the soup of water mixed with heavy metals. Dissolved salts, toxic and human waste, garbage juice, germs worms, microbes, spores, pathogens, otra, arsonic and so on all of which make a cocktail in one tumbler or bottle.

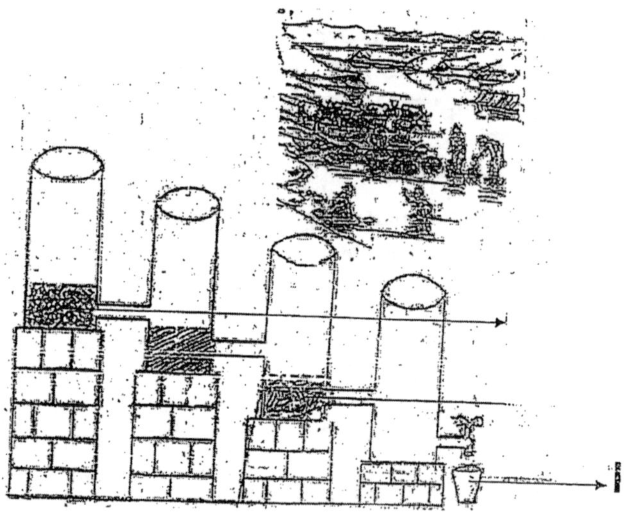

Drumstick seeds can clarify water up to 99.5%.

Flocculation is the first step in water purification and the plant's seeds provide an alternative to the commonly used Alum salts and synthetic polymers. These chemicals harm both the environment and health. Alum salts have been linked to Alzheimer's disease.

Drumstick seed powder was used by ancient Romans, Greeks, Egyptians as a flocculent in purifying water. Recent research proves that banana skin absorbs heavy metals and lime skin absorbs dissolved salts.

Charcoal and ash of prosopis juliflora Tamil-Vellikathan (thorny bush).

Stalk bits of water hyacinth and drumstick (Murunga) seeds whole or dried and powered.

Potable water a teaspoon of jeera (Jeera Gam) and dried ginger powder is added to it every day after cleaning the receiver.

Chapter - 74

Ryfo Water Purifying System

(Cost estimate in India - ₹ 3000)

Water supplied in the city of Chennai whether it is Metro water or ground water, is highly contaminated and sometimes oily and pungent. Frequently, different types of worms, larvae and suspended particles are seen floating in the water making it unfit even for bathing or washing clothes, not to speak of using the water for drinking and cooking. Heavy metals are found beyond the approved limit. Of late, ARSENIC (A slow killer) has been found in several locations.

Ryfo system of purifying such water is based on traditional technology combined scientific knowledge. The water drawn from this system is good enough for bathing and washing clothes. Where drinking water is very scarce and drought conditions prevail causing dehydrating and increasing water borne diseases, the water drawn may be filtered through a cloth and then boiled for human consumption. Even sea water can be made suitable for human consumption.

The Ryfo system can be built on house tops (open terrace), in the balcony or down in the garden. It may be built out of brick and mortar, or empty oil and tar (bitumen) drums or large plastic containers may be used.

Purifying materials used are as follows. In Tank:

1. Black rock stone (query waste) and activated carbon or charcoal or coal and sieved wood ash; in tank.

2. Sieved and washed river sand along with quick or slacked lime, that is ordinary mineral calcium called CHUNAM, plus jaggery (half a kilogram or sugar quarter kilogram) and in tank.

3. Drumsticks and twigs of the same tree, crushed, plus gooseberry twigs crushed along with neem leaves and twigs and papaya leaves or water hyacinth stalks, crushed are used.

The materials put in should fill only a third of each drum, leaving two-thirds space to hold water.

The mounted height of Drum 1 is 2 feet, Drum 2 is 1.5 feet, Drum 3 is 1 feet and the fourth drum collector is half a foot from drum base level. All the four drums or tanks must be closed with transparent plastic sheet as direct sunlight kills several harmful bacteria.

All that is put into the three tanks or drums, including sugar or jaggery, may be changed (renewed) once in 3 months.

The drums or tanks need not be connected by a plumber. Only plastic nozzles need to be fixed in the drums. The nozzles may be connected by a 30 or 40 centimetre long cycle tube and tied properly to prevent leakage. Cycle tube connection not only makes the setting easy and less expensive but also facilitates the separating and cleaning of the tanks or drums once in 3 months with bleaching powder.

If drums are used, they may be painted black outside and inside.

Aid from the rich countries for the poor countries has fallen significantly since 1992. The situation is particularly alarming in Africa. Over the past 3 years, the total World Bank/ International Development Association (IDA) funding for agriculture and rural development for sub-Saharan Africa has amounted to less than US$ 250 million per year, or less than US$ 1.40 per undernourished person in the region – a mega drop in the ocean of needs.

Delhi will run out of groundwater within the next 14 years if immediate action is not taken to combat the situation, states a study during 2000 by Tearfund, an international aid agency.

Chapter - 75

The Poor Should Get Water Free

Human Development Report 2006

A Global Action Plan under G8 leadership is urgently needed to resolve a growing water and sanitation crisis that causes nearly two million child deaths every year, says the 2006 Human Development Report, released on 9 November 2006, Across much of the developing world, unclean water is an immeasurably greater threat to human security than violent conflict, according to the Report, entitled Beyond Scarcity, Power, Poverty and The Global Water Crisis.

Each year, the authors report, 1.8 million children die from diarrhea that could be prevented with a glass of clean water and a toilet. 443 million school days are lost to water related illness, and almost 50 percent of all people in developing countries are suffering at any given time from a health problem caused by a lack of water and sanitation. To add to these human costs, the crisis in water and sanitation holds back economic growth, with sub-Saharan Africa losing five percent of GDP annually, far more than the region receives in aid.

The report adds; make water a human right – and mean it. Every one should have at least 20 litres of clean water per day and the poor should get it for free.

Cherrapunji in Assam, where it used to rain all 24 hours a day and all the 365 days in a year is now drought struck and people are purchasing water. This is a case in point to show that erratic climatic changes are affecting the globe and depleting the water table.

Fall of the water table is a global phenomen which takes place not only because of inadequate rainfall but because of so many other factors as well, including global warming, the green house effect, deforestation and increasing demand of water by people, farms and factories.

Chapter - 76

Drinking Water From Air

If this plant is installed near a water body, fresh water or sea water canal, due to the water borne atmosphere the yield of fresh water will be much more. The machine will be more productive.

Srinivas K. Reddy reported some years back that in Jalimudi Village in East Godavari district of Andhra Pradesh, India, water is obtained from atmospheric air. There trial runs were able to produce about 1000 litres of potable water every day and several tests showed that it was fit for human consumption. This watermaker is manufactured by Water Maker (India) under the direction of Meher Bhandara.

The news item further reported that the unit works on electricity and follows refrigeration technology to condense water from air. Then air is directed by a blower, passed through filters and a refrigerant and circulated leading to condensation of water into a net receiving tank. Naturally, the productivity of the unit will depend on the humidity and temperature around the unit, 70 to 75 degrees humidity and a temperature between 25 and 35 degrees Celsius is said to yield good results. The higher the humidity and temperature the more the yield of water.

The cost of the machine is ₹ 3 lakhs. All this information was reported in the Hindu, with the photo and is explained here with permission Srinivas Reddy and the Hindu.

From what is reported is looks as though the machine will be quite productive in warm countries with a sea coast or on the banks of sea water canals.

The report does not mention about the working and maintenance cost and if it can be worked by a diesel engine or by solar power ₹ 3 lakhs is too much money for the purpose. However, if technically and economically feasible it can be installed in tanks, colleges, clubs, hostels, hospitals and large business undertakings. If retailed to domestic consumers in plastic bottles and sachets, government tax on ink, paper, labelling and transportation cost will make the price prohibitive for the end user.

Nevertheless, the machine can be better than the Reverse Osmosis (RO) machine from the cost and wastage point of view.

Chapter - 77

Recommendation

1. Further extraction of groundwater should be discouraged.
2. Use of metallic hand pumps and wind mills connected to metallic borewells should not be promoted.
3. The open system of providing water and lifting water should be encouraged all over the world, specially in developing countries.
4. Sea water must be exploited as far as possible and desalinated using solar energy.
5. Brackish water, where available, should also be desalinated in solar stills. Villagers may use the Ryfo Pit still (sea chapter 104).
6. National Governments should take sea water to nearby coastal villages with sluice and lock and filter arrangements and regulate the flow of sea into the land.
7. People should help themselves to dig narrow open canals and lead the water from one village to another.
8. Safety outlets may be provided by the people in every storage tank for surplus water to over flow into dry wastelands.
9. NGOs should help to build village community overhead tanks and windmills with canvas bucket lifters.
10. National and International Governments should build saucer lifters and install mini generators.
11. They should encourage multinationals to promote and maintain these system as a service to society.
12. Ryan Foundation solar still must be popularised and every home should have one still which may be made of bamboo, transparent plastic sheets and basins or trays made of clay or wood as far as possible. The pit model explained in page is excellent for villages.

13. More and more plant trees and weeds that grow in sea water and saline soil must be identified and promoted to generate livelihood activities for the rural poor.
14. Fish and fish based industries such as fish meal, fish gum, fish oil, dried fish, chicken and cattle feed and fertilisers etc., must be promoted. When plenty of inedible fish is used as a fertilisers soil salinity is reduced and fertility increased.
15. Inland water ways transport arrangements for quick and cheap transport of fish and other rural produce must be promoted.
16. Governments must spend on water and not on warfare.
17. Governments should promote and not provide water and sanitation facilities.
18. Rain water must be harvested in every home.

Development is About Building Human Capabilities

Chapter - 78

Where There is No Fuel

(For cooking and distilling sea water

Make

Your pellets or briquettes)

Introduction

No doubt, land, labour, capital and management are the basic factors of production but water, fuel, transport and communication are also important, if not equally important. The world at large is suffering from shortage of safe water and cheap fuel. While the sea and the oceans may be exploited for water, by desalination or distillation, alternatives have to be found for cheap and easily available fuel which the poor can afford. (UN report on India date 18.05.2013 laments that 230 million go hungry daily in India. The most feasible alternatives are solid pellets and briquettes compressed out of all sorts of foliage, sawdust, coconut waste, paper waste, and so on without polluting the atmosphere and damaging the environment. This chapter explains how to do it.

Steam Distillation cannot be Done only With Direct Solar Heat. Parabolic Reflectors, Magnifying Glasses, Foliage Briquettes and Pellets, Fish Oil and Inedible Oil From Pods of Plants and Trees are Required to Boil Water For Steam Distillation.

Process 1

Gather anything that burns when set fire to, such as foliage, coconut pith, fiber, waste paper, tailor's waste, water hyacinth (leaves, stalk and roots), acacia, sea weeds, reeds, wood shavings, sawdust, banana skin, chicken waste and so on dry them well to small particles in a "chekku" or a "gani" (See diagrams of common village grinders (in chapter nos 204 to 209) for a field test report of briquettes (see chapter 206).

For foliage, sweep and collect what is available near and around your settlement. Further, make arrangements with educational and other institutions having a large compound, which usually sweep, heap and burn the collection.

A pulveriser or a flour mill, a paper shredder, or a cow, or donkey-drawn grinding device may be used (Chapter 204). Store the ground matter in a dry condition.

Process 2

Take a galvanised iron (GI) pipe 3 feet long and 1 inch in diameter, oil the inside of the pipe with fish oil obtained by steam distilling dead fish or unwanted fish parts. (How to distill is explained in Chapter 202.) You may also use any other water oil. Then mix the ground matter with the fish gelatine (glue) as explained and make a paste. Fill the paste into the GI pipe and ram the ground matter. A metal rod may be used to compress it. Then dry the tube in the hot sun or place it behind pot (touching the pot) where something is cooking and leave it like that for or days to dry well. Thereafter, push the inside matter out of the tube and it will come out like a stick or in broken parts. Thus, you have made your pellets to burn with a flame, as cooking fuel, or to steam distill sea water. (Leaves of the tree Acacia holosericea produce high calories of heat (see chapter 210).

Process 3

Take over all the inedible fish from the fishermen in coastal areas; alternatively introduce tilapia in fresh water wells, ponds and lakes in your village. Tilapia is one of the fastest multiplying fish and within six months, without any care, you will find thousands of them. They are edible but are better for industrial purposes. (For breeding, see Chapter 212, for cage culture in community wells, (see Chapter 148) Put the dead fish in a village made pressure cooker explained (in chapter 247) and heat the drum. To close the drum, use a cut piece of car or lorry tube and tie the sheet fast to serve as a stopper with a tape around the drum, as shown in chapter 202. When you remove the rubber sheet, smoke, steam and smell will escape. Keep away from the drum when the smell and steam escape.

Then tilt the drum gently and separate the top layer of oil. Below the layer of oil you will find fish gelatins (glue) appearing like frozen ghee. Scoop out the glue, put it in a frying pan and turn it a while with a wooden ladle. What you churn will turn into fine quality glue. This is the glue to be mixed with the ground matter to make a paste, which is filled into the pipe.

Process 4

If fish is not available, use raw assava starch (Chapters 100 and 101) or use tamarind seed powder starch, as explained in chapter 240. In fact, any edible cereal, including rice, and non-edible pods or beans can be boiled to get sticky starch. Another way is to mix a cup of jaggery or sugar in water in a small vessel. The mixture gives good quality glue.

At the bottom of the drum, after removing the oil and the glue there will be a thick layer of fish residue, which is excellent as fish feed and feed for ducks and pigs. Some of it may be thrown into the water, where Tilapia multiplies. The residue makes a fine fertiliser too. (See chapter 245.)

Process 5

Chapter 291 explains a RYFO ram and Chapter 99 explains a block press. These handy or convenient machines ram and compress the foliage matter into the dye to make briquettes, which are broken, dried and used as fuel on a commercial scale, especially to boil sea water for steam distillation. The residual ash can also be used as a fertiliser for garden plants.

Process 6

For boiling the fish in the drum, one may use solar energy with reflectors (Chapter 186) and for tracking the sun, (see Chapter 189.) Waste oil burners are explained in (Chapter 176) where necessary magnifying glasses will have to be used for enhancing the heat of the sun. With powerful magnifying glasses, the heat can be raised to 300 to 400 degrees Celsius.

> A world managed by men has transformed the traditional food producers of Africa, namely women, into the sade victims of famin and shortages – Womens World-ISIS.
>
> We are now applying a measure of GNP which counts cutting down a tree as growth but gives no value to the standing trees. Economic growth is only a measure, people are the ultimate end. Growth must be woven around people, not people around growth.

> Almost four centuries ago, Franacis Bacon prophetically said, that "he will not apply new remedies must expect new evils." Bacon's timely wisdom should not be forgotten. We must move quickly, we must make sure that future generations will not pay the price for our unforgivable neglect and delays. We must act before poverty spreads further. Time is knocking at our door to revisit development thinking.

A new set of tools are required now because the old ones have just failed to work anymore. The wrong illusion that, sooner or later, the benefits of centralised development will flow down to the poor and the remote corners of the world should be abandoned forthwith. Instead, a model of development based on people's resourcefulness and ingenuity to achieve efficiency and social equity should be seriously followed and encouraged. Poverty cannot be eradicated without economic growth and economic growth cannot be brought about if millions are still enslaved in step poverty. We cannot afford to risk another decade of stagnation through inaction.

Note 1: Every other family in villages (why, not in cities too?) must make briquettes for its own use and for the market. Briquettes and pellets must be made hard enough to be packed in used cement bags for the market and transported on donkeys or hand-drawn carts.

Water supply and sanitation problems are no longer technical in nature. They are pathetically political by design. This is exactly what Hans Alderrs, Minister of Housing, Social Planning and Environment in the Netherlands, has been focusing attention upon. Alder's statement is the bare truth and the whole truth. When politics becomes "poly-tricks" it cannot be otherwise.

The Simple Truth is that Third World Governments have Neither the will Nor the Ability and Resources to Tackle the Devastating Degradation of Land, Water and Forest Resources.

Chapter - 79

Facts About Water

1. In parts of Chile there has not been a drop of rain for four centuries. Source Severn Trent Water Ltd. Information Service (quoted by Tamil Nadu Pollution Control Board, July 1991).
2. Water shortages have combined with soil erosion and desertification to degrade farm and grazing lands and reduce food production.
3. Instead of digging deeper and deeper and sucking the earth to leave it dry and sick and then declaring that potable water is getting more and more scarce governments should take sea water into the land by open cut man made canals and also connect them to dry rivers to help people, the rural poor in particular, to help themselves with adequate and safe water without making them depend on the government pipelines and supply system which always expensive and troublesome.
4. According to the World Bank and the UNEP, 67 percent of Africa's non-desert lands, 34 percent of Asia's non-desert lands and 20 percent of Latin America's non-desert lands may turn into non-productive desert in the near future. As many as Six million hectares world wide become desert annually and an additional 21 million lose their productive potential each year.
5. Americans draw about 400 billion gallons of water each day for residential, industrial and agricultural purposes. This figure translates into 1600 gallons for each person each day. Prolifigate use of water in beginning to show in many parts of the country. Henry La Roae of the US Geological Survey, predicts, that the supply of underground drinking water in southwest Florida will be completely exhausted in 10 years (Zero population Growth November 1991).

6. Health depends on water and sustainable development depends on health. Goods and services cannot be produced by families, farms and factories without water. The national income, standard of living and export earnings of any country are bound to fall if there isn't adequate usable water.

There is Still no International water law in Force.

7. According to UN figures, 20 million more Africans lacked safe water, and 30 millions more Africans lacked adequate sanitation at the end of International Drinking Water Supply and Sanitation Decade than at the beginning!

Africa recovery (UN) September 1991.

8. One-tenth of China's territory is below sea level and population, agriculture and industries are concentrated in these areas.
9. While Global Water recharge has remained almost constant over the centuries withdrawals are believed to have grown more than 35 fold during the past three centuries and are projected to increase by an additional 30 to 35 by 2000.
10. The degree of acidity is expressed in a scale of numbers known as the pH value (pH stands for potential of hydrogen) and is a measure of the activity of hydrogen. Pure water with pH 7 is neutral. pH ranges from 1 to 14 the lower the pH the higher the acidity.
11. Looking back at the International Drinking Water Supply and Sanitation Decade (*Idwssd*) we are all disappointed that the primary goal of providing safe drinking water for all by 1990 was not achieved. Over one billion people remain without access to good quality potable water while nearly two billion people have no sanitary means of disposing of human waste. G. Arthur Brown, Chairman of Steering Committee for Cooperative Action for the International Drinking Water Supply and Sanitation Decade in UNDP Source, December 1991.
12. Dams have displaced eleven million people in India. Of these, 8.5 million people have still not been resettled and remain landless. Narmada Dam scheme in India is already 8 years behind schedule. It may be made economically unviable by the build up of silt even before it is finished.

    - Oxfam News Summer, 1992.

13. Cherrapunji in Maghalaya (India), where it used to rain all 24 hours of the day and all 365 days of the year, is now totally drought ridden and the people are walking several miles to fetch a pot of unhealthy water.
14. Water, pollution and scarcities are not god made problems but problems made by politicians and bureaucrats for their advantage.
15. Development of water resources is the surest and shortest path towards poverty alleviation and improvement of the quality of life of the people.
16. The proposition that the non-monsoon flow of rivers can be increased appreciably by afforestation is not true either scientifically or on the basis of available flow data.
17. Moreover, pumping of ground water requires energy, which is woefully short. Many tube wells can operate for no more than Two to Four hours a day for want of electricity.
18. The cooum orginates at Kavery Pakkam in North Arcot district and was designed by the British to carry the surplus waters of the local tank at the Kesavaram Anicut (Dam). The canal branches out into two, one to link the Poondi reservoir and the other as the extension of the cooum. Within the city limits of Madras (Chennai) the cooum stretches 17.6 kilometres.
19. Improving basic water and sanitation services is an important first step in alleviating poverty, protecting the environment and promoting human development.
20. Nearly 70 percent of the world's population lives in or within 80 kilometres of a sea coast says the report. "The World Bank and the Environment Fiscal 1992."
21. A study conducted by the Indian Market Research Bureau for the UNICEF and published by the National Drinking Water Mission of the government of India in 1990 indicates that faced with a choice between a hand pump and an open well, people often opt for the well water for drinking and cooking.
22. In India, the tug of war between the states and the center states over water sources and supply highlights political manoeuvring and characterise water management policies which have not produced any good result for the past 2 decades. On the contrary, due to bad water management in several areas in the city of Madras (Chennai) such as Minjur and Mylapore saline water has intruded and made once potable water of the area now unpotable

and in some places the water table has fallen from to feet and in other places the surface land is visibly sinking. The problem of sea water intrusion was observed by some UNDP agencies and the Metro water board as early as 1984. It is reported that in the Mylapore and Minjur areas sea water intruded into the coast by some 6 kilometres in 1983, 5 kilometres in 1984 and 9 kilometres in 1987. Perhaps it reached 15 kilometres in 1992 and this is the evil of over extraction of ground water. But our planners and politicians seem to go by the Keynes Philosophy that we will all be dead in the long run and therefore such problems should not worry us.

23. While planning for water we have to balance what is technically sound, environmentally friendly, legally permissible, economically viable, socially acceptable and politically possible.
24. In Titlagarh, which is reeling under acute water crisis, the water table plummets to 330 feet during the summer months when the mercury touches 50 °C. Despite the Tel river at a distance of 8 kilometres there is no piped water supply here and water is being sold to the poor at ₹ 2 to 4 a pitcher.
25. According to a report which was published in the Hindu on 16.5.1993 in Odisha, of the 2840 villages in Kalahandi, 754 habitations are waterless. In Bolangir, of the 2,741 villages, 311 are waterless habitations and in Koraput, out of 6,294 villages, 4,056 habitations are waterless. More than 800 drinking water wells in these districts have dried up. About 3,300 tubewells in these districts have become defunct and 1750 tubewells yield water with high iron content. Besides, 30 percent of the tubewells in the area are not functioning.

Chapter - 80

Where There is No Pressure Cooker

(makes fish glue to bind foliage briquettes)

When something is to be steam cooked under pressure and a pressure cooker is not available with the rural poor, two vessels and a rubber sheet may be used, as shown in the diagram. A small vessel is placed inside a big vessel on top of a small pillar (stand) as shown in the diagram. Two or three small solid building blocks or one big hollow block or an old tin may be used as a pillar stand. The big vessel must be at least double the size of the small vessel.

The material to be steam cooked is put inside the small vessel and the small vessel is not closed. Water is poured into the big vessel up to three-fourths the height of the small vessel. Since the small vessel is loaded with the cooking material, it will not lift due to buoyancy.

An old lorry tube (automobile tyre tube) is cut open to a size of half a metre by half a metre and this rubber sheet is used to close the big vessel and tied air tight. Two to three pieces of rubber may be joined if necessary.

Tilapia Fish Multiplies Fast in Fresh Water and Milk Fish Multiplies Fast in Sea Water. All Parts of Any Fish May Be used in This Cooker to Get Oil, Glue and Fertiliser.

Chapter - 81

Glue Out of Fish Waste

Take all unwanted parts of edible fish and the whole of edible fish. If you included salted or dried fish, wash and remove the salt first. Keep the collected parts, soaked in water at least for a whole day and wash them at least four times with drinking water. Then chop and mince the contents and wash again in potable water twice or thrice.

Transfer contents to a wooden board, chop and mince them including the bones, and put the mixture into a suitable vessel, don't put a lid on the vessel so that the unpleasant smell escapes while boiling. For boiling, place the vessel with the contents inside a simple double walled drum (take a big drum, place a brick in the middle of the bottom of it, put a smaller drum on the brick and fill water between the walls of the two drums and place the vessel in the middle of the small drum) close the big drum with a lid and keep over a fire it and the steam will circulate and cook the contents. For diagram of a double walled drum (see Handout 287.) Once or twice remove the lid and stir the contents with a ladle and cook until the contents becomes Jelly like. Allow the cooked matter to settle down and then skim and remove the top layer of oil. Filter and separate the liquid in the middle, that is between the oily top skimmed and removed and the cooked matter below. Filter this liquid in a plastic or wire mesh or thin cloth (bandage cloth or mosquito curtain cloth may be used). As you heat this liquid outside the drum on a low fire it will thicken and give you a fine glue. Keep on heating to get the desired consistency.

The solid portion left after the first filtering may be boiled again inside the drum for a second filtering and still again for a third filtering. The final residue makes an excellent fertiliser in both wet and dry form and it is odourless.

The liquid glue thus prepared will remain in liquid form always but must be kept in a closed container. It has a permanent usage life. On wood this glue penetrates deep and dries slowly and if made into a thick paste it is very strong but not water resistant. It has pale yellow colour and no odour. As it does not solidify when unexposed to air it can be packed, stored and marketed in cheap plastic sachets (like the milk sachets).

The Ryan Foundation survival by Sea Water movement wants women in coastal fishing family settlements to prepare this common purpose glue and market it for consumer and industrial use. The glue is excellent for binding foliage briquettes for domestic cooking and sun dried bricks for low cost housing.

> This book moves countries of the world to innovate using locally sourced resources with a focus on job creation, building social capital and generating multiple cash flows by stimulating entrepreneurship and business model innovation.

Chapter - 82

Home Made Briquettes

The poor in remote villages don't get cooking gas, electricity or kerosene and even if they do they cannot afford them to boil water every day to get safe distilled water. Ryan Foundation solution is to use foliage briquettes made by each family in their own huts or outside.

All sorts of dried leaves fallen from plants and trees, shrubs and weeds, coconut tree droppings, husk and fonds etc., including water hyacinth and rail cactil are collected, dried, pound and granule. The granules are mixed with country glue or ragi or any edible grain starch boiled to liquid state and filled into a mould. The handy mould available in most village homes called in Tamil, Murruku Acchi may be oiled with cheap castor oil readily available in villages, pressed to harden and left through the night. When the mould is opened the next morning the compressed briquette will fall out ready to be used as cooking fuel. They may also be rammed inside a metal pipe. The best binder is fish glue. When non-edible fish and unwanted fish parts are boiled, oil floats on top. Below the oil is good quality glue and at the bottom there is fish feed, pig meal and fertilisers. (for diagrams and detailed explanation contact Ryan Foundation). Only actuals are charged. Many briquettes may be made in advance and stored to dry for ready use.

The method explained above can be made feasible only if sea water is made available in every village.

The Poor will Remain Poor If They Depend on the Technology of the Rich.

No Metal Parts, For Using Donkeys (See Chapter No. 41 Buffalo's May Also Be Used. The Residue May Be Given To The Animals as Feed.

Chapter - 83

Donkey driven Wooden Grinder

1. ABCD is one block of hard wood and 'V' shaped PHQ is scooped out of this block.
2. Conical EFG is cut out from another block.

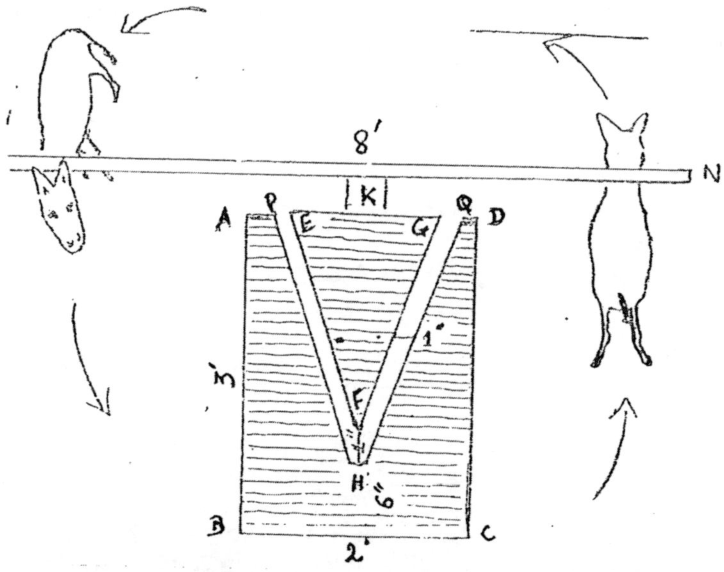

3. K is a solid piece of block 4 inch x 4 inch fixed to the centre of EG with four L brackets.
4. MN can be either a wooden post or a bamboo pole and the top side of K is fixed to this long arm as shown.
5. At H is a metal cup buried and screwed and FH will be like a pivot to meet point F. The bold should just fit into the cup at H. Thus, F – H will form a bearing which may be greased now and again.

6. The gap between PHQ and EFG should be uniformly about one inch.
7. The material to be ground may be filled into the cavity between FE and QG.
8. The two arms MK and NK may be turned by two donkeys or cows and the material inside the 'V' cavity will be ground.
9. BC may be buried in the ground to a required depth to make MN sit on the donkeys. If horses are used to turn the grinder, BC may have to be raised on a platform.
10. To remove the ground material from the grinder two people may lift the arms at M and N and take the grinder EFG out of PHQ.
11. The surface PH, QH, EF, and GF may be made rough with a chisel if edibles are to be ground and one inch nails driven fully into them on the grinding surface if non-edible are to be ground.

> It is erroneous to believe, as the US does that the environment friendly technologies are contrary to the competitive spirit of business development. Gro Bruntland, Prime Minister of Norway at the Earth Summit at Reo-de-Janeiro.

Chapter - 84

Hand Made Briquettes

(out of dried leaves and starch as glue

Diagram one is the traditional Indian Villages system to powder lime-stones and sea shells for building construction. The circular pit is about 3 metres in diameter and the gap between the inner and the outer circles is about 25 centimetres. The circle walls are bricks, cement plastered. In

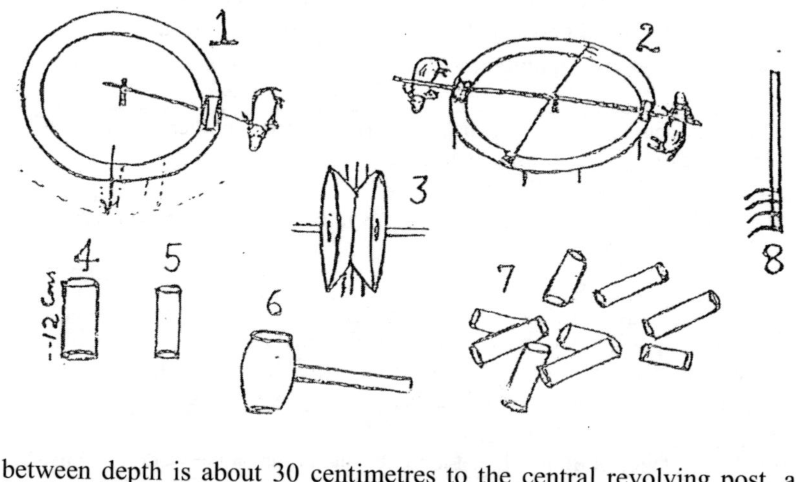

between depth is about 30 centimetres to the central revolving post, a casuriana, bamboo or eucalyptus pole is attached and passed through the middle hole of the solid one piece stone roller seen inside the pit. The roller may be 75 centimeres in diameter. The pole end is tied to the neck of a buffalo and when the animal goes round and round, the stone roller grinds. The circular brick construction should not rise more than 5 or 6 centimetres above ground level. Diagram 8 is the rake to mix the mass while grinding as shown in position in diagram 2.

Diagram 2 – Poles are extended on either sides and two buffaloes are used. The second grinder is not made of stone but it is the rim of a motor lorry wheel. One centimeter metal plates may be welded to it as shown in diagram 3 to serve as cutters while grinding.

In this bio-mass grinder the raw material used may be sawdust, coir waster, groundnut shells, rice husk, pine needles, wood shaving, bagasse, corn cobs, shells, barks and all sorts of dried leaves, including water hyacinth, which is available in abundance.

Diagram 4 and diagram 5 is a metal tube of about 6 centimetres in diameter and five is a solid metal roller to just fit into the tube and work like a piston. Bamboo tubes and wooden rollers may also be used.

Cassava starch is prepared as explained in (chapter 101). The ground mixture is removed, mixed with the starch and the paste is fitted into Tube 4, resting one end on a flat floor. Then Roller 4 is inserted at the other end and hammered. A wooden hammer may be used. When the paste is compressed and hard, it may be ejected using the roller and dried in the sun. The tube may be lubricated with oil or sieved ash before use. Finished briquettes are shown in diagram 7 ready for use or sale.

> It makes no sense to mount a campaign against deforestation when alternate fuels or resources are not made available.
>
> The North as well as the rich in the South will have to change production and consumption patterns.

> United States based environmental expert, Jennifer Turner told reporters at the US Consulate General in Chennai on 13.05.2015 that India has coal fired plants and one-third of her coal requirements are imported. Due to stringent licensing coal mines are unable to expand in India.
>
> Mr. Turner, Director of the China Environmental Forum at the Woodrow Wilson Centre said that in the US 45 percent of water is used in nuclear and coal plants to cool them.
>
> Shortage of coal and water is acute in several countries, especially in India. India should solve these problems quickly. Water shortage problem can be solved by exploiting the sea as resolved in the Rio summit.

Quality and Productivity Much Depend on the Dryness of the Burning Matter and The Binder Used. Fish Glue is Better Than Starch From Pods of Plants. Jaggery Mixed With Mineral Lime (Chunam) Binds Well.

Chapter - 85

Foliage Briquettes – Field Test Report

The foliage is first sun dried for about 3 days to bring down the moisture level to 15 percent from about 50 percent. If they are spread out on metal sheets or tiled roof tops the moisture level further drops by about 5 percent. Longer drying makes the leaves brittle. They are crushed and shred more or less to a particle size of one centimetre. After particled, cassava starch is added and compressed by hand, or in moulds and dried again. The harder the compression longer the burning capacity.

If compressed in a Cinva Ram or Ryfo Ram, the calorific value can be about 3,500 to 4,000 kilocalories per kilogram and if compressed in a car jack as explained by Ryan Foundation, it can take a calorific value of 4,000 to 4,500 kilocalories for kilogram or about the same as charcoal. But in common rural production it is easy to give a calorific value of 3,200 kilocalories per kilogram to serve as excellent and priceless domestic fuel for rural homes. The smoke and ash produced by foliage briquettes are little but this of course depends on the type of leaves (foliage) used and the percentage of moisture the finished briquettes contain. If the briquettes are moulded in a hydraulic press or in an injection mouldling machine with cassava starch as a binder they are as good as coal. The finished briquettes are again dried on tin sheets and corrugated roofing sheets. Corrugated sheets dry them quicker than flat sheets.

> 80 million people enter the work force every day while already millions are unemployed or underemployed.

Water hyacinth, eucalyptus leaves, locally available shrubs, and so on are all mixed to make good briquettes and briquettes made this way are excellent for boiling sea water in the RYFO, four pots still to get healthy drinking water. Indeed, foliage briquettes are ideal replacements for firewood as cooking fuel. They can be made by villagers entirely with local raw materials and without investment or dependence on outside help and transportation. Rural technology can be appropriate only when it totally eliminates transportation costs.

Desalination is a proven technology experiencing solid growth. As of Dec 1995, a total of 11,066 desalinating units had been installed or contracted for world wide, with a collective capacity of 7.4 billion cubic metres per year Sandra Postal Technology Review April 1977, Massachussetts Institute of Technology. (Author's comment, but any day distilled water is cheaper, safer and close at hand and can be consumed without restrictions.)

A study published by the Population Action International (PAI) in USA warns that by the year 2025 every third person in the world will suffer acute water shortage.

More than 335 million people live in countries defined as water stressed in 1990. A water stressed country is defined as one that has less than about 1,700 cubic metres of renewable water per year per person. Water scarcity is defined as availability of less than 1,000 cubic metre of water per person, per year.

The study report is called Sustained Water : Population and the Future Renewable Water Supplies, present figures are projected to increase eight fold by 2025 to between 2.8 billion and 3.3 billion people by 2025. The difference between the two figures reflect how different rates of population growth could affect water availability, even over a period of just 35 years.

Chapter - 86

Domestic Ball Grinder for Granuling Foliage Paddy Straw, Etc

Every home needs a grinder and several types of hand operated grinders are used in several countries. There are also animal and pedal operated grinders besides those driven by wind or electric power. Not all of them are appropriate to given situations as some are too costly and some are too

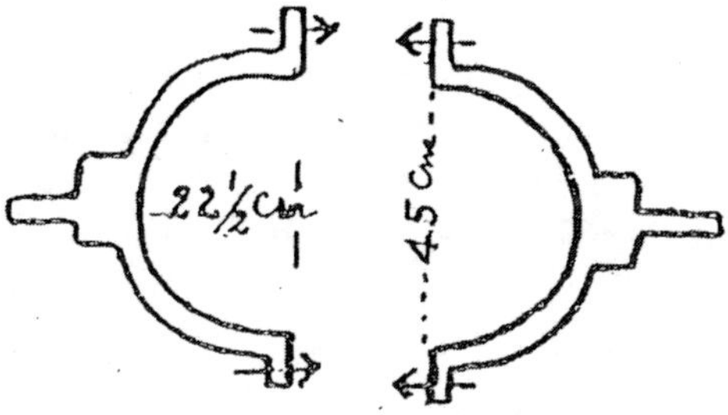

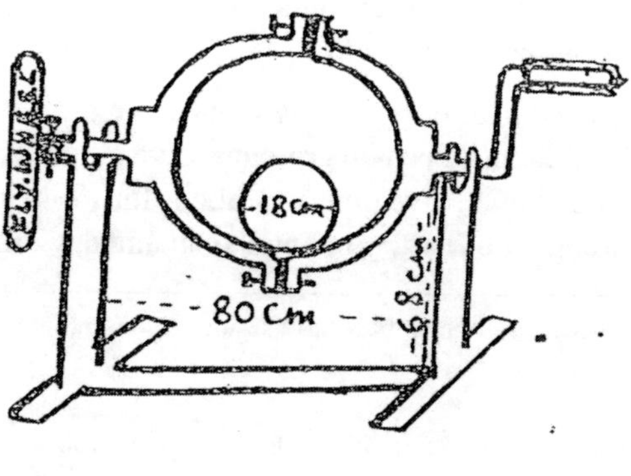

big for the rural homes while many types don't suit the climatic conditions. The Dynapad, for instance, is too big for a small hut and cannot be used in the open during rains and in the winter cold.

The Domestic Ball Grinder invented by the Ryan Foundation can, by and large be accepted as cheap, trouble free, convenient and very suitable to almost all Third World countries and also for urban homes.

The grinder consists basically of three large parts – the wooden or metal stand and two semicircular bowls with arms as shown in the diagram. The bowl like parts are identically equal and may be cast in aluminium having 45 centimetre diameter. When joined and fitted with bolts, there should not be a gap of crevice and holes may be provided in the casting as shown in the diagram with four arrows. If necessary, a rubber, leather, plastic or aluminium washer may be used to fit them tight. A detachable flywheel may be fixed at the axle end with a cotter pin as shown.

For loading, the two bowls should be removed apart from the stand, the grains put in along with an iron ball or hard stone ball of 18 centimetres diameter fitted and mounted on the stand as shown. The two limbs of the stand where the axle of the globe rests may be fitted with bush bearings. Ball bearings are not necessary as they will cost more and maintenance may be a problem.

When the handle is turned (and even a child can turn it without strain), the iron ball will crash the grains and within 15 minutes, a kilo of grains will be powdered. With the addition of water, wet grinding can be done. The ball grinder is an ideal device to grind wheat, maida, rice and so on and the same grinder can be used also for extracting oil from seeds and nuts on a domestic scale.

The two bowls may be sand cast and machined in villages but this cannot be done in all villages and in all Third World countries and then they may be cast in industrial centres.

It is not necessary for every family to have a unit. A village can have about six units and families can do grinding service for the community for a small payment and thus, save time and labour for women and save them also from the drudgery of hard grinding day in and day out.

Exploitation of water, it has now become a serious issue. India is threatened with a difficult and tumultuous water future.

Water must contain some protein, carbohydrate, calcium, iron, vitamins, enzymes, minerals, sodium, potassium and so on. To iodise water add about 50 milligrams of potassium iodate to 10 litres to distilled water.

So many years after the world becoming MODERN water scarcity has become an acute global problem. Hundreds of countries of the world are drought ridden or water stressed. Humanity does not and cannot put back even half the water it draws from the ground. Almost the whole of Africa is practically dry and states like California in USA are suffering from severe drought for the past several years. Brazil is craving for water. Our earth is in a vacuum sans water and can suck the earth at any time as it happened in Gujarat in the first decade of the present century. Small earthquakes will expedite the sucking and calamities after calamities will follow.

May Be Common Village Property to be used By All Families Living in the Village

Chapter - 87

Ojaj Planetary-Mixer

OJAJ sturdily build and attractively finished mixing machines are regarded as the most up-to-date equipment to handle and mixing process of Liquids – Creams – Pastes

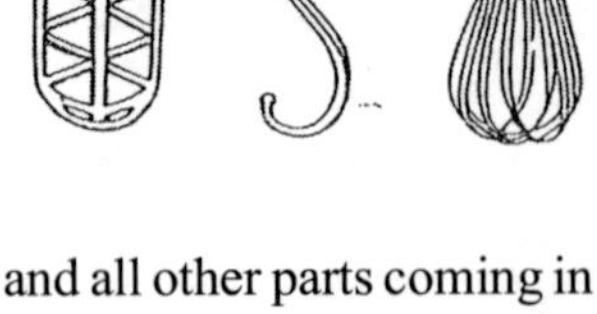

The bowl, agitator and all other parts coming in contact with the product are of stainless steel.

To attain the highest efficiency, the Planetary Mixers are provided with three different and churners to suit various products which are to be processed.

There are Seven Models to choose from :-

Model	Container Size	Capacity	H.P.	Approx. Speeds
F-10	Dia 9" x 9" Ht.	10 kgs	1/2	30, 60, 90
F-20	„ 11" x 12" „	20 „	1/2	30, 55, 100
F-50	„ 14" x 18" „	50 „	1	20, 45, 90
F-100	„ 20" x 10" „	100 „	3	15, 40, 80
F-150	„ 22½" × 22½	150 „	5	12, 35, 70
F-250	„ 27" x 27" „	250 „	7½	10, 30, 60
F-500	„ 36" x 36" „	500 „	10	10, 25, 50

Some of the applications of Ojaj Mixing Machine in various industries are as follows:

Pharmaceutical Industry: For Ointments, Granulate Masses, Syrups, Suppository Masses, Simple Emulsion etc.

Cosmetic Industry: For Toilet Creams, Milks, Fards, Hair Conditioners, Toothpastes, Shaving Creams, Brilliantines.

Food Industry: For Mayonnaise, Vegetables, Fruit Juices, Condensed Milk etc.

Confectionery: For Almond Paste, Chocolates, Pastries, Biscuits etc.,

Textile and Rubber Industries: For Atex, Vegetable Oil Emulsions, Sizing etc.

Chemical, Paint and Ink Industries: For Glues, Chemical Products, Colours Mastics etc.

An OJAJ Mixing Machine will save your money by :

Cutting Down Labour Costs, *Shaving Time in Operation,* Obtaining High Quality Product.

Agents in Chennai: M/s Bansons Engineering Works, 61-A, Seethamma Rd, Alwarpet Chennai – 18.

Ojaj Engineers

2nd Floor, Bharat House, 104 Mumbai Samachar Marg (Apolo Street) Mumbai – 400023.

Communicated by the Appropriate Technology Information, Chennai.

Chapter - 88

The Nepal Type Grinder / Crusher

This technology can be commonly seen in Nepal and a slight variation of it, high enough to be turned by a camel, in some African countries. It is not common in India as the horizontal arm type. There is a camel-drawn one in the refugee camp of Jalaloqsi in Somalia which is often shown to foreign visitors.

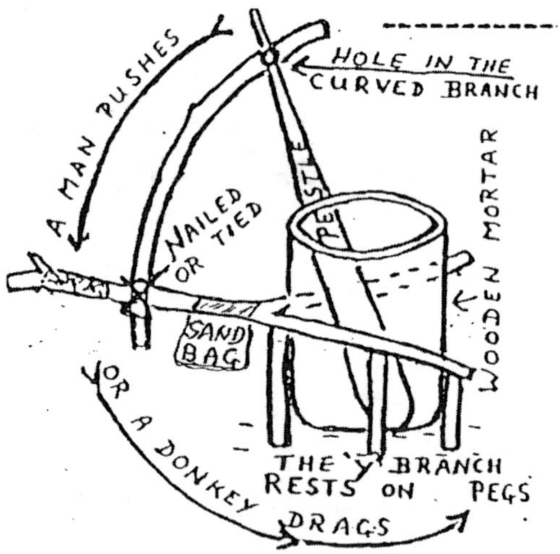

Sans Metal Parts

The Nepal type, called Chiuri, is very highly appropriate in Nepal because of the hills and valleys and tribal settlements of the country, where plenty of wood branches of required sizes are available.

What is seen in the diagram consists of six main parts all made of wood :

1. A hard mortar carved out of a tree trunk of required size.
2. A long pestle, also shaped out of a branch of a hard wood.
3. The Y-shaped – branch, broad enough to hold the wooden mortar as shown, between the 'V' portion of the 'Y'.
4. A bent piece of branch to connect the pestle and the turning arm 'Y'.
5. Six 60 centimetres long pegs to be driven all around the mortar to hold it in position and to make the 'V' part of the 'Y' branch rest on them.
6. A gunny or fiber bag filled with sand to make the 'Y' branch turner press the seeds or weeds during the crushing operation.

   (Waste oil may be applied at all contact parts to reduce friction during movement).

This press can either be turned by a man or woman or by donkeys or mules. If a camel is used, the construction must be high, so that the portion where the sand bag hangs rests comfortably on the upper neck of the camel (between the head and the hump) so that the wood does not hurt the camel.

However, the best way to operate this press is to harness two or three donkeys to turn the tail of the 'Y' branch. This is where the lost animal power can be restored for human benefits.

A billion cubic feet of sea water pumped into this or any well will join the ground water stream and the water in the well will not rise above sea level.

Invited by the Government of India and the Government of Gujarat, the author is seen with a team of officials making arrangements to connect an 80 feet deep dry well in Valsad district near the coastline to the sea, with concreate pipes to get the sea water into the well to raise the ground water table. He explained to the water managers that sea water is heavier than ground water, it will go deep down the earth and push the ground water up and therefore there is no question of salinating the soil or the existing ground water under normal condition.

Warm and water starved Countries Must Go all out For This Tree Which is a Gift of Nature For People in Remote Drought Areas.

Chapter - 89

Acacia Holosericea

(Tamil Nadu-means ears of a deer)

This tree, which is leguminous and drought resistant with a high calorific value grows in different types of soils and climatic conditions. It is small in size, ornamental, and fast growing and it belongs to Australia but thrives in all countries where it has been introduced and propagated. The nitrogen loaded nodules of the tree enrich the soil and therefore it is ideal for wasteland reclamation. It bears large phyllodes, which are nearly 25 centimetres long and 10 cm broad, they are usually covered densely with fine hair.

The trunk has a girth of about 30 centimetres and branches starting from a height of 50 to 60 centimetres from the ground level. Only till the roots catch the soil the saplings are to be watered and thereafter nature and annual rainfall take care of it. Small bright yellow flowers are clustered in prominent spikes about six centimetres long.

The climate most suited for cultivation ranges from 30 to 40 degrees Celsius but the tree can survive dry spells for 8 to 9 months in a year with scanty rainfall of about 125 millimetres annual. Narrow hairless pods, nearly six centimetres long are coiled in dense clusters and contain shiny black seeds.

The wood of tree gives high quality charcoal and the tree has a natural self regeneration property. Branches grow upward and the leaves are 12 to 24 centimetres long and five centimetres to eight centimetres broad (almost like mango leaves) and cattle don't browse them. As the leaves have insecticidal properties white ants don't attack the wood. The tree has proved excellent as an avenue tree inside institutions like schools, colleges and hospitals and on either side of small roads. Crops of mature seeds may be formed within 2 years of planting. The tree grows well along with Eucalyptus Camaldulensis.

The Auroville township in Puducherry, South India, introduced the tree in its campus a decade ago and the tree has been propagated from there to several parts of South India by the Social Forestry Department, which has realised the potentialities of the tree.

No Watering is Required If This Tree is Near a Water Body Such as Ponds, Lakes or Sea Water Canals.

The branches of the tree are good for making roofs for low cost huts after making them into shingles. The twigs are excellent for making briquettes.

From a distance the tree looks similar to an oleander tree without flowers and the leaves are broader than the oleander leaves All Third World countries should propagate this tree to save fuel wood crisis, to enrich the soil and reclaim wastelands. It propagates by seeds, scarified by mechanical abrasion or immersed for one minute in boiling (100 degree Celsius water. The life of the tree varies from 5 to 15 years.

It makes no sense to mount a campaign against deforestation without making available alternative fuels and resources.

Cacti and loes rise up to fifteen feet in height while thorny bushes cover the ground in Androy, the ancestral land of one of the 18 noted tribes of Madagascar (Malagasey). The Tandroy people are often referred to as "people of the thorns." They survive by eating the drought the resistant cassava which can be introduced in drought prone areas.

Selvador Persica (Piloo) has similar properties.

Some Trees Fix Nitrogen in the Soil. A Water Harvesting Pond and a Biogas Plant in Every Farm Will Help To Improve Enormously The Productivity and Profitability of Dry Land Farming. In Addition, This Will Help to Mitigate Climatic Change.

Chapter - 90

Vasaka Provides Employment

Flourish well near sea water canals, ponds and pools.

Botanical Name : Adhatoda Zeylanica Medic syn A. Vasica Nees

Vasaka is otherwise called Adhatoda or Adusa.

Family Name : Acanthaceae

Indian Names :

Hindi : Vasika, Adusa, Bansa,

Sanskrit : Vasaka,

Bengali : Vasaka, Kannada : Adusoge, Gujarati : Ardusu or Aldusco,

Malayalam : Atalotakam, Marathi : Adulsa, Telugu : Adasaramu,

Tamil : Arathorai or Adaorai,

In and around Delhi the plant is popularly known as Piya-bansa.

The international trade name is Vasaka.

Vasaka is about 2 to 3 centimetres tall, an evergreen shrub with large leaves. It grows uncared for all over India, Burma, Sri Lanka, Pakistan, Bangladesh and several African countries and other countries with a tropical climate. Nepal has plenty of it.

Stalks of the spikes are shorter than the leaves and dense flowers grow in the short spikes. Capsules have four seeds, Petals of the flowers are white with purple shades and fruits are capsular.

Uses :

1. The leaves contain an essential oil and people can be employed in steam distilling the oil.
2. The oil may also be sold for burning lights, for cooking and for lubrication of wheels and pumps.
3. The leaves contain an alkaloid called VASICINE, which acts as an expectorant. As the alkaloid is to be chemically extracted, fresh leaves can be sold to chemical factories collecting and packing the leaves will give employment for women and children.
4. The leaves are excellent manure in both green and dried form. They can be dried, pound or ground by one of the methods shown in this book along with other leaves having fertiliser and insecticide properties and sold.
5. The leaves yield a yellow dye which may be extracted and sold to handloom weavers or those engaged in fiber industries.
6. Parts of this plant are not attacked by insects or fungi and therefore can be used in the framework of shingles.
7. As the plants are not eaten by cattle it can be grown to reclaim soil and as field and plant guards.
8. The leaves can be collected and sold to wholesale fruit markets to ripen fruits and to preserve them (as commonly done in India).
9. Villagers boil the leaf of this plant, take the essence of it and give drops of it to those suffering from chest cold. A larger does is said to irritate the stomach lining.
10. Sticks can be used for making building slabs and this work can employ several teenage boys and girls.
11. All the parts of this neglected plant can be used in making pellets and briquettes for steam distillation.

See Author's Book on Plants and Trees That Generate Employment.

Chapter - 91

Breeding Tilapia

It takes nearly 4 months for Tilapia fry to mature. The male fish initiates the breeding process. It goes to the bottom of the tank or lake and by waving its tail back and forth makes a circle to invite the female to get to that spot and lay eggs. When the eggs are laid the male fertilises them and then the female takes the fertilised eggs into its mouth for protection. The eggs get life and become fry after about 10 weeks. They come out of the mother's mouth and swim around and get back into the mouth for safety. In about 8 weeks, they detach themselves from the mother and fend for themselves.

Usually Tilapia reproduces four times a year under suitable conditions and three times under unfavourable conditions. Over 1,000 eggs are laid out each time. Therefore the Tilapia fish get overcrowded within a year. And then there are too many fish, due to inadequate water space and inadequate food for all of them, they perish and float on the surface. Usually, the whole lot of them die about the same time and very often the death is due to lack of space or oxygen, underfeeding or overfeeding.

It is necessary to watch and regulate the feed and growth of Tilapia in a pond. Small frys may be sold away to those who want them and the full grown ones may be harvested for the kitchen.

A good way of regulating is to net the big ones and make industrial glue out of them (See Chapter 150). Another way is to dry them in the sun, remove the oil by pressing and use the granules as fertilisers. The oil may be used in kitchen lamps or for making organic paint.

In Kerala, the Philippines and some other countries, they throw inedible fish or edible fish that is spoilt in the kitchen gardens which makes the garden yield more fruits and vegetables. Raw fish will give a bad smell and attract bees and mosquitoes. It is better therefore to bury them in a pit and after a couple of months, dig the sand from the pit and use it as manure in gardens.

Pulicat lake, the second largest lagoon of India is a shallow 461 square kilometres lake whose maximum with is about 19 km. the lake is partly in Tamil Nadu and partly in Andhra Pradesh hosts 52 fishing villages around it and on the islands in it. Studies show that the lake was about 3.8 metres deep before the Dutch arrived in 1610 and made Pulicat their chief East Coast settlement. The lake has been silting up at a rate of about 1 metre per century and what is left now is 0.8 metre. According to present projections, the lake will turn into a solid marsh within 50 to 60 years and die an ecological death if action is not taken to reduce the rate of siltation.

Iraq and Syria have been on the warpath for the past several years opposing the plan of Turkey to construct dams to store surplus water of the Tigris and Euphrates. Iraq has a row already with Iran regarding the sharing of the Skattal Arab waterway, Ethiopia, Egypt and Sudan are all fighting for their share of water from the Nile.

In Central Asia too high risk conflicts are going on over the rivers Ama Darya and Syr Darya and the pure water of the Areal Sea.

Among Zambia, Mozambique, M, Zimbabwe and Sengal the Chaba River is a deep cause of tension, and Senegal are both trying to get more control over the river.

Chapter - 92

Compost Made Easy

(Better environment begins at home)

Converting domestic garbage into compost or briquettes to distil sea water.

Many individuals, institutions and registered foundations have come out with new and easy ways of disposing of kitchen and dining table wastes. While all these recommended methods are good, the Ryfo method seems very feasible, most economical and convenient. Many homes in several countries are now taking to the Ryfo method.

The system is very simple. A corrugated roofing sheet painted black on either side is setting a slanting position near the rear entrance to the kitchen. Preferably the sheet may be made of metal. Kitchen and dining table waste in wet or dry form are straight away spread on the corrugated sheet to dry. On the following day, glass, tin, hard plastic which will not burn and cannot be disposed of easily, are separated to fill land fills and sent away or used to fill pits, potholes and low lying areas nearby. Foliage, paddy straw, paper waste and so on are added to the matter to be composed. There are organisations in Chennai, such as, Kuppathotticom, which collect solid wastes and send them to faraway landfills. The advantage with the RYFO system is that the housewife or house maid need not keep the rubbish stored till the next morning but can spread out on the sheet then and there before the rubbish gathers flies and bacteria and gives out a stench.

After 2 days, when the rubbish is soft and brittle, it is crushed to granules by the many methods shown in this book. The granule particles are mixed with easily available and cheap binder (glue) and the semi paste is filled into a 1 inch thick metal water pipe of a metre long and rammed and the loaded pipe is exposed to the sun. After a day or two, when the rammed matter is pushed out, it comes out in the form of several pellets ready for the kitchen stove. (mixed with a binder the rubbish may also be pressed in a Cinva Ram – explained in this book to get briquettes as big as building blocks). The metal pipe should be oiled inside before loading. Fish glue and fish oil are the best to make these briquettes and they are explained in this book.

By this clean method the household refuse and waste are composed hygienically and also used as cooking fuel or as an alternative source of energy.

> The bare foot revolution started in some countries by the end of the last century and by several NGOs has miserably failed. Only the self help revolution will bring relief to people and help their progress. God Helps Those who Help Themselves.

> Ryan Foundation is a relief agency working with and for the poor to help them help themselves.

Chapter - 93

Salt cum Safe Water System

Process : Sea water is poured through the inlet into the screw type evaporation pan. The big lid which looks almost conical, and the sea water container both take the heat of the sun and/or the heat applied at the bottom

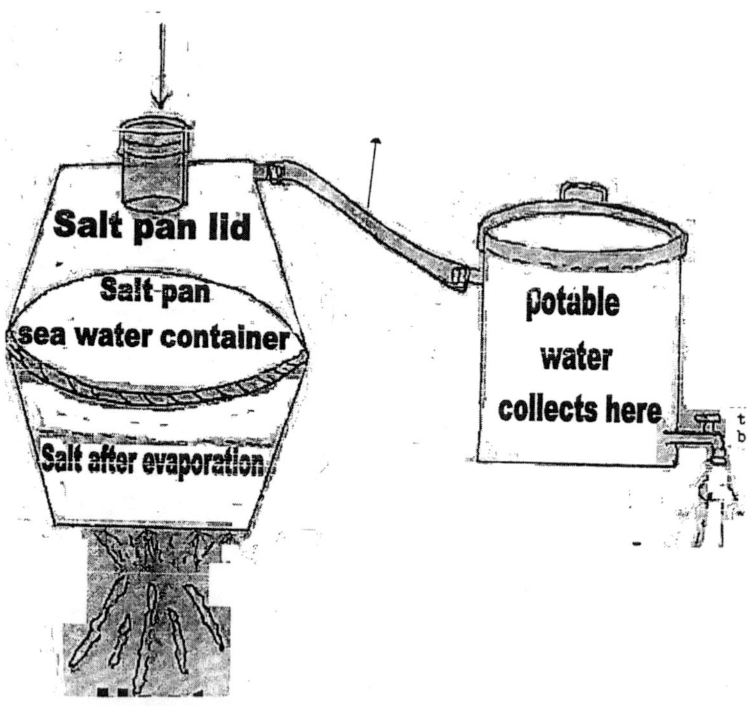

by burning foliage briquettes. The vapourised water rises the top of the lid enters the hose pipe and the steam passes into the potable water receiver and is condensed into pure water by the overnight cool breeze.

Experts studying the economics of salt production say that by the year 2020 salt production must be doubled to meet the requirements of the growing dyeing and pharmaceutical industries. Currently India produces 25 million tonnes, of which 6 million tonnes are used as edible salt and 12 million are used as industrial salt.

After 2 or 3 days of heating the residual salt in the big container is dissolved in some distilled water and evaporated in the open sun to get clean salt.

Chapter - 94

Protozoa

(Microscopic animal life found in water (along with other impurities and heavy metals)

RO machines eliminate salt, suspended particles and some evil bacteria but not the very many harmful heavy metals which sea water and ground water in many locations contain. People must be given safe water without heavy metals such as arsenic, mercury, alumunium, fluoride, iron particles and many other metals. Excessive arsenic leads to lung cancer, ulcer in the intestines and kidney damage.

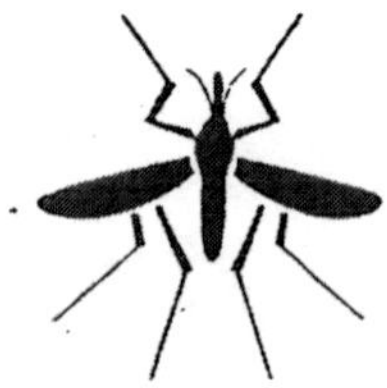

Ro does not remove toxic waste, garbage, juice, germs, worms, spores, pathogens or ova and it has a very high or very low pH level, much beyond the limits permitted by the World Health Organisation (WHO). Pure water as H_2O does not exist in nature.

Poor countries and some big cities in Africa, Asia and Latin America must spend upto 20 percent of their income on buying water – Source FAO, OECD, World Bank, UNESCO and Worldwatch Institute reported in D + C 5/6/94.

When water is so scarce universally why should about 40 percent of treated potable water be used to flush our toilets all over the world?

Chapter - 95

Sea Water for Prevention of Mosquito Menace

With reference to the new reports, Waterways to be cleaned round the clock (The Hindu dated February 22), any amount of cleaning of the Inland waterways will not mitigate the mosquito menace as long as domestic and institutional sewage continue to be drained into the waterways.

An easy and quick way of arresting the spread of mosquitoes is to let sea water flow through the waterways, as the British did at the Buckingham canal to prevent breeding the mosquitoes. Moreover if the inland waterways are connected and sea water is made available for residents, the same can be used as a safe resource for drinking and cooking by boiling and distilling it. Only some common kitchen ingredients would have to be added to restore the lost nutrients.

The sea provides several varieties of edible vegetables such as, Bladder Wrack, Hijiki, Arame, Dulse, Nori, Kombu, Wakame, Sea Palm, Irish Miss of these halophytes have superior nutritional content – rich in protein, antioxidants, carbohydrates, vitamins minerals and in particular Carotenes.

Chapter - 96

Ryfo Sea Water Generator

(A concept to be tried out)

RYFO SEA WATER GENERATOR (a concept to be tried out)

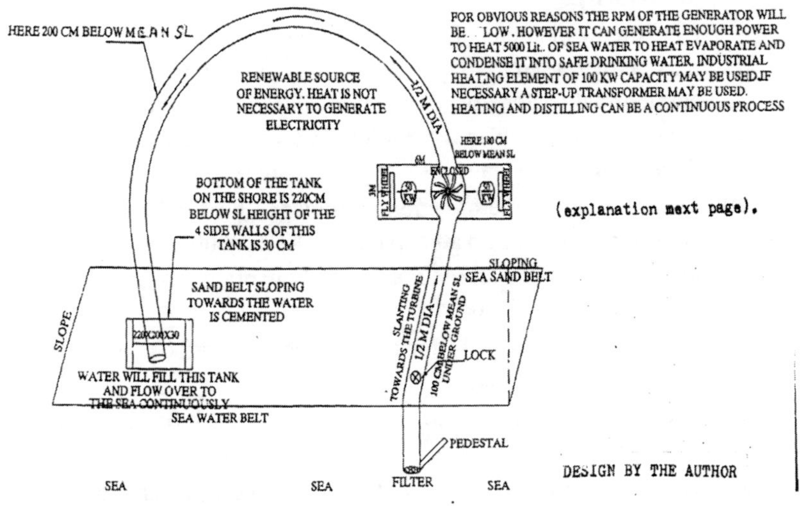

Explanation

1. Where the sea water enters, the pipe, should extend three metres into the sea and at the mouth have a stainless steel or fiber glass filter like a cap, fastened to stay in position. A pedestal may support the pipe.
2. The long pipe leading the water from the entry pipe to the turbine is buried 100 centimetres below ground and should slant slightly towards the turbine.

3. The turbine pit should be 6 metres x 3 metres and the axle of the turbine should be wall to wall on a pin and bush bearing.
4. The generation on either side should be fixed above axle or even higher.
5. At both ends of the axle there should be heavy flywheels.
6. The entire turbine must be covered inside a stainless steel case except at the water entry and exit points.
7. At the entry point the pipe should be 100 centimetres below mean sea level and at the exit point 220 centimetres below mean sea level, in other words, the bottom of the concrete tank on the high sandy slope at the outflow point should be 220 centimetres below mean sea level.
8. The entire pipe system should be as long as possible. However, 1,000 m may be the standard length, the more the length, more is the velocity.
9. Half metre diameter dia pipe may be the standard. However, it may be increased to metre if the unit is to produce electricity on a commercial scale. But it is better to have several units all along the shore connected to a common grid than to make the unit large. Small is better manageable.
10. This unit is to generate enough power to heat industrial heating elements of 100 kilowatt capacity to heat 5,000 litre of sea water in a stainless steel container every hour. If necessary a step-up transformer may be used.
11. In addition, parabolic reflectors may be used to focus solar energy on the sea water tank.
12. The distance between the entry point and the exit point may be 100 metre.
13. The distance between the left arm and the right arm of the U-shaped pipe should be at least 100 metres in the middle portion. It may be a semi-circle.
14. Opposite to the turbine, another turbine with generators may be installed on the out flowing pipe to double or almost double the electric current produced.
15. The concrete tank at the outflow point should be 220 x 220 centimetres on a sandy slope of the coastline.
16. The bottom of this tank should be 220 centimetres below mean sea level.

17. The distance from the outlet tank on the slope of the sand to the normal reaching point of the waves toward the land should be at least two metres. The slope should be a cemented ramp.
18. The tank into which the turbine and generators are installed must have walls on all four sides to prevent people, children in particular, and cattle from falling in.
19. There should also be a roof above to prevent rain water from filling the tank.
20. A lock may be provided at the entry point to stop the flow for repairs and maintenance.
21. A lock may be provided at the entry point to stop the flow for repairs and maintenance.
21. Heat is not necessary to generate electricity.
22. There may be two units side by side and commissioned on alternate days.
23. The maintenance cost of this project is minimal.
24. Strictly speaking this is a renewable source of energy from the sea and can be promoted on the coastal belt all over the world. It will solve the problem of energy to boil and distil sea water to get safe water.
25. Estimated cost of one unit of the project in India is about ₹ 10 lakhs.
26. Each unit can provide safe drinking water for at least 10,000 families every day continuously.

Over The Past Ten Years The Dutch Water Table Has Fallen By 75 Cm And A Lot of Vegetable That Depends On Ground Water Has Died Out – Environmental News From The Netherlands, March 2000.

The water in the tank can also be siphoned out of the tank to fall on a sloping cemented floor and flow into the sea.

Note: For large scale production of electicity only the length, breadth, depth and the size of the turbine and the generator are to be increased, in which case the diameter connecting the inlet and outlet points will make a big semi-circle. The length of the diameter will size increase.

This book explains thousands of income generating activities that can be promoted to give a living for people and raw materials for industries by bringing in sea water into the land by man-made canals and channels.

Chapter - 97

Palmyrah Tree Kattumaran

How to make it

1. Cut five fully grown Palmyrah trees.
2. Cut off the top side and the root side and make all the five trunks of uniform size, as shown in the diagram.

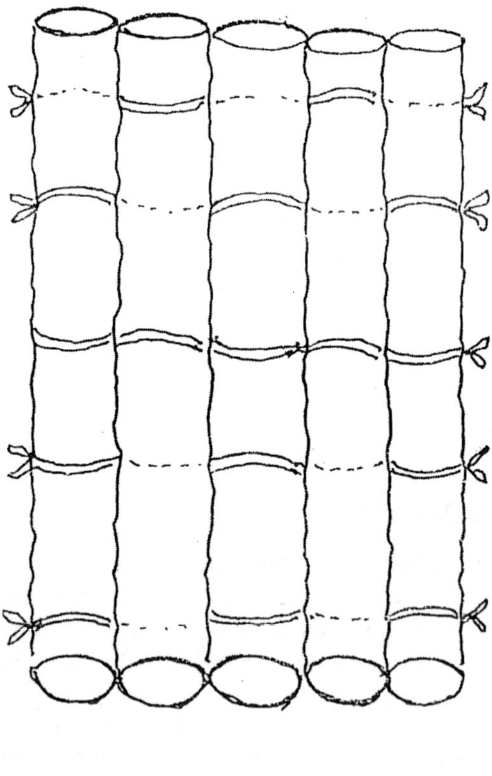

3. Tie up the trunks one with the other and all of them together with old (used) electric wires or sisal or coconut fiber ropes not thicker than a man's thumb.
4. Get a tin of tar (bitumen) from the Highways Department, heat the tar to a semi-liquid and paint all the five trunks with the molten tar on all sides.
5. Give a thick coat of tar paint on the two cut sides of all the trunks.
6. Dry what you have made near the coastline for three or four days and then push it into the sea.
7. Now the Kattumaran is ready for fishermen to go out into the sea to catch fish.
8. Throughout the coastal belt of peninsular India there are lakhs and lakhs of uncared palmyrah trees. Most of them are on lands which belong to the government. You have to take the permission of the Collector of the District to cut them. The forest department can have then cut to size and give it to the fishermen families.

This Kattumaran is ideal to be used in the Buckhingham and other sea water canals, in rivers and in big lakes.

Chapter - 98

Ryan Foundation Rice Pot Still

Kill Two Birds With One Stone

Water treatment facilities do not have the know-how to remove pesticide traces. Tests that monitor water quality often cannot detect them. So the water we get at home is not really safe. Water purifying equipments cannot tackle pesticide traces and heavy metals like chromium, arsenic, selenium, mercury, iron, aluminium or cadmium. These affect our body, reduce immunity to diseases and even lead to cancer as some water experts say.

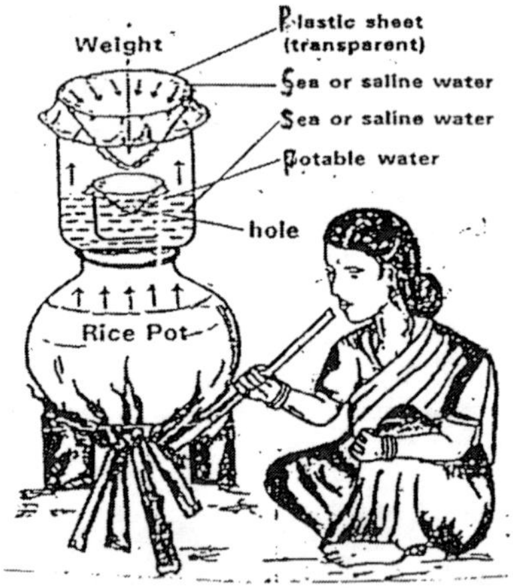

This is a low investment, low technology contraption to reach safe water to the unreached and the rural poor (almost no investment).

Remove the plastic sheet the next day for best results. This is how the poor can provide themselves with safe drinking water, though in small quantities.

The rural poor should always use RYFO foliage briquettes and pellets for cooking or boiling water and they should make the briquettes in their front yard or backyard. (RYFO Ram makes excellent briquettes) See the RYFO pamphlet "Where There Is No Cooking Fuel." Where cheap fuel is easily available, the rice pot can be eliminated.

Less Water Evaporates and Condenses Fast

Water evaporates at room temperature and the vapours condense (change into liquid state) when they cool dawn. The condensed liquid is absolutely tasteless. Therefore, filter half a litre of the steam-heated sea water in the big container and mix it with 10 litres of the tasteless condensed water in the small container to restore the lost taste and nutrients such as carbon dioxide, hypochlorite and hydrated lime lost during distillation.

Sea water must be taken into the land by man-made canals, as done in Dubai, in Spain, in peninsular India and in Kutch, in Gujarat and many other sectors of the world, and made easily accessible to the rural poor (Ref : RYFO Sea Water Tanks in chapter 223).

Sea water is better than no water. Saline agriculture is better than chemical agriculture. Humanity can survive by exploiting the sea and the saline tree, as demonstrated by Isreal, Arizona and some Gulf countries. Sea is the resource of the poor.

Place a thin cloth bundle as big as a table tennis ball containing bits of fresh banana skin, drumstick (murunga) seeds, lime skin, a pod of onion, a pod of garlic, grains each of rice, dhal and ragi and so on to restore the lost nutrients and taste in the distilled water. Pure distilled water is lethal if consumed for long. Change the bundle once a week.

Note: The inverted plastic cone may be made of steel, clay or fiber glass and used as lid. A screw type lid will prevent steam from escaping and yield more safe water.

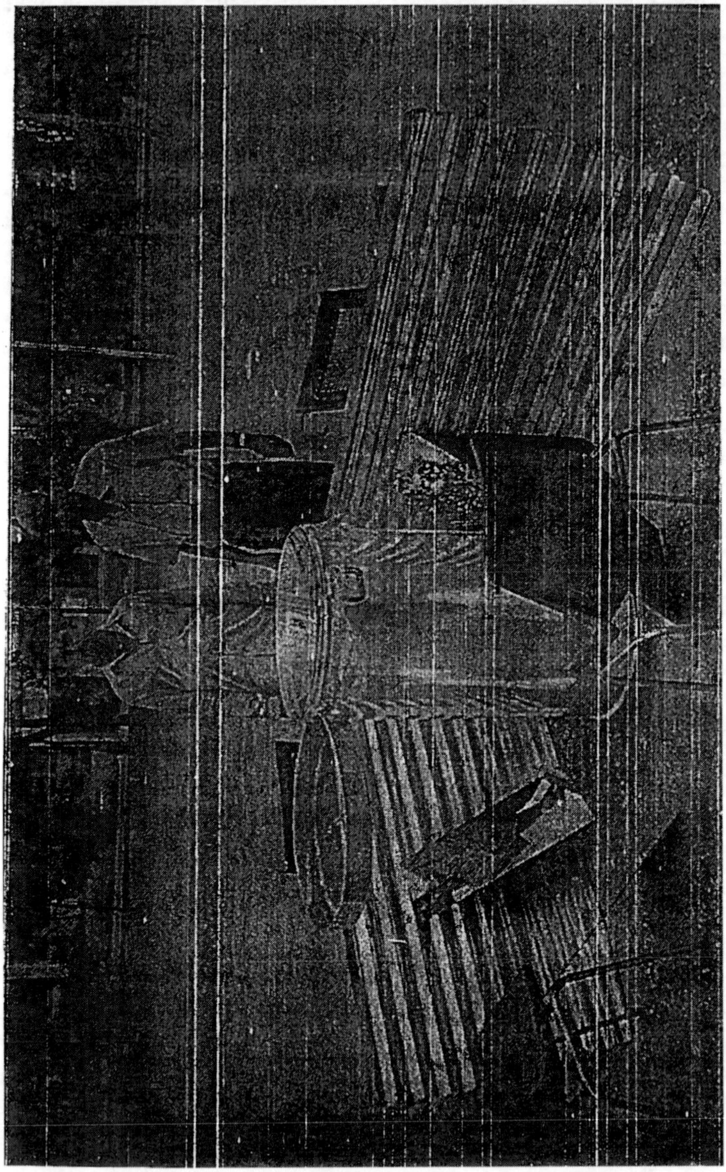

Six Years after adopting and pursuing the goals of the Rajiv Gandhi Watershed Mission, large parts of Madhya Pradesh are realing under the impact of one of the worst droughts in recent years.

Chapter - 99

Ryfo Water Tank Another Device for Distributing Water to the People

Water, both for drinking and for washing, remains a major problem in almost all Third World countries. The taps are blocked or broken, the hand pumps installed by the UNICEF and so many other Aid Agencies are not maintained and in many places children have filled them with sand and stones. Some wind mills stand here and there, like skeletons of giants for

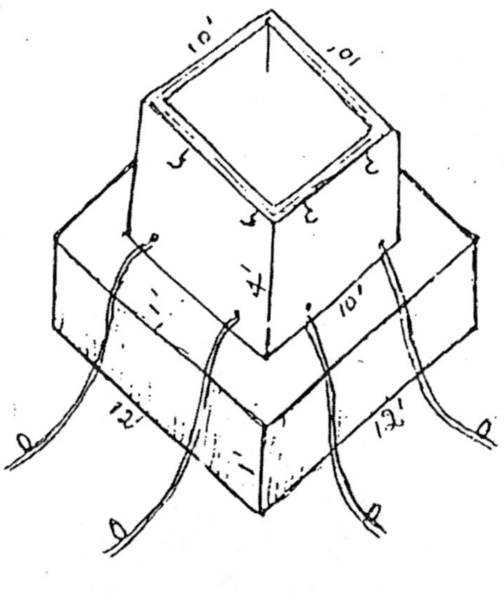

want of spare parts. Birds build their nests and washermen dry their clothes and water remains an eternal problem. It is common knowledge that a good majority of Third World people do not have water close to their homes and have to walk several miles to fetch water.

The Ryfo water tank built out of mud bricks, clay and plaster is a reasonable solution to this problem, provided drinking water is available in the village. Thus, water tank can be given to the people since it does not involve much expenditure and has trouble free maintenance.

As seen in the diagram it is a simple tub like construction (cistern) raised about a foot from ground level. A standard tank can be of inner size 10 x 10 x 4 ft (height). The dimensions may be bigger if necessary but the height may be restricted to 4 feet so that water can be poured into it without difficulty.

Flexible PVC tubes may be fixed to leak proof holes made at the bottom of the tank so that, when filled, water will flow out automatically through the tubes. Curved hooks may be fixed on all sides along the upper end of the tub as shown in the diagram and rope or wire loops tied to the delivery end of the PVC tubes as shown. The delivery end will always be lifted and hung on the hook on the top end of the tank and when removed and brought down to the bucket or vessel on the ground, water will flow into it.

Metal barrels or wooden barrels mounted on small carts with two wheels and drawn by a man or by an animal may fetch water from the nearest available source and have the tub filled every day. If there is a well in the middle of the village the tub may be built adjoining the well.

Chapter - 100

Another Model of Sea Canal Lock

For the past several years Ryan Foundation International which innovates low technology to solve problems created by high technology has been advocating unpaved sea water canals across drought ridden and barren lands of the countries of the world.

As world population expands by a projected 2.6 million people over the next 30 years, and as consumption levels spiral upwards, water problems are bound to intensify with best dam site already developed and many rivers and ground water resources already overtapped, opportunities to solve these problems lie only in exploiting the sea.

Picture shows a strong and ideal lock (sluice) to regulate the flow of sea water into barren lands which is inexpensive and can be easily constructed along coast lines anywhere in the world by governments or NGOs and operated and maintained by the village community without any training and without physical strain.

As soon as water is provided to the thirsty and starving rural poor they will at once take to domestic desalination by the RYFO stills, to canal transportation, saline agriculture, acqua culture, salt making and to several cottage and village industries based on saline plants and trees and their leaves, nuts, fruits, fiber, and many other derivatives from them.

Only the sea and the saline tree can save humanity from peril. Sea water is better than no water. Saline agriculture is safer than chemical agriculture.

Where water, food and employment opportunities are scarce, they must be promoted by innovative methods. Ryan Foundation "Six Tanks Still" is the best answer for large scale desalination for cultivation and this method has now been declared feasible by national and international governments. Efforts by any agencies, government and non government are under way now to commission this system. The poor will remain poor if they depend on water supply in pipelines from far away places.

Chapter - 101

Sea-Water Tank (another device)

Several projects and proposals of Ryan Foundation International require easy and short distance access to sea water. The Suez Canal or the Panama Canal are man-made sea water canals for ships to pass. But the water of these canals used in several other ways as well. The Panama Canal across the isthmus of Panama which links Atlantic and Pacific oceans is a huge undertaking which was completed in 1914 at a cost of about 340 million dollars an enormous sum in those days. However, to meet the RYFO requirement, is not required. They can be met with a very small investment, which even villages can afford for their own benefit.

All that is required to provide sea water for people living not far from the coastal belt (80% of the world's population lives within 100 km from the coastal belt) is to give them water in three cubic metre tanks build one kilometre away from the coastline and connect the brick and mortar tank to the sea by brick and mortar pipes (without using cement as even the big building of olden days were built).

Usually, the level of land near the coastline is about 5 feet above sea level. The connection may therefore be given 5 ½ to 6 feet below sea level, with a lock arrangement inside the tank. Such as arrangement will always keep the water in the tank to the level of the sea. The rise and fall of the sea level does not matter as it will not badly overflow the tank. When the tides are very high, the tank will of course overflow a little. The overflowing water may be channelised for saline agriculture or let into a man-made lagoon or led back into the sea.

The open tank may be fitted with crossbeams or crossbars in a check pattern and closed with slabs so that dust and dirt do not fall into the tank.

Craft cultivation is cultivation in a small piece of arable land close to one's home and such cultivation should be encouraged. It is akin to roof gardening and organic gardening.

From one tank water may be taken to another similar tank in the next village by underground pipes or closed gutters and from the second village to the third and so on. If the inlet pipe or gutter from the sea is about 2 feet broad, about 10 to 15 tanks in about to 15 villages can be connected to the chain and all of them will maintain the level of water to the level of the sea, even if 500 pipelines draw the water from these tanks to different locations for desalination, saline agriculture and fish ponds. The supply will be perennial and the sea will not go empty.

Hundreds or thousands of such tanks may be built all along the coastline and connected and interconnected to make abundant sea water available to people everywhere. Sea water is better than no water and the sea has a world of wealth.

Where necessary, overhead water tanks may be built above or beside the ground level tanks to lift the water up using windmills with canvas buckets, Persian wheels or animal turned water lifters. The appropriate technology methods have been published widely by Ryan Foundation.

There is a common tendency among people, particularly rural people and agriculture specialists, to say that sea water will pollute the ground with salinity. But by the piped method suggested above by the Ryan Foundation, there is no question of salinity spreading into the soil and there is no room for complaint or conflict on this issue.

Some villages may have mortar paved tanks in their villages of the size 2 metres x 2 meters and 1 metre deep to rear sea fish. The urban rich may draw the sea water into their compound and have their own swimming pools, fish and duck ponds and so on.

Under the most optimistic assumption – including nearly a 10 percent increase in water supply from the completion of a joint project with Sudan – Egypt appears headed for a crisis. In a land where rain is scarce, 55 million people depend almost entirely on the waters of the Nile, none of which originate within the nation's boundaries. Its population leaps by 1 million every 8 months, and with virtually no additional sources to tap, Egypt has no choice but to squeeze more out of its existing supply by upgrading and better managing its irrigation systems.

Newly built housing colonies can have the small RYFO desalinating system on top of the terrace and supplies absolutely clean and healthy water for the flat owners free of charge after recovering the initial cost, which is minimal.

Large institutions such as hospitals, hostels, schools, colleges, churches, temples, mosques and residential settlements can have centralised large scale RYFO plants one, two or three to meet their in-house requirements and also to supply safe water to their neighbourhood.

Governments and local bodies should quickly swing into action to get these sea water tanks build and connected. Where they are unwilling or lethargic, the people can take the matter into their own hands and build tanks with contributed or donated money and connect the ground level tanks with flexible hose pipe buried underground. A kerchief like cloth may be tied at either end of the hose to serve as filters. Even if these hose pipes are damaged by some construction workers, there is nothing much to loose or regret. The hose pipes may be repaired and restored or replaced and it will cost very little. Laying of such small hose pipes to draw water from the sea does not infringe any government law or rights of people of the area. However, it may be done with the consent and cooperation of the neighbourhood.

Hundreds of plants, trees, weeds and shrubs have been identified that grow well in sea water and these can be promoted extensively in rural areas for reforestation and restoration of the environment. Reforestation will provide fuel wood for cooking and rural women will be saved the time consuming and back breaking work of gathering fuel wood.

When plenty of fuel wood is made available close at hand, it can also be used to boil and desalinate sea water to get healthy potable water by the very many cottage scale methods published by Ryan Foundation and thereby

This book moves countries of the world to innovate using locally sourced resources with a focus on job creation, building social capital and generating multiple cash flows by stimulating entrepreneurship and business model innovation.

women will also be relieved of their burden of trekking long distances and carrying water on their heads and shoulders to meet the family needs.

By culturing sea fish in their family ponds, they will get plenty of animal protein for better nourishment and health. Fish oil, fish glue, fish fed and fertiliser may also be made use of in livelihood activities.

Villagers can also wash their cows and goats with sea water, keeping them clean and healthy and their surroundings cool and fine.

Public health has been defined as ensuring the conditions in which people can be healthy thus, easily accessible, safe potable water is an imperative to ensure good public health, while accessibility to safe potable water is a basic human right and right of every individual. The WHO, UNICEF, World Bank and the UNEP will be making significant contributions to public health as well as income generation and overall progress of the rural masses by providing potable water for them by the RYFO six tank method which has been declared feasible by water experts as well as NGOs and national and international governments. It costs only half million US dollars to desalinate water for 50,000 people. Villagers can maintain the system without any external aid or help.

Raising the Water Table – Gujarat

Gujarat used to be the land of lakes, which in the 1930s met 78% of the irrigation needs. The picture is totally different today. In May this year, the Chief Minister of Gujarat stated that the water table in his state had fallen by 1000 feet and the drought condition was precarious. However on the 26th of May, Vandana Siva, international scientist, wrote in the Hindu that the ground water in Gujarat had fallen to a depth of 1500 to 1800 feet, leaving shallow acquifiers and surface storage systems dry.

Chapter - 102

Water is an important Factor of Production

The outdated economics of Adam Smith and Marshal has to be supplemented and re-written. The economic concept that land, labour and capital are the basic factors of production and development is now irrelevant, inadequate and out-dated. Today's factors of production for global development and human well-being are not only land, labour and capital but also water, energy and environment (infrastructure).

Given water energy and infrastructure, it is possible for common people to acquire with their own efforts land, labour and capital but given land, labour, and capital alone, it will not be possible for them to acquire water, energy and the environment, especially when water supply and energy are managed (mismanaged) by governments.

Therefore, people must be given easy access to water energy and a good infrastructure for their livelihood and economic progress. There has to be a shift in emphasis on the factors of production and priority must be given both to drinking water and to water for agriculture and industries. Land, labour and capital must take the backseat.

> Richard Leakey, who heads Public Service Reforms in the European Economic Community reports in Courier issue 182, 50 percent of Kenyans live in poverty. Absolute poverty doubles every 20 years. Urban poverty has grown because urban infrastructure has deteriorated.

To put it differently, the three great liberators of production and development are not limited to land, labour and capital but include drinking water, cooking fuel and infrastructure (good environment). Only these three additional factors release people, especially the villagers, who form the bulk of the masses of the world, from the tragic necessity of mortgage their future by destroying their soil (land) by cutting trees and gathering fuel wood and by running ever greater hazards of disease in themselves and their children in search of water.

How can degraded land, demoralised labour and devalued capital bring about production and distribution under the order of the day when they are threatening the health, existence and chances of survival of people in this planet?

Our economic accounting systems are deeply flawed as they have no accounts for environmental capital. No business could survive without a capital account. Neither can this planet. The United Nations, the World Bank, the OECD (Organisation for Economic Co-operation and Development) and the Global Agencies should bring in new international standards of national economic accounting before the end of 2020 or earlier. These reforms will instill a concern for environmental resources in governments, multinationals and national and international industrial and commercial undertaking and then they will consider both environment and economics factors in decision making (i.e. Environment includes water, energy and infrastructure).

> World population which has just passed the 6 billion mark will stand at 8 or 9 billion by the middle of this century and will continue to grow and suffer shortages of food and water if the sea and the saline tree are not exploited.

As early as 1992 the World Commission on Environment and Development, reconvened in London, recorded that the developing countries debt now exceeds $ 1.3 thousand billion and debt service exceeds net development assistance by a factor of four. There has been for much is the past decade a net transfer of resources from developing to industrialised nations, which has been correctly described as nothing less than a blood transfusion from the sick to the healthy. This is an evil economic concept based on land, labour, and capital, without taking into account the environment, water, energy and social cost.

> Good governance is fundamental for the success and sustainability of any national poverty alleviation strategy. Evidence shows that countries that are able to develop and maintain transparent government institutions are more effective both in combating poverty and in improving the living standards of their citizens – IFAD / UN – Nov 2000 No. 8.

Again, the book published by Unesco a decade ago entitle "Environmentally Sustainable Economic Development Building on Brundtland," warns that economic activity can no longer proceed under the banner of "business as usual" arguing that it is no longer tenable to make economic growth the unquestioned objective of economic development policy. The traditional concept of growth which the authors call "throughput growth," with its reliance on an ever-increasing throughput of energy and other natural materials, cannot be sustained and must yield to an imaginative pursuit of economic ends that are less resource intensive. This argument has serious implications regarding the urgent political decisions needed to change current social, economic and environmental policies.

According to the World Bank and the UNEP, 67 percent of Africa's, 34 percent of Asia's, and 20 percent of Latin America's non-desert lands may turn into non-productive desert in the near future.

Sea water can be desalinated and distilled for human consumption and used for flushing toilets and cleaning roads, for saline agriculture, cultivating halophytes (salt loving trees) for promoting sea fish, sea weeds and reeds and also house boats and floating markets. Further, to make dry areas humid, to make trees and plants flourish, birds to sing and bees to buzz.

As many as 6 million hectares world wide become desert annually and an additional 21 million lose their productive potential each year. This being so, mere land can no longer be considered as an important factor of production or development.

Americans use about 400 billion gallons of water each day for residential, industrial and agricultural purposes. This figure translates into 1,600 gallons per person per day. Profligate use of water is beginning to show in many parts of the country. Henry La Rose of the US Geological Survey predicts that the supply of underground drinking water in South-West Florida will be completely exhausted in 10 years (in "Zero Population Growth Nov 1991"). This dismal prediction is due to the fact that good water is not considered a finite factor of production and development and is not regulated for the survival of future generations.

The UN report on the "World Social Situation 1989" which blasts the so called adjustment policies imposed.

Now is the time to take up community driven plans to use village assets like ponds, trees, kitchen, gardens, animals and birds to regenerate rural ecology and environment.

By 2030, the global middle class could top 5 billion, up from 2 billion today. These people will inevitably demand higher living standards, as well as greater freedom, dignity and justice. Why should they settle for less?

By the "International Monetary Fund (IMF)" and the World Bank for their devastating effects on the African people points out that the recipes are particularly cruel in the employment income, food, nutrition, health and education sectors. The report, which was submitted to the 10th session of the UN Commission on Social Development, states that the IMF and the World Bank programmes have completely failed in their goal of restoring conditions for sustainable economic growth. The policies, says the report, have neglected the social, political and human aspects of the development crisis and make it even more difficult for Africa to achieve food self sufficiency.

The report further points out that by benefiting wealthy commercial farmers (who have control over land, labour and capital) the IMF policies have not only side-stepped the problem of alleviating rural poverty, they have also made the adjustment of Africa's priority goal of food self-sufficiency even more illusory.

Tamil Nadu Government wants to interconnect Mahanadi, Godavari, Krishna, Pennar, Palar, Cauvery, Gundar, Pambo, Achankovil and Naippar rivers to distribute drought in the country and increase the number of poor people.

Even in the few cases where some measures of economic growth has been achieved the report notes that countries have become ever more indebted and dependant on external resources.

According to recent IMF statistics, the ration of debt to exports of goods and services deteriorated a great deal and is still deteriorating. Somalia, Mozambique and Sudan had ratios of over 100 percent.

Such a sad picture is presented mainly because the IMF did not consider water, fuel and other factors of production for the present and the future generations.

The problem facing nations has been very succinctly expressed by Robert MC Namara in the following words. "The basic problem of poverty and growth in the developing world can be stated very simply. The growth is not equitably reaching the poor. And the poor are not significantly contributing to growth. What we can do is to begin to move now to ensure that absolute poverty, utter degradation – is ended. But can it be ended when water, fuel, infrastructure and the factors of production of the poor are not provided to them ? "water" in particular !

Plants and trees shed leaves which dry and become foliage. Foliage with other burning matters crushed and made into briquettes. Briquettes are used to boil and steam distil sea water into safe drinking water. Residual impure salt is dissolved in distilled water and evaporated to get clean white salt as a by product.

Let our planners and policy makers realise that the main engine of production and development is, and has always been, self help and hard work and not aid, land, labour or capital alone. No amount of land, labour and capital, government statutes or legal measures can substitute for the people themselves taking up the responsibility, especially when it comes to finding basic needs of life such as water, fuel, food and shelter. Governments should provide the poor with water, full shelter, security and safe environment, which are the real factors of production for the poor and the world is mostly inhabited by poor people.

The book is all about water, which is posing the gravest threat to mankind. It explains several cottage scale and commercial scale methods based on appropriate low technologies to reach water to the unreached. Easy access to safe water is the entry point for sustainable development.

When I was a Student of Commerce I was Told. Money is a Matter of Functions Four A Medium, A Measure, A Standard, A Store. But in Life I Find That Money is a Matter of Objective Four – A Benz, A Mansion, Luxuries and More.

Chapter - 103

The Malmo Declaration

The Malmo Declaration of 2000 says it all

'At the dawn of this new century, we have at our disposal the human and material resources to achieve sustainable development, not as an abstract concept but as a concrete reality. The unprecedented developments in production and information technologies, the emergence of a younger generation with a clear sense of optimism, solidarity and values, women increasingly aware and with an enhanced and active role in society – all point to the emergence of a new consciousness. We can decrease poverty by half by 2015 without degrading the environment, we can ensure environmental security through early warning we can better integrate environmental consideration in economic policy, we can better co-ordinate legal instruments and we can realise a vision of a world without slums. We commit ourselves to realising this common vision.'

Alas : it is already middle of 2014 and poverty has not decreased in any way. In fact, in real terms it has increased and inflation is killing the poor.

Only doles, freebees, bribes and scams have increased. What is worse is the fact that our judiciary is moving slower than a snail and delays convicting culprits or lets them go scot free. On an average our judiciary takes over ten years to settle a simple and uncomplicated case of crime, murder or rape.

The MALMO Declaration must be reconsidered and revised and reworded the Declaration.

> The earth, our home, is beginning to look more and more like an immense pile of filth – Pope in encyclical 2015.

Chapter - 104

Freshwater Stress and Scarcity in Africa by 2025

An area is experiencing water stress when annual water supplies drop below 1700 cubic metres per person. When annual water supplies drop below 1,000 cubic metres m^3 per person, the population faces water scarcity.

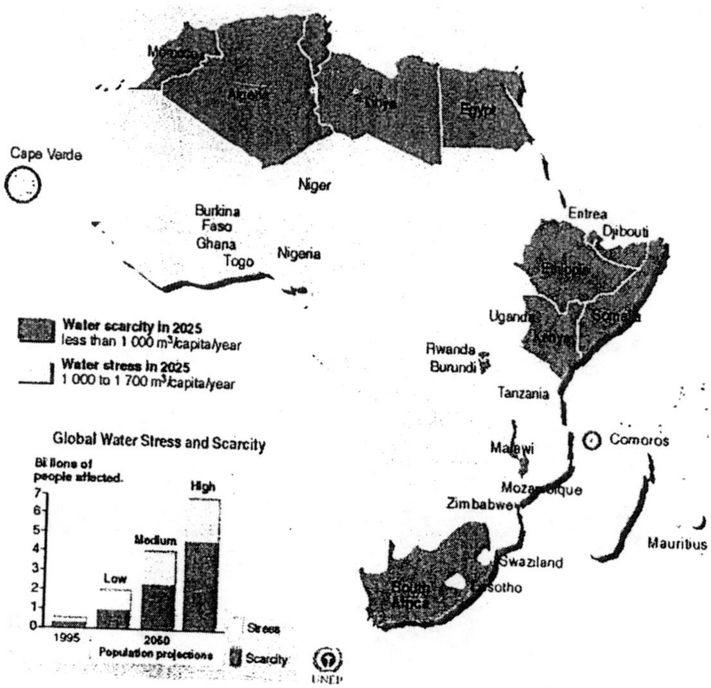

Source: United Nations Economic Commission for Africa (UNECA) Addis Ababa. Global Enironment Outlook 2000 (GEO) UNEP. Earthscan, London, 1999, Population Action International.

One of Africa's most pressing natural resource issues is water. In 2003 Unep was asked to take the lead in establishing the African Ministerial Council on Water.

Egypt Threaten, Military Reprisals Against The Ongoing 2014 Ethiopian Construction of A Large Dam on The Blue Nile.

World Bank reports speaks about the failure of existing economic strategies to combat poverty and calls for new, socially oriented pro-poor approach.

Chapter - 105

Appeal to All Governments

1. Water is critically scarce, don't make it worse.
2. Don't ask people to manage water when they don't find it.
3. Prevent water wars within and between stares and neighbours.
4. Don't centralise water supply you will reach the unreached.
5. Give people water, not Coca Cola, Alcohol, beer and sarayam.
6. No more bore pumps, they suck air, arsenic, selenium, sewage, rust and so on.
7. Forty percent of potable water goes into toilets stop it-use seawater as in dig sea water canals Hong Kong like railway lines – stop the construction of mega dams. It is easy and economical to distil sea water.
8. It is a human right to have free and safer water ; don't tax water.
9. Appeal to the rulers of the world to provide sea water for sustainable development. Seawater is better than no water. Saline agriculture is safer than chemical agriculture.

Governments are Really For The Rich and Those who want To Become Rich By Competition. Corruption, Theft and Violence. The Minority Rich Control The Majority Poor with Their Power and Plenty.

Chapter - 106

Demand and Uses

Demand for water in India is expected to rise drastically to about 833 billion cubic metres (BCM) in 2025 and 899 BCM in 2050. At present, water demand stands at 712 BCM.

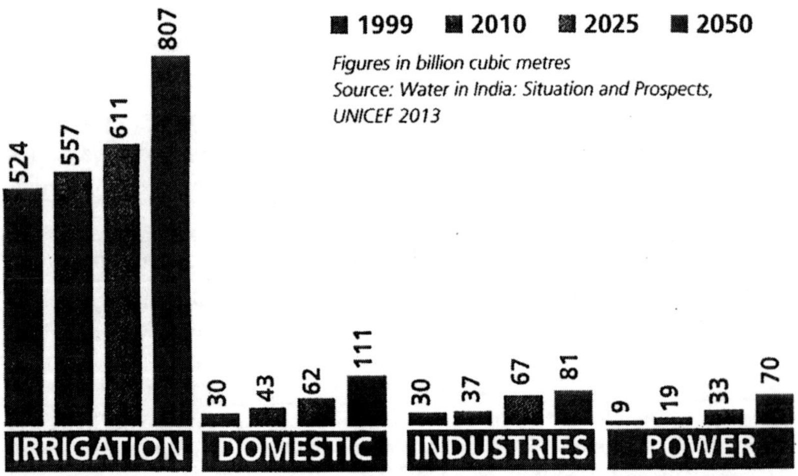

POPULATION TROUBLES

Water is getting scarce because of growing population

Years	*Population (in millions)*	*Per capita average annual availability (m^3/year)*
2001	1,029	1,816
2011	1,210	1,545
2025	1,394 (projected)	1,340
2050	1,640 (projected)	1,140

Source: Water in India: Situation and Prospects-2013, UNICEF.

Chapter - 107

Wastage of Water

Rainfall Wastage

Sixty five percent Rainwater run off goes into the sea, which is a major wastage considering that over 70 per cent of country's farming is rain-fed. The runoff also causes soil erosion, river flooding and siltation of water bodies.

Poor Sanitation

6.4 percent of India's gross domestic product (328,500 crore) worth of economic opportunities were lost in 2006 because of poor sanitation. Poor sanitation results in water contamination that leads to widespread diseases and wastage of public resource.

Depleting Groundwater

210 billion cubic metres is the amount of groundwater extracted in India annually, which is the highest in the world. Groundwater today provides for more than 60 percent of net irrigated area. As a result, over 60 percent of districts in the country are facing problems relating to poor quality and shortage of groundwater, says the mid-term appraisal of the 11th Five Year Plan.

Industrial Waste

Ninety percent of waste water discharged in rivers after treatment – 38,254 million litres – do not meet environment norms. The sewage water should be treated so that it is fit for bathing. World Bank estimates that the current industrial water use in India is about 13 percent of the total freshwater withdrawal in the country and the water demand for industrial uses and energy production will grow at a rate of 4.2 percent annually.

Pollution

Seventy five percent of available water resources in the country do not meet the World Health Organisation water quality standards, because of pollution caused by rapid industrialisation, urbanisation, poor farming and irrigation practices.

Urbanisation

Seventy percent of India's water pollution happens due to waste from cities than house only 36 percent of India's population, says a report of the Central Pollution Control Board. Owing to poor infrastructure, cities are able to treat only 31 percent of the waste, while the remaining is dumped untreated. The report also points out that only 38 percent of the population in the cities have access to sanitation and 78 percent to clean drinking water.

> Only that rural technology which can engage the whole family can be said to be absolutely appropriate.

Chapter - 108

Are Water Wars Inevitable?

Mr Ismail Serogelding, Vice President of the World Bank presented his report to the 20th session of the International Seminar on Planetary Emergencies in Sicily, Italy. According to the report which was released on 21st August 1995 in Rome, atleast 40 percent of the world's population lives in housing that has neither plumbing nor potable water. Eighty countries are not really able to provide potable water for their people and due to this there is no development, not to speak of sustainable development. Water scarcity is leading these countries to great health and environmental problems. To aggravate this problem the world population is steadily rising in spite of efforts to control it.

World Bank and the Food and Agricultural Organisation (FAO) of the United Nations indicate that the consumption of water by and large doubles once in 20 years. For the past 20 years water above ground level and below ground is getting more and more scarce. There is a fear now that the already existing international cold war for water will worsen into a hot war.

Adel Dorwish writes in the UNEP Journal, Our Planet, No. 3, 1995, When President Anwar Sadat signed the peace treaty with Isreal, he said that Egypt will never go to war again, except to protect its water resources. King Hussein of Jordon has said that he will never again go to war with Isreal, except over water. United Nations Secretary General Boutros Boutros Ghali has warned bluntly that water will cause the next war in the area (the Middle East). All this confirms that water is the main issue for stirring quarrel, enemity and hatred among nations. Even in the Holy Bible water quarrels and enemity have been recorded during the time of Abraham and Isaac in Genesis chapter 26 (GNV).

Isaac's servants dug a well in the valley and found water. The shepherds of Gerar quarreled with Isaac's shepherds and said, This water belongs to us so Isaac named the well "Quarrel Isaac's" servants dug another well, and there was a quarrel about that one also; so he named it "Enemity." He moved away from there and dug another well. There was no dispute about this one, so he named it "Freedom". He said, Now the Lord has given us freedom to live in the land, and we will be prosperous here.

Peter Gleick of the Pacific Institute whose researches on environment and security pointed out in the seminar at Sicily that atleast 40 percent of the world's population lives closer to 250 major sources of potable water which are the object of hatred and frequent conflicts. Mr. Seragelding clarified, "it is not a coincidence that conflict over the control of these resources is even today, one of the biggest obstacles to peace treaties between Isreal and its Arab neighbours. He warned, that the availability of water per head in West Asia and North Africa will be down by 80 percent by 2025. He also mentioned that there will be a 30 percent fall in Egypt, 40 percent in Nigeria and 50 percent in Kenya.

According to the FAO, water scarcity is largely due to rising demand and mismanagement. FAO blames tourism, luxury hotels and wrong life style for the worsening of water shortage. This global agency estimates that 15,000 cubic metres of water can supply for 3 years 100 nomads and 450 livetock and 100 urban families but can supply 100 luxury hotels only for 55 days.

The threatening problem in the future in several countries will not be want of arable land but want of water to cultivate the available arable land. This problem is already felt severely in India, Pakistan, Bangladesh and several African countries, particularly in the northen part of the African continent. On an average agriculture demands about 60 percent of the water available in a country but many countries don't even have a quarter of this.

While the average per capita consumption of water is 2.5 litre per day for a person in a developing country, it is 400 litres per day in developed countries. Ryan Foundation water experts hold the view that a person anywhere in the world must have at least 100 litres of water per head per day to meet personal needs, including bathing, washing clothes and so on to keep clean and healthy.

Tapping of river waters during the monsoon and storing it in artificial lakes, as done in Russia, is one solution to the water problem but this has not been emulated by other countries. In fact, many countries don't even deepen existing natural lakes to hold more rain water. What is worse, in countries like India, Governments fill up existing natural lakes and catchment areas with debris and the city garbage collections to build on them residential colonies or commercial complexes. This is particularly so in the city of Chennai which is already suffering from gross water shortage.

Development Education Exchange Papers of the FAO, September, 1995, says, referring to food scarcity, Typically, the vulnerable include, farmers on marginal lands, pastoralites and fisher folk with diminishing resource base and other members of the informal economy. Food insecurity now grips more landless, female, elderly, urban and African people than it did in 1970s.

The United Nations Research Institute for Social Development (UNRISD) states in its report to the UNRISD / UNDP International Seminar on Economic Restructuring and Social Policy (New York) 11 – 13 January 1995). "The social and economic consequences of the free market reform have been drastic. In general, the primary incomes of the poor are down, the number of people living in poverty is up, and social income – access to public services – has decreased. Targeted interventions meant to protect the poor and vulnerable groups from the worst aspects of adjustment never reach the poor for their livelihood and survival" the same report adds "the world bank has said that its work must be judged by the extent to which poverty has been alleviated and continues" "We must move quickly, we must make sure that future generations will not pay the high price for our neglect and delays. We must act before poverty spreads further." This warning points to human laziness and lethargy which if not arrested will overtake the poverty prevailing now and the situation will go entirely out of control.

Problems dog plans for a joint Jordanian – Israeli canal from the Red Sea to the Dead Sea, which would use the 100 metre drop between the two seas to generate electricity for desalination. This plan will succeed only if all the surrounding countries co-operate to share the benefits taking heed from the Biblical reference quoted alone new water sources must be created instead of generating new quarrels and enemity.

While inaugurating the 65th Annual Session of the National Academy of Science, India (NASI) on 3rd November 1995 at Trirupati, Prof. M.G.K. Menon, International Scientist, called upon scientists everywhere to come out and with a strategy on desalination and on use of brackish water for cultivation and said that people in this fast drying planet have no the other option. Emphasis may also be given to tapping and storing flood waters during the monsoon and impromoting artificial lakes.

Global summits, World Bank resolutions, statements and national parliamentary promises are not going to improve the water situation unless vigorous follow up action is taken. To help the poor out of the death rap water scarcity and the death trap starvation they must be given sea water canals, like railway lines, running on all direction; fish, saline agriculture, desalination contraptions, and a moisture borne atmosphere to bring about a total revolution in their rural economy and lifestyle.

> 100 million suffer drought
>
> The number of people being fed by the United Nations World Food Programme (WFP) due to drought has more than quadrupled in the last 4 years, according to the agency. It reported in August that more than 100 million people in over 20 countries are suffering the effects of drought, the number of affected people WFP now feels has soared to 16 million this year, from 3 million in 1996.
>
> "We have seen an incredible increase in drought-Induced hardship over the last decade, said Francesco Strippoli, WFP Senior Humanitarian Advisor. Today we're literally working in every part of the world to provide emergency food aid not to people who have lost their crops or livelihoods but due to drought.
>
> The Horn of Africa remains in the grip of severe drought, with more than 16 million people facing severe food shortages in recorded history. With a national maize requirement estimated at 3.21 million tonnes, the country will need to import some 1.4 million tonnes over the next year, according to WPP.

Chapter - 109

Water Crisis

The Controller and Auditor General of India (CAG) reported in July 2014 that about 90 percent of the illegal colonies in Delhi had not functional sewer lines and atleast 197 out of the 895 colonies had no water lines that they spent over Rs 3029 Crores on roads, drainage and water connections.

Chapter - 110

The Multipurpose Ox Cart

To Transport water from Village to Village to Supply Water to the Unreached, in Cans and Drums

Transporting goods is an acute problem in many Third World countries, especially in rural areas. The picture shows a multipurpose ox cart which can be completely assembled by villagers with locally available raw materials. Bamboos, Casurinas or straight branches of village trees can be used to assemble the cart.

The total length of the cart may be about 3 metres, breadth about a metre and height also about a metre. All the joints may be tied with fiber ropes or builder's building wire or old (discarded) electric wires. Nails and screws are not necessary. When constructed, the weight of the cart will be about 500 kilograms and the cart can easily take a load of about 500 kgs. upto one tonne.

The load bearing box (container) in the middle may be made either with bamboo or casuarina or with thick wooden planks or tin sheets. But the ideal will be to tie up a fibre netting. The fiber rope may be first tied to make a net container and then fixed onto the frame of the cart.

The most important part of the cart is the wheel, which may be made out of discarded automobile tyres (of private cars). Scooter tyres may be used for donkey drawn carts.

These simple village made multipurpose carts can be drawn by any draught animal and by human beings as well. Women and children can drag them without difficulty and with rubber tyres, it is easy to pull them off the road and across the fields and repairs and maintenance can be done by the owner himself without having to go to a workshop and spend money on replacements.

Chapter - 111

Solar Cooking in Your Kitchen

The best of German and Swiss technology at Indian Prices

Brought to you by:

Gadhia Solar Energy System Pvt. Ltd.

Scheffler Community Solar Cooker

The cooker consists of three key components:

- A Primary reflector
- A Secondary reflector
- The tracking mechanism

The primary reflector is a 7 m^2 parabolic dish fitted with a set of acrylic mirrors. They reflect and concentrate the sunlight falling on the dish onto the secondary reflector.

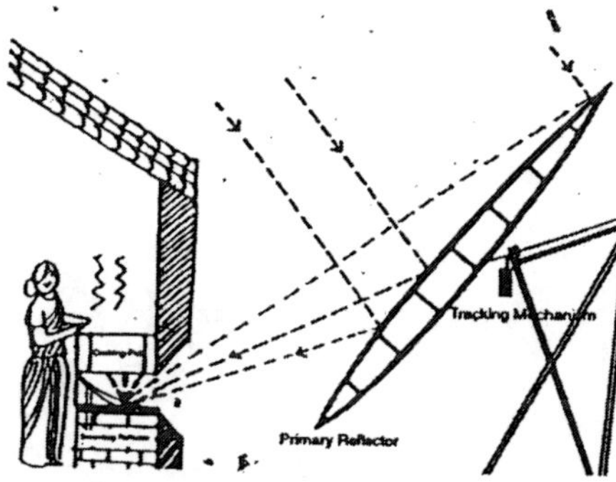

The concentrated light is then deflected off the secondary reflector onto the cooking vessel inside the kitchen. Temperature of the concentrated sunlight at the focus can reach up to 450 °C! This can cook for 60-80 persons daily. The primary reflector only needs to be reset once every morning and it automatically tracks the sun's movement throughout the day!

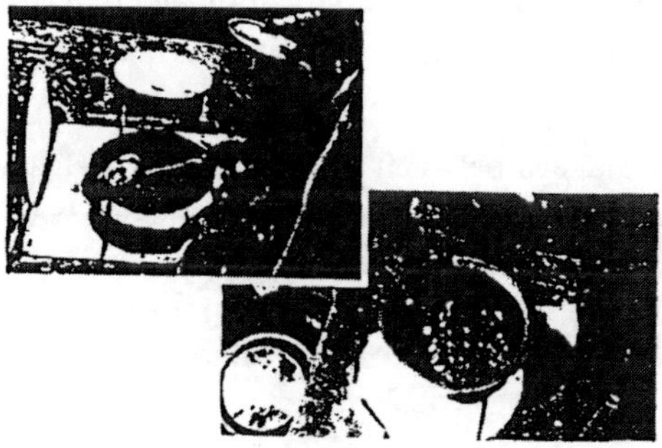

Boil or fry anything from vegetables to rice to purees! Besides making chappatis and dosas, you can even bake your own bread, biscuits, cakes and pizzas!!

Chapter - 112

Water for Food

Several estimates warn that the population of the world could increase to 10 billion by 2050. Ninety five (95) percent of the increase has been predicted to be in the Third World countries (poor countries) and two-thirds of the total growth is predicted to take place in the cities. Therefore, the urban poor need special attention.

Its Governments and NGOs must take action now (through its already late) to bring about agricultural diversification and intensification in order to enhance the income and job opportunities of farm agriculture through ecologically and economically sustainable farm and non-farm activities.

Growing imbalances between our ability to produce food from the deteriorating and diminishing state of usable land and water supply of and expanding biotic and abiotic stresses pose a risk serious famine. We have sufficient and reliable statistics now that in per capita terms, land, water and forest availability is shrinking and fertilisers are in short supply.

For many countries the best way to face this problem is to follow the example of Israel. In 1948, that country was a barren desert. Even today vast parts are barren, sandy and salty. But very enterprising Israelites have reclaimed and are still reclaiming land, vitalising revitalising the soil and producting food and fruits. The deserts are turning into gardens and orchards.

Israel has an average annual rainfall of 30 centimetres only. The country recycles water, desalinates sea water and promotes saline agricultural and halophytes. People there produce agricultural produce and sell them to less industrious people who live in fertile countries in Europe and elsewhere. As India's eminent jurist Nani Palkiwala put it, "Israel is to India what quality is to quantity. The area of Israel is 27,817 square kilometres less than one of India's 3.3 million square kilometres while the population is only 5.3 million (less than half the population of Mumbai or Delhi), as against our 900 million. But its exports for the last year totalled US$ 23 billion against India's US$ 22 billion.

In India, the average yield of rice from about 40 million hectares (mha) is only a little over tonnes per hectare. This can easily be doubled by cultivating fields and farms desalinated sea water using wave, wind and solar energy.

Land use statistics (LUS) reveal that in real terms the actual physical land area served by the public canals rose from 15.805 (mha) in 1985 to 16.067 mha in 1990, or by 0.262 mha. During these years ₹ 11,107 crores was spent on this sector. Thus, the average cost of irrigating one additional hectare of land during these years was ₹ 4.26 lakhs. As irrigation by canals is, on an average, only ₹ 50 as productive as land irrigated by wells, the expenditure in this sector should be taken to be equivalent to ₹ 8.52 lakhs per hectare of equivalent irrigation by ground water. This is pouring the tax payer's money down the drain.

In 1986, Prime Minister Rajiv Gandhi publicly stated that investments in big irrigation projects had ceased to yield returns. In his words "We need some definite thrusts from the projects that we started after 1970. Perhaps we can safely say that almost no benefit has come to the people from these projects. For 16 years we have powered money out. The people have got nothing back no irrigation, no water, no increase in production, and no help in their daily life."

To put it differently and strikingly, in spite of investing about ₹ 53,900 Crores between 1950 and 1997, India's agriculture sector and rural economy are in very bad shape and are going from bad to worse.

We can vitalise this sector and make it productive and job oriented only by exploiting the sea for desalination and saline agriculture and promoting halophytes and sea water weeds. Let the world be inspired by Israel, Saudi Arabia and Dubai. (for further reading, "water management in Homes and villages" by the author, (200 pages.)

> The Horn of Africa remains in the grip of severe drought, with more than 16 million people facing severe food shortages. Kenya alone is suffering from one of it's worst droughts in recorded history. With the national maize requirement estimated at 3.21 million tonnes the country will need to import some 1.4 million tonnes over the next year, according to WFP.

Chapter - 113

Plenty of Alcohol But No Safe Water to Drink

India has several million coconut and palmyrah trees from which villagers extract illicit alcohol (TODDY or KALLU). In almost every village plenty of illicit liquor is available but not a glass of safe drinking water.

The men folk spend money on toddy while their wife and children starve for food and thirst for water.

Rajaji introduced total prohibition of alcohol first in the Salem District in the Chennai Presidency and later on throughout the state of Madras, now called Tamil Nadu. The state was progressing without alcohol and women and children were happy.

Chief Minister of Tamil Nadu abolished prohibition and now more illicit alcohol is flowing in Tamil Nadu than plain and safe drinking water. Women and children are the victims.

What is worse, the government is selling millions of bottles of Indian made foreign alcohol when there is no clean water for people to drink. It sells water to the people knowing fully well that water is their property, given by God and not by their government.

Chapter - 114

Water and International Conflict

Demand for safe water in our planet doubles once in 21 years and the population doubles once in 30 years. Nature does not replenish even half the quantum of water we draw out from the earth, globally speaking, our water table is falling by about a foot every year and by a meter in some areas. Cherapunji in Assam where it used to rain all through the day and year is now drought ridden. Due to the EI Nino effect and climatic change almost all over the world rivers are running half their capacity during most part of the year. In India, the holy Ganga (Ganges), a highly polluted river does not reach the sea. Any number of climatic change conferences such as the one which took place in Denmark in December 2009 cannot stop or mitigate the EI Nino phenomenon. During 2009 EI Nino had intensified draught across Latin America, Guatemala and Ecuador.

Egypt has threatened to attack Ethiopia because Ethiopia is taking too much from the Nile while Syria says Turkey is stealing water from the Euphrates. Under these explosive circumstances we necessarily have to turn to sea.

Many of the wars of this century were about oil but wars of next century will be about water, says Ismail Seagelding, of the World Bank.

It is said that water is getting expensive as milk or oil in some countries. We must ask ourselves if this is the kind of economic growth we want, if this is the Welfare economics of the Noble Laureate of India, Amarthya Sen. Emphatically, the solution to this grave problem is to take sea water into deserts, parched lands and to dry and draught areas. Sea and deserts have to be harnessed in patches and stages for human welfare in many parts of the world.

Focusing more on India, fast depleting ground water could result in the collapse of the Indian agriculture, affecting one fourth of its grain harvest, warns the international "Think Tank" World Watch Institute.

Mr David Seckler and his colleague of the International Irrigation Management Institute in Sri Lanka has been quoted by Washington based "Think Tank" in its annual state of the world report saying that water tables are falling at one to three metres per year almost everywhere in India.

The rain in Tamil Nadu and peninsular India during the end of 1998, was more than usual and unexpected. Yet Tamil Nadu pumped out more groundwater during the period. The normal withdrawal during the period was between 230 and 235 tme ft tonnes of metrifeet. During the end of 1998 however, it touched 240 tme ft., thus depleting the storage at Mettur Dam despite the slightly increased supply of water from Karnataka, the neighbouring state.

Digging and sucking the ground deeper and deeper for water is like killing the goose for the golden egg. At some point this house of cards will collapse. When it does, India's grain harvest could fall by as much as 25% warned Seckler. A severe wave of famine is round the corner.

The situation in Pakistan is even grimmer, the World Watch Institute said, adding that grain harvest area per person in Pakistan was expected to fall from 0.31 hectares in 1950 to 0.08 hectares in 2000 and 0.03 hectares in 2050.

Water is unevenly distributed. Some countries such as Canada, Norway and Scotland, have surplus water, but other regions such as the Middle East, Africa, Central Asia, and Singapore are desperately short of water. Ironically, these latter regions will have a leaping population growth during the next 50 years and their children will be thirsty, hungry and undernourished.

Egypt has a severe problem 96.5 percent of the country is arid desert and water supply from the Aswan Dam is limited by the 1959 Nile agreement with Sudan. To meet the needs of the growing population, Egypt plans to reclaim land in the eastern and western deserts, but after the year 2000, the major constraint to development will be lack of water, not land.

Meanwhile, the country faces serious water quality problems caused by the increased use of fertilisers, the increased soil salinity, the discharge of untreated sewage into the Nile and the irrigation system and the use of water for bathing and washing animals.

Consider the Aral Sea which is fast drying up because of hydroelectric schemes and irrigation, leaving the people of five countries, namely Uzbekistan, Kazakistan, Tajikistan, Turkmenistan and Krygyzstan without water. The problem is unresolved. Moreover, the water of the Aral is getting more and more polluted.

United Nations hydrologists and metrologists met at the UK Institute of Hydrology during early December, 1998 and agreed that urgent action needed to prevent much of the planet from dying of thirst and starvation.

Water scarcity is now slowly spreading to capital rich USA which tries to solve water problems in other countries. Even the Big Brother is in trouble. The USA is now trying to buy water from Canada. A business company in New Foundland called McCurdy Enterprises is trying to export the water from Spring field Gisborne Lake in the Canadian province, but the government of Canada and environmentalists are against making water an interstate-trading commodity on environmental grounds. Environmentalists are absolutely right in their stand. What alternative does America have?

The model country, Singapore, is now buying water from Malaysia and Indonesia. Their contract with Indonesia is for the supply of water for several years, perhaps until Indonesia also runs short of water.

Food production depends upon the availability of water. Development Education Exchange papers of the FAO, September 1995, says, referring to food insecurity.

Typically, the vulnerable include farmers on marginal lands, with diminishing resource base and other members of the informal economy. Food insecurity now grips more landless, female, elderly, urban and African people than it did in the 1970s.

Only more water can produce more food. Sea water can produce surplus food and bring down the prices of grains, fruits and vegetables.

Global summits, World Bank resolutions and national parliamentary promises are not going to improve the water problem. Big dams only damn the neighbourhood and people's livelihood. To help the poor out of their death traps, they must be provided seawater (canals), fish, saline, agriculture, desalination contraptions (published Books for change) and a moisture borne atmosphere to bring about a revolution in their rural economy and lifestyle. There can be no life without water. There is no alternative for water.

In a positive move to help water starved people the world over, the Federal Republic of Germany is planning to send man-made floating islands. Each island is 140 metre long 55 metre wide and 13 meter high. Whichever coast they drop they hope to make available within a few hours healthy fresh water and also electricity. These floating island, basically made of steel, will carry on board a desalination plant electric power driven by turbines.

This problem-solving project was proposed in a study by a team of experts from STEAG, an industrial firm in Essen, at the request of the Bonn government. According to the German view the rising demand for water for drinking and other domestic purposes can only be met in the future through the desalination of sea water. Even today, about 10 to 12 million cubic metres are manufactured in this way daily. By the year 2000, the worldwide demand could be around 4000 million cubic metre.

If governments cannot give the poor, healthy portable water, let them give at least sea water. There is a world of wealth in sea water and thousands of livelihood activities based on marine resources and saline agriculture. Sea water, sea fish, desert sands and saline agriculture can rehabilitate humanity and feed 10 times the present population of our planet and give them the purchasing power to purchase the food that is available thereby wiping out poverty for years to come.

> Once the ten most important lakes of Chennai were Vysarpadi, Perumbur, Paravallur, Madavaram, Chetpet, Spur Tank, Nungambakkam, Mylapore, Mambalam, Kottur and Kalikundram. None of these exist in today's concrete jungle. Where have they gone?

Chapter - 115

Ryfo Alternative Cement

RYFO Alternative Cement (to join and long then bamboo pipes and plaster cracks and leaks)

Excellent formula for making hand compressed adobe (sun dried) bricks which need no firing and no machines.

1. Six bonds of clay, dried and hand sieved to 40 mesh.
2. Two bonds of surki (surki is burnt clay powdered) and sieved to 40 mesh.
3. One bond of quick lime or slaked lime (quick lime is sea or river shells roasted and powdered). Slaked lime is natural lime stone roasted and powdered.
4. One bond of dried and powdered water hyacinth, paddy straw or bamboo leaves. They may be hand pound or stone roller ground and sieved.
5. One bond of waste paper (papier mache) in semi-liquid state.
6. One bond of tamarind seed powder, cassava powder, or yam powder boiled in equal quantity of water by volume and added to the mortar mixture in liquid state.
7. Quarter bond of Jaggery dissolved to an almost semi-liquid consistency.
8. One-eighth bond of myrobalan (Kadukai in Tamil if available, mixed in water and added).
9. Alternatively, if available one-eighth bond of pulp of varuna. The varuna tree is called Barna in Hindi. Varuna in Sanskrit. Nirmathalam in Malayalam and Mavilingam in Tamil. It grows well throughout India, especially near rivers and rivulets. When the pulp of this fruit is mixed with mortar a strong cement is obtained. The pulp has been using in the building industry since time immemorial.

Chapter - 116

Tax and Subsidies

Development economists know that when the price of water and fuel is increased, on the one hand, the price of everything goes up, including cost of food, healthcare and education, and on the other, the standard of living of the common people goes down very low.

A country can see real progress and prosperity only when domestic and agricultural fuel and water are made available gratis to the common people and the rural poor. Luxury items and those that bring about deterioration in health, education and moral standards may be heavily taxed but not water and fuel consumed by the common people and the poor for their survival. In the farm sector, small holders will collapse if they are to pay a price for water and fuel. If small holders a country are not able to break even, the economy of that country will break down. This is not logic or economics but plain common sense.

Many countries, including European countries, heavily taxon fuel and water and then give heavy support price for farm produce. With taxes and support prices administration expenses and corruption always increase. Indeed corruption has become an acute problem now in 2014 and the European Economic Commission (EEC) has detected large scale misuse of support prices and subsidies on farm produce. Specific cases have been brought to light where output has been exaggerated and support money has been enhanced in some countries. Investigation by the EEC is now going on to expose these fraudulent practices. It makes no sense to tax and subsidise. Even a two years old child knows that if she receives two then sweets and gives away two she will have nothing left for herself. She also knows that when somebody gives her two sweets locks away one, she is being deprived, cheated or exploited.

The successful Tapchan wave energy plant in Norway is drawing world attention. The waves are led into a furrow in the rocks to fill a basin and from there the sea water is led into the sea through a turbine. This technology must be copied and introduced wherever possible.

Contaminated Water

Contaminated water supplies cause more than million deaths per year and a far higher number of cases of illnese. The health of more than 400 million people largely women and children, is threatened by homes filled with smoke from cooking on wood fires. Tropical countries are estimated to be losing 0.5 percent to 1.5 percent of their GNP per year through soil erosion. Deforestation, lose of biodiversity and poor management of soild and hazardous wastes accelerate atmospheric change, endager land and water reserves and constitute a serious health risk.

Chapter - 117

Sand from the Sea

About Eight lakh cubic metres of sand was being removed (dredged) from the late 1990s Diwards annually for several years from the mouth of the Ennore Estuary near the city of Chennai. The sea sand so removed was not used to fill and raise pits and pothols and low lying areas in the city and so more and more water stagnates in the city and does not flow into the canals and open gutters cut across the city. Nor was the sand removed from the estuary thrown in catchment areas to be washed by rains for two or three years and made fit for the construction industry as in Dubai and other desert countries. India is wasting her natural resources and spending money on machine-made and imported materials. Poor countries will remain poor if they don't exploit the abundantly available natural resources. Appropriate technology should be the motto of poor and developing countries.

All over the world we need to employ heavy machine-made dredgers to remove sand from the bottom of the sea and to raise the land level above the sea level. Such a move will arrest the rising sea level and many countries of the world such as Bangladesh and the Maldeve Islands which are being fast eaten up by the sea can be save.

Countries like Singapore and the Island Republic of Seychelles are already retrieving land from the sea. In Mumbai, the author of this book lived in the Backbay reclemation in a big Bombay building called Bank House which was built on the land reclaimed from the sea. Next to the Bank House there was the Secretariat of the government of Maharashtra (Bombay) the government of the Stanvac building, a five star hotel and many other high rise buildings. Many governments try to annex the land of neighbouring countries instead of retrieving their own land gone from the sea.

All along the important road called the Marine Lines in Mumbai big boulders and tetropads are placed to prevent the sea from eroding the land. Mumbai is a narrow strip of island but where the land area is big and broad sea water cannals, lakes and lagoons may be dug as the British rulers dug the Buckingham Canal, connecting Andhra Pradesh and Chennai to provide work for the jobless and also to provide water for distillation, desalination and promoting saline agriculture.

The canal took all the flooding rain water, cleaned up the roads and city gutters and provided water for farmers to wash their cattle.

The rains are getting more and more scare and the groundwater table is falling every year by about a metre. Man is the plan and design of the almighty.

This very fast multiplying fresh water fish is God's gift for the poor in many tropical countries. It tolerates water which is little saline. The dead fish yields oil, glue and fertilisers for income generation.

The Case of Tilapia

The Tilapia species OREOCHROMIS NILOTICUS continues to be the focus of development in most of the tropical countries and the poor people of the poor countries themselves want the fish. If millions of people survive today, especially in Africa, it is not because of their governments or scientists but because of Tilapia. As one who has lived and worked with the poorest of the poor in several poor countries, Africa in particular, I can say for sure, that Tilpia helps millions of people to keep their body and soul together.

In several countries NGO's have promoted and popularised Tilapia preserve, Tilapia fertilisers, Tilapia gum (glue), Tilapia chicken feed and so on and generated employment to thousands of people. It is indeed a very fast multiplying and sturdy fish, which can be used as raw material for several livelihood projects.

Therefore, the food and agricultural organisation (FAO) of the United Nations and ICLARM (International Centre for Living Aquatic Resources Management) in the Philippines have been promoting Tilapia. Financed by UNDP, the ICLARM released in 1992 what it calls the "Super Strain Pink Tilapia for propagation to farmers in the Philippines after several years of research and crossbreeding. The developed fish is larger and stronger than what is commonly farmed in many parts of Asia and Africa. Unlike other varieties of Tilapia the new pink variety prospers in flooded rice fields and therefore farmers can introduce the variety in their paddy fields for additional food and income.

The Super Strain has been evolved not by debatable hormonal science or genetic engineering methods which are disputed by several scientists and economists now, but merely by crossbreeding one variety with another, starting from the African Nile Tilapia.

The popular Nile Tilapia can now be crossbred to produce all male progeny to avert inbreeding problems and excessive multiplication. The Nile Tilapia is now farmed extensively in Asia and the combined annual production in Taiwan and the Philippines alone is over 60,000 tonnes. The FAO journal CERES, often highlights this miracle fish, the life saver of the poor.

With the UNDPs financial support and encouragement the latest Super Strain has been produced by crossing four wild varieties of Nile Tilapia with four Philippines strains in different permutations and combinations. This pink strain has survived all conditions differing quality of water and even rough handling. It is surprising that the government of Tamil Nadu have banned the propagation of Tilapia for reasons known only to them. Damn the fish and damn the poor seems to be their policy.

According to the Korean, everybody has a right to water as a social good and no one may claim the sole right to it.

Over the past 10 years the Dutch water table has fallen by 75 centimeters and a lot of vegetation that depends on ground water has died out. Environmental News from the Netherlands, March 2000.

Chapter - 118

Fish is Getting Scarce

Globally about 200 million people are economically dependent on fishing and fish based industries. Sea food provides over 16% of the total consumption of animal protein within Asia and a billion people are dependent on fish as their primary source of protein.

Jobs of over 10 million people could be at risk in the next two decades if the over mechanised fishing fleets continue to operate to double the catch with their installed capacity.

Fish brings in the badly needed foreign exchange to many countries. For instance, Chile in South America earns about US$ 400 million every year by exporting fish and fish products.

To prevent further scarcity of fish and dwindling catch, the best solution will be to take sea water canals into the land and promote sea water lagoons, ponds and pools for the poor to cultivate fish to meet their family needs and have something for the market.

When this is done the fish population will increase and get highly localised. Every village must be encouraged to have its own fish pond, drawing water from the nearby sea water canal dug by the people and for the people, encouraged and helped by the government.

When plenty of fish is made available several fish based industries (oil, fertiliser, glue, etc.) can be promoted to enable people to stay in their own villages instead of migrating to cities to live in slums and squatter settlements in search of jobs that aren't there.

Sea water lagoons can save the world from the increasing food crisis and the impending fish crisis.

Sea water is better than no water. Saline agriculture is better than chemical agriculture. Sea water canals can also be used for washing, saline agriculture and water transport.

There has to be a sea change in our approaches to development, especially when we talk about sustainable development. The sea and the saline tree have the answer.

As already stated in the earlier pages, sea water must be brought into the land by man-made sea water canals. The water must fill the dry rivers, ponds, pools, lagoons, swimming pools and catchment areas. This is the only way to replenish our fish stock and save the poor from malnutrition and starvation.

> A hungry person needs food today and not tomorrow. The Roman philosopher Seneca said 2000 years ago, "Hungry people listen not to reason, nor care for justice nor are bent by any prayers."

Chapter - 119

Common Uses for Tamarind Kernel Powder

It is good for binding briquettes for boiling sea water for distillation.

1. The powder is good as a sizing material in the textile, tanning, paper, board and jute industries. It contains POLYSACCHARIDES. It is also used to clarify and concentrate latex in the rubber industry.
2. The hard and brown skin (testa) is a source of good tannin and can be sold to the tanning industry. Tannin can also be extracted on a rural scale under the open sky.
3. The roasted kernel powder when ground and used in the confectionary industry can reduce costs, as the kernel contains 70 percent carbohydrates, 20 percent protein, 3 percent crude fiber and 3 percent ash.
4. Raw tamarind seeds are crushed (or roasted and crushed) and mixed with cattle feed, chicken and duck feed.
5. The glue is also mixed with clay, surki (burnt clay powder), and mineral lime (chunam) to add to the binding property of sun dried bricks. When the gum is boiled and made thick, the bricks bind quicker and stronger. Of necessary, some of the gum that oozes from trees like the murunga (drumstick) tree may also be added in semi-liquid form.
6. The tamarind kernel powder contains 7 percent to 8 percent fatty oil which can be extracted and used for soap making. The oil is rich in LINOLEIC acid (2 percent) and OLEIC acid (19.6 percent). It also contains LINOLEIC acid (2.8%) and BEHENIC acid (9 percent), all finding are used in soap manufacturing and other industries.

7. About 5 lakh tonnes of tamarind fruit is produced in India every year which gives 1.5 tonnes of seed that can yield 10,000 kilograms of oil per annum.

> ONLY WHEN THE LAST TREE IS CUT, THE LAST RIVER DRIES, THE LAST FISH DIES WILL WE REALIZE THAT WE CANNOT EAT MONEY.
>
> E.S.Cheltenham in his book, BATTLE FOR WATER : EARTH'S MOST PRECIUOS RESOURCE state that in Asia, approximately 86 percent of fresh water is used is for agriculture, 8 percent for industry and 6 percent for domestic purposes.
>
> Sudan is the largest country of Africa suffering without water.

Chapter - 120

Oil Distiller With Village Materials

A simple and inexpensive way of distilling oil from the leaves, nuts, pods, barks and roots of oil-bearing plants and shrubs for steam distilation of sea water.

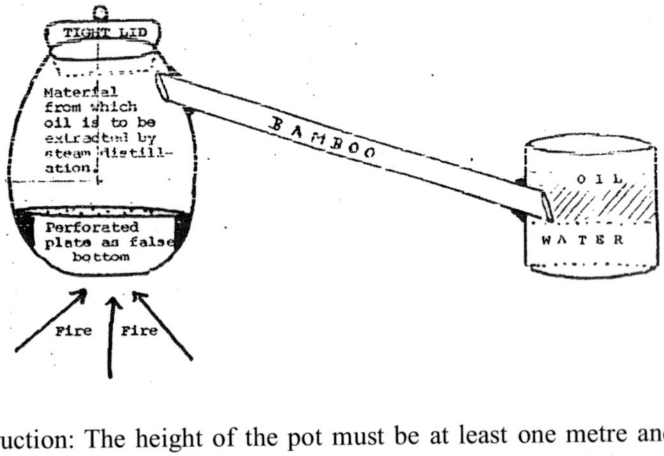

Construction: The height of the pot must be at least one metre and breath half a metre. The lid must be tight and leak proof. A wire belt may be tied round the pot and the lid connected and tied to it with a binding wire as shown. The earthern pot must be as strong as possible, made of good clay, surki (burnt clay powder), mineral calcium (chungam), some jaggery and plaster of paris (natural gypsum stones burnt and powdered). If cement is easily available a little may be added. The bamboo must be two metres long and smooth inside. The two joints must be tight and plastered as shown.

How to use : Mount the pot on three stone with a fire underneath. Fill water upto the perforated false bottom. The steam will carry the oil and flow out to the water bath and condensation will take place inside the bamboo. To hasten condensation the bamboo may be wrapped with a wet cloth the pot may be lodged inside a tank or small chamber built of bricks and placed on the ground 1.25 metre of the size 1.25 metre x 1.25 metre to prevent accidents in case the pot breaks due to excess pressure inside. But this should not happen as the excess steam will escape through the bamboo. (if metal wire is not available to be used as a belt around the pot a strong fiber rope may be used).

All types of nuts, barks, seeds, leaves or roots that have to be steam distilled for oil may be put into the pot. The oil in the steam will condense and float on the water in the bath as shown and can be siphoned off. In every village about a dozen families may distil the oil and make it available for all those families who supply the raw materials. Kilo of oil may be sold for 30 paisa (Indian) and a kilo will be adequate to cook a days full meal for the family. This distiller will provide rural employment to save trees from being cut, and keep the surroundings clean.

Chapter - 121

Doorway Pot Cooler

Cooled Water for Drinking During Summer

The poor, who cannot afford a refrigerator and people who do not have electricity connection in their houses may follow this method to make drinking water, lime juice, buttermilk or coconut water, cool in summer. The device may be used, also to preserve eatables in containers under low temperature.

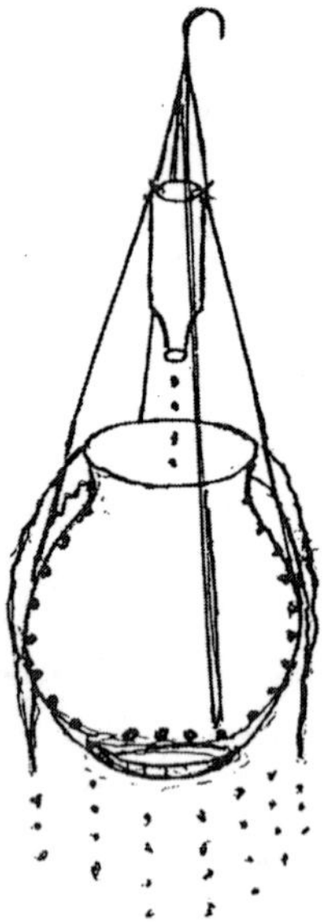

All that is required is a large size mud pot, a pot rest made of straw, and a bottle. A big plastic bottle, with the bottom removed is preferable. About a hundred holes may be made all around the pot each bot enough for the little finger to enter. The pot rest or a used scooter tyre, may be tied to four ropes and hung from a nail or hook. The pot sits on the ring held by the ropes on the four sides.

The bottom of the plastic bottle is removed and the bottle is tied to hang upside down as shown in the diagram.

A saucer-like mud lid, without any hole covers the pot, a thick towel covers the whole pot and the lid, excepting the bottom of the pot. The towel ends hang along the sides of the pot.

Water or drink to be cooled is put into five or six bottles, corked and placed inside the bottle. Then, cold water, may be sea or saline water, is poured on the towel and filled into the bottle through the opened-out bottom. The inverted bottle has a tight lid with a small hole for water drops to trickle down in quick succession. This keeps the towel and the pot always wet. Some water which seeps into the pot through the holes will drain out through the holes in the bottom of the pot. The inverted bottle is filled with more water as and when necessary.

Drinking water kept even in a plain mud pot becomes cool but under this system the water becomes much cooler, even during a very hot summer the water keeps very cool. This system is commonly followed in Indian villages and even by tribals. The pot may also be hung from a tree branch out in the open.

> The Yellow River in China which flows through the country's most important agricultural areas, dried up for 226 days in 1998 reports Environmental News from Netherlands, March 2000, No. 1.

Chapter - 122

Salicornia in Kutch

Salicornia is a useful sea water shrub which multiplies very fast. The Thaper Group of industries has selected the Kutch region of Gujarat in India to cultivate it and sell the produce and products made out of it.

Salicornia farming started in 1979 in Arizona, USA, initiated by the University of Arizona and R and D is still being carried out.

Rahul Chaturvedi who has been associated with the project in Kutch is trying to get about a dozen motorised borewells to flow the ocean water into the farm of 160 acres, which has been brought under cultivation already. Presently, about 13 experts are doing R and D in the farm and the results are very encouraging up to now.

It is said that about 25,000 square kilometer in Gujarat and about 6,25,250 Square kilometer in other parts of the world can be brought under Salicornia farming.

Salicornia is a good fodder, for cattle, says, Vardhman Jivdaya, a researcher in the field, and that edible oil can be extracted out of the seeds. It can also be used as raw material in paper and board factories and for making bricks for small buildings. Scientists say that the plant absorbs carbon dioxide from the atmosphere.

Salicornia farming has started also in Rajasthan. Project manager Vivek Padamanabhan at Luni says, salicornia seeds were imported from Arizona after it proved successful in Saudi Arabia and propagation was first tried in Bhangesar in the Pali district of Rajasthan. So the seeds are already available in India and need not be imported to promote new farms.

Salicornia needs a cold atmosphere for the initial 100 days. So in India sowing should be done in November or December. The bush bears flowers in May and seeds are ready in June – July.

Belarpur Industries is collaborating with the Halophyte Enterprises for the Luni project. Gordon Amberbian of this company is himself working there with specially designed tractors.

The yield in this project is about 15 metric tonnes of seed from 1 hectare. The seed has 28 to 30 percent oil by weight. The quality of the oil is rich and can be compared with saffola oil.

After the oil is extracted the residue has about 42 percent protein in it, which is used as Maji Maharo of Jodhpur has also taken to this farming and many farmers of Rajasthan are showing much interest. Pakistan and Oman have also initiated Salicornia farming.

Salicornia leaves are used as spinach in some Western countries Reported by Tridiv Vaidu in the Gujarat magazine Chitralekha – Annual Issue 1996. Translated into English by Miss Ullas M. Kapadia of Vastrapur, Ahmedabad, for RYFO.

Chapter - 123

Avenue Trees Already Available in Chennai

S.No.	*Species*
1.	Samanea Saman
2.	Mimusops elengi
3.	Delonix regia
4.	Peltophorum Pterocarpum
5.	Cassia Siamea
6.	Cassia Merginata
7.	Cassia Javanica
8.	Syzygium Cumni
9.	Derris Indica
10.	Tabeluia Rosea
11.	Lannea Coromandelica
12.	Albizia Lebbeck
13.	Thespesia Populnea
14.	Hibiscus Tiliaceus
15.	Tamarindus Indica
16.	Sterculia Foetida
17.	Polyalthia Longifolia
18.	Polyalthia Pendula
19.	Ficus Religiosa
20.	Ficus Bengalensis
21.	Ficus Retusa
22.	Cassia Fistula
23.	Parkia Biglandulosa
24.	Swietenia Mahagoni
25.	Couroupita Guianensis

Most of these trees can be found in warm tropical countries. They need watering only for a month or two but require a water borne atmosphere. Foliage of these trees make excellent briquettes to steam distil sea water.

Can be taken up on a large scale

S.No.	*Species*
1.	Butea monosoerma
2.	Erythrina Indica
3.	Calophyllum Inophyllum
4.	Alstonia Scholaris
5.	Bambax Ceiba
6.	Dalbergia Sissoo
7.	Spatho dea campanula
8.	Ailanthus Excelsa
9.	Albizia Procera
10.	Artocarpus Heterophyllus
11.	Bauhina Purpurea
12.	Bauhina Variegata
13.	Lagerstroemia Flas – regikal
14.	Mangifera Indica
15.	Michelia Champaca
16.	Millingtonia Hortensis
17.	Tecomella Undulata
18.	Terminalia Arjuna
19.	Terminalia Bellirica
20.	Terminalia Tomentosa
21.	Pterocarpus Marsupium
22.	Pterocarpus Santalinus
23.	Anogeissus Latifolia
24.	Bambusa Vulgaris
25.	Bamningtonia Acutangula
26.	Caesalpinia Coriaria
27.	Cochlospermum Gossypium
28.	Crataeva Religiosa
29.	Commiphora Caudates
30.	Diospyros Melanoxylon
31.	Bassia Latifolia
32.	Elaeodendron Glaucum
33.	Feronia ELephantum
34.	Aegle Marmelos
35.	Filicium Decipiens
36.	Gmelina Arborea
37.	Hardwickia Binata
38.	Holopteka Integrifolia
39.	Kigelia Pinnata
40.	Melia Azrdatach
41.	Melia Dubia
42.	Salevadora Persica
43.	Sapindus Emarginatus

Around 40 percent of the precious water used for irrigation is lost on it's way to the fields. The amount of irrigated agricultural land per head of population has been declining due to water shortages since 1980. The water table is falling steadily in important farming regions such as the American Midwest and the Punjab in India.

Chapter - 124

Sea Water Kitchen Gardens

Plants and trees that cannot stand saline soil or brackish water can be raised by this method innovated by the Ryan Foundation. Experiments conducted in some African countries have proved very successful.

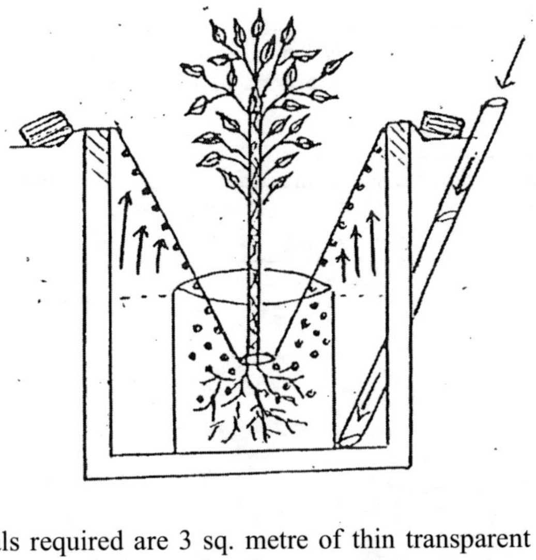

Materials required are 3 sq. metre of thin transparent plastic sheet, 200 hand-comprssed and sun-dried bricks, a 20 litre garden or kitchen mud pot, RYFO mortar (no cement) and a 6 cm. dia hallow bamboo about 120 cm. long.

First, a one metre X one metre pit is dug and the floor and sides are lined with bricks and plastered with mortar. While building a side wall the bamboo is placed slanting as shown in the diagram. The mud pot is placed in the middle of the pit as shown and filled with good earth mixed with organic manure. (RYFO manure is recommended).

In the middle of the plastic sheet a circular hole of 10 cm. diameter is cut. The pit is covered with the sheet and the central portion of the sheet taken down into the mud in the middle pot as shown. Weights are placed on the colour of the sheet around the pit as shown and sand heaped around the pit on the plastic to prevent air entering the pit.

The required vegetable, say the quick growing okara (lady's finger) is sowed or planted in the middle of the central pot as shown.

Sea water is poured through the top end of the bamboo and the mouth of the bamboo is plugged.

During day the sea water evaporates, hits the sides of the plastic sheet, gets condensed and rolls down the sheet to the roots of the plant. Thus, the plant gets something like pure rain water and flourishes. The method can be adopted also to growing trees in the middle of deserts provided man-made sea water canals are taken into the deserts. The sea and the tree holds the key for human progress.

> Fused out tubes of tube lights, metal ends removed may be used in the place of bamboos. Table and kitchen waste may be mashed and put around the plant as manure.

> We can expect growing pressure-points around water, food and energy scarcity as countries progress. By 2030, almost half of the world's population will live in regions of high water stress or shortage.

Chapter - 125

Tyres Multiply Fish

It sounds strange but it is true and it is happening in Japanese Islands for the past several years.

There is a convention or unwritten law in Japan that worn out and unwanted tyres of automobiles of all sizes must be thrown into the sea.

The local administration areas of Japan, called prefectures, have marked areas in the sea shore where old tyres are to be thrown. Periodically, they arrange for steam boats to lift these tyres and dump them about 10 kilometres from the shore.

Big fish lay eggs inside the tyres and the fry stay there and grow up till they are fit enough to dodge and escape from the big fish which chase them to swallow them.

This system has made Japan the leading fish producing and consuming country in the world, better even than Norway, known for deep sea fishing. Why don't all coastal countries follow this system to multiply the fish population and provide plenty of fish (animal portein) to the people ?

Vegetables grow on boats

Similarly there are hundreds of unmechanised boats in that country of the size of about 100 ft × 50 ft × 4 ft filled with soil from the earth and vegetables, fruits and flowers are grown in these floating boats.

These floating gardens are tied with long ropes about 1000 feet long and tied to posts and pillars on the coast. Once in 3 or 4 days they are drawn towards the shore and fresh water is sprayed on them. The sea water-borne moist atmosphere is healthy for their growth. During windy days these boats are towed to less windy sides or neighbouring islands.

Fish in tyres and vegetables, spinach and weeds can be grown and multiplied even in sea water canals, lakes, ponds, swimming pools and catchment areas to grow more food for the people and reduce transportation distances and thereby also reduce the cost of fish and vegetables.

Half the world is starving or undernourished. These methods have to be resorted to in order to save the poor and malnourished children.

In a Lighter Vein

Question	:	What are the states that grow food to feed people in India?
Answer	:	Andhra State, Punjab State, Tamil Nadu State, Kerala State and the United States.
Alas	:	Even these states, which were producing surplus food, are now importing food grains due to water scarcity and soil nutrition deficiency added to power scarcity. Will India copy Japan to increase production of fish and vegetalles? It is a million dollar question.

Chapter - 126

Halophytes Must Be Exploited

“India has been meeting its growing demand for agroforestry products such as plywood, timber, pulp and paper through imports, while the domestic industries have been under-performing. According to the National Research Centre for Agroforestry (NCRA), the country imported 6 million cubic metres of timber and round logs and spent, 18,000 crore last year.

India does not even grow halophytes (sea water plants and trees) and drought trees which require little or no care. India cuts more trees than what nature can replenish. Money allocated for trees encourage corruption and fraud and contributes to inflation.

> The book HALOPH by J. Arensen published by the office of the ARID LAND STUDIES of the University of Arizona lists 1,560 species of HALOPHRTES (salt loving plants and trees), many of which may be cultivated with sea water or with a mixture of sea and sewage water.

Chapter - 127

Dhuandhar Falls

One of the most powerful waterfalls of India DHUANDHAR in Jabalpur of Madhya Pradesh is a sight to be seen. River Narmada crashes on the marble rocks of BADA GHAT to create a cascade of water.

The water overflows during rainy season and almost dries up during summer causing draught in Jabalpur.

RYFO SOLUTION: When the water overflows it must be led into a large and deep swimming pool. When the pool is full a thin layer of castor oil must be sprayed on it tubes filled with air must be thrown on the water to float to further reduce evaporation.

Water from these pools may be drawn during peak summer when there is water scarcity. Several pools of this type will solve the water problem in Jabalpur.

All over the world surplus water during rains may be stored and preserved thus to ease water scarcity during peak summer.

Millions of gallons of rain water goes into the sea now world over and this must be stopped.

Dust floating in the air will settle down on the tubes and oil will not mix and dirty the water. The water in the tank need not be filtered, treated, heated or chemicalised.

In the middle of large lakes and catchment areas huge tanks may be built of the size 100 x 100 x 20 and rain water around the tank in the catchment areas may be pumped into the tanks by wind mills or diesel engines. The oil and tube will keep the water clean and the supply of water to the people can be regular. This appropriate technology will be several times cheaper and easy to commission than the RO process which does not reach the poor in remote areas. Moreover the villagers or local communities themselves can maintain the system without any external help or aid.

Good management is in finding sensible alternatives.

One of the most attractive tourist spots of Kerala has the lake VEMBANAD. The lake is suffering from chemical run off from paddy fields, waste from households, over fishing, sand mining and lack of drinking water. Isn't there a way to save this and similar beautiful back water lakes.

Chapter - 128

Man-made Lakes (Case Study)

The world has thousands of man-made lakes. About 100 of them are considered very big and great. The year book 2010 lists 30 of the greatest and remarkable lakes. Some of them are listed below:

1.	Owen Falls – in Uganda –	204,800 million cubic metres	completed 1954
2.	High Aswan (Sadd-el-Aali) in Egypt	168,000 m.c.m	completed 1970
3.	Upper Wainganga – in India	50,700 m.c.m	
4.	Sao Felix – Brazil	50,600 m.c.m	
5.	Sammenxia – in China	35,400 m.c.m	1960
6.	Glen Canyon – in Arizona	33,304 m.c.m	1964

In Tamil Nadu, the last state in the India peninsula, there is the Veeranam lake, the largest man-made lake of Tamil Nadu dug during the region of King Paranthaka of the Chola Dynasty in the 19th Century A.D.

The last Sri Lanka king Wickrama Rajsinghe also had several deep tanks dug in 1806.

The Pulicat lake, again in peninsular India, which is the second largest lagoon in the country is a shallow 441 sq. km. lake, the maximum width of which is 19 km. The lake, partly in Tamil Nadu and partly in the adjoining state, Andhra Pradesh host 52 fishing villages around it and on the islands in it. Studies show that the lake was about 3.8 metres deep before the Dutch arrived in 1610 and made Pulicat their chief East Coast settlement. The lake has been silting up at the rate of about one metre per century and what is left now is 0.8 metres. According to present projections, the lake will turn into a small marsh within 50 – 60 years and die an ecological death if action is not taken to reduce the rate of siltation.

We need more and more of extensive man-made lakes in India and other countries to hold the surplus rain water now going into the sea. Such lakes could hold an enormous amount of river water which can be drawn through pipes and supplied to towns and cities.

Well built open cut man-made canals should connect the rivers which carry surplus water to man-made canals for storage. As stated earlier a thin layer of castrol oil sprayed on the water in the lakes will retard evaporation and make water available during summer or drought.

Tamil Nadu Government in South India wants to interconnect Mahanadi, Godavari, Krishna, Pennar, Palar, Cauvery, Gundar, Pambo, Achankovil and Naippar rivers. It is a sure way of distributing drought. Sir Viswesveraya who built several dams in India, Prime Ministers Nehru, Indira Gandhi and very many water experts in India and abroad have given up the idea of interlinking rivers. If done, water will not only flow into dry and semi-dry rivers and millions of people will be deprived of the water they are now getting. Much water will go underground but the ground water will not come up in any good measure.

Chapter - 129

Karachi Facing Drought (Case Study)

Karachi (Pakistan), a city of roughly 18 million people is growing fast in population and the water shortage in the city is acute. The city gets most of its water, about 550 million gallons per day from the Indus river and another 100 million gallons from the Hub dam that is supplied water from the adjoining Buluchistan province. But in recent years drought has dwindled the supply and only half the needs of the city is met which is about 12 million gallons per day. More water is supplied by tankers than by pipes. The tankers go round the city Two or Three times a week and fill the underground cisterns.

Criminals siphon water from various points and have built their own private arrangements to supply and sell water. Reports say that the Karachi Water and Sewage Board is over staffed and payment for water supplied is not collected regularly. The net result is, the water board is sinking while Karachi is drying.

Karachi must steam distil sea water on a cottage scale and on a commercial scale using solar and briquette renewable energy with parabolic magnifying glasses to increase the heat as being done now in Saudi Arabia according to the innovation published by the Ryan Foundation in India.

> Some governments want to nationalise the sun, wind and wave energy which are god given natural free resources. They should never be priced or traded with or taxed. They are gifts of god to millions who are living and billions who are yet to be borne. People should be encouraged and helped to draw any amount of energy from them for their survival and well being.

The root cause of millions of people being unemployed, underemployed, underpaid or living below the poverty line is scarcity of water even to wash, bathe and meet basic needs and proximity to the available scanty water. In many rural and urban homes family members have to hunt for water and stand in queue for half a day to fetch a small pot of water. Sea water is better than no water. Sea is an important resource of the poor. Governments should exploit the sea for the welfare of the poor, Pakistan is not a poor country but a poorly managed country.

Chapter - 130

Ground Water Depletion in USA (Case Study)

The Colorado river basin in the US has lost nearly 65 cubic km. of fresh water over a period of 14 years which is double the volume of America's largest reservoir, the lake MEAD of Nevada.

About 50 cubic km. of the loss was from ground water which is more than three quarters of the total volume now available for use. For the past 15 years the Colorado basin region has been experiencing severe drought.

The loss came to light from observation by NASA's gravity recovery and climate experiment (GRACE) mission.

California is battling against severe drought for the past 3 years 2012, 2013, 2014 and the drought may continue for some years. The main sources of water for California are snow melt from the Sierra Nevada, imported water from the Colorado river and ground water. All the three sources are yielding less and less water. Therefore, for the past six years, the Orange County has been using highly purified waste water to replenish groundwater resources. Our sources of supply are literally drying up says Michael Markus, general manager of the Orange County water district.

The Fountain Valley recycling plant produces 70 million gallons a day, turning residential waste water from dishwashers, showers, washing machines and toilets into drinkable water.

But recycling used dirty water from thousands of households is also a problem. The collected water goes through three stages of production; filteration through a series of tiny straws to remove bacteria, reverse osmosis to remove dissolved salts, and chemicals and exposure to ultraviolet light with hydrogen peroxide. By the time it leaves the purifying plant it is distilled water sans nutrients needed by the human body.

Pure distilled water is lethal. It has no vitamins, minerals or enzimes. These lost nutrients may be restored in homes that use the water by soaking in a pot of water, some organic matter such as common grains and vegetables readily available in the kitchen. They may be bundled in a cloth as big as a tennis ball and put into the pot to stay for 2 or 3 days. See chapter 174.

Water must contain some protein, carbohydrate, calcium, iron, vitamins, enzymes, minerals, sodium, potassium and so on, to iodise water add about 50 milligrams of potassium iodate to 10 litres of distilled water.

Bibliography

1. Pumps without water Felix Ryan, Ryan Foundation 1998.
2. Survival by Sea Water Felix Ryan, Ryan Foundation 1990.
3. Water Management in Homes and Villages Felix Ryan 1993, Books for change 19, Richmond Road, Bengaluru, India [Felix] Ryan, Ryan Foundation 1991.
4. How to Convert Sea water into Drinking Water Felix Ryan, Ryan Foundation 1998 Felish Ryan 1993, Books for change 19, Richmond Road, Bangalore, India
5. Sustainable Safe Water for all Felix Ryan, in press.
6. Ground Water S. Ramakrishnan, H34/F TNHB, Colony, Tiruvalluvar Nagar, Chennai – 41.
7. Fresh water for India's Children and Nature, Ashok Nigam et al : 1998, UNICEF and WWF.
8. Water Sanitation and Health and MOreso, Tinker et al : Earthscan Paperback.
9. Riskful Confusion of Drought and Man Induced Water Scarcity, J. Bandyopadhyay, 1989, AMBIO 18 (5) 284 – 92.
10. Alternative Sources of Drinking Water. The Mahiti Water Experience, Dinesh Kumar et al Pacific Institute San Francisco.
11. Systems Approach to Water Management, Biswas Asit, Mcgraw Hill (London).
12. Water Resources Management in India, Achievements in Perspectives, Guyle Moigne et al : the World Bank.
13. Water in Crisis, P.H. Glieck, Oxford University Press.
14. Technology Mission on Drinking Water, Government of India, 1986.
15. Water for Third World Development, A.K. Biswas, 1976.
16. Rainwater Harvesting, Dept. of Rural Development, Min. of Agriculture, G.O.I, 1989.
17. Ground Water Resources in India, Central Ground Water Board, G.O.I. Faridabad, Haryana.
18. International Food Policy Research Inst. - (IFPRIO 1994).
19. Water Resource Development in India, A.N. Khosla - Paper presented to UN Water Conference, Mardel Plata.
20. Ground Water Availability for Drinking in Gujarat, Economic and Political Weekly, March 26, 1996, New Delhi.
21. Alternative Water Allocation Mechanisms, Economic and Political Weekly, March 30, 1996, New Delhi.

22. Montreal International Forum, Montreal Centre, June, 1900.
23. Sustainable Financing of Water and Sanitation Proceedings of 22nd WEDC Conference – A Nigam.
24. Sea Water for Survival Felix Ryan Proceeding of WEDC Conference – Reaching the Unreached, Sept 1996.
25. Dividing the Waters, Food Security, Postel S. World watch paper 132, WW Institute, Washington, D.C.
26. Conserving Water, The Untapped Alternative. World Watch – paper 67.
27. Freshwater Management in the Arid Region, Vani, Tatu, Patel and Others, IRMA, 1997.
28. Traditional Drinking Water System in Rajasthan. M.S. Rathore, 1991-92.
29. Rural Water Supply and Sanitation on Indian Water Management Sector. Rehoej, Glennie, Abeyratna and Sjorslev – prepared for World Bank by G.O.I.
30. Peoples Initiative in solving water crisis, Eco. and Political Weekly 29 (34).
31. Water Supply, Sanitation and Environmental Sustainability. The Financing Challenge, I. Serageldin, 1995, World Bank.
32. Sustainable Development of Ground Water Resources. T.Shah, 1992, Eco and Pol. Weekly, 27 (10-11).
33. User Friendly Irrigation Designs, N. Sengupta, 1993, Sage Publications.
34. World Fresh Water Resources, P.H. Gleick (Ed) Oxford University Press.
35. Sustainable Fresh Water Supply for Chennai Centre. AMMM Nurugappa Chettiar Research, UNICEF – WWF.
36. Looking back to think Ahead, Tata Research Institute, New Delhi, 1997.
37. UNDP (1990), Global Consultation on Safe Water and Sanitation for 1990s – Safe Water 2000, New Delhi, Sept. 1990.
38. UNICEF (1979), People, Water, Sanitation – 1995, 45 / 46, UNICEF, Geneva.
39. UN 1980, International Drinking Water and Sanitation Decade – Report of Secretary General (A/35/367).
40. U.N. Earth Summit Agenda 21, Programme Action from Rio.
41. UNCTAD 1988, Our Common Future Recommendations related to water.
42. Managing Water Scarcity for Water Security – A Review, FAO, April 1997.
43. The Cost of Inaction – World Bank, 1995.
44. The Internationl Drinking Water Supply and Sanitation Review WHO.

More and more artificial deep lakes with high rising bunds must be dug and built as was done by the last Sinhala King Vikram Rajasinghe of Sri Lanka in 1806, like the Veeranam lake considered to be one of the biggest lakes of Tamil Nadu, built during the reign of King Paranthaka of the Chola dynasty in the 19th century.

Releasing the world bank's draft report on India's water economy, John Briscoe, country director of Brazil and former senior water advisor for South Asia said that the country (India) would have to change from water development to management. He added that unless India makes drastic changes, it will have neither the money to maintain and build new infrastructure nor water for its economy and people.

Index

C

L

M

Q

R

S

SURVIVAL
BY
SEA WATER

SURVIVAL BY SEA WATER

(VOLUME - 1)

Dr. F. A. Ryan

GYAN

© Author

All rights reserved. No part of this work may be reproduced, stored, adapted, or transmitted in any form or by any means, electronic, mechanical, photocopying, micro-filming, recording or otherwise, or translated in any language, without the prior written permission of the copyright owner and the publisher. The book is sold subject to the condition that it shall not, by way of trade or otherwise, be lent, resold, hired out, or otherwise circulated without the prior publisher's written consent in any form of binding or cover other than that in which it is published.

The views and opinions expressed in this book are author(s) own and the facts reported by them have been verified to the extent possible, and the publishers are not in any way liable for the same.

Published, 2021

Published by

Gyan Publishing House
5, Ansari Road
Daryaganj, New Delhi-110002
Phone: 011-47034999, 9811692060
www.gyanbooks.com
E-mail: books@gyanbooks.com

Printed at: G. Print Process, Delhi.

Cataloging in Publication Data—DK
Courtesy: D.K. Agencies (P) Ltd. <docinfo@dkagencies.com>

Ryan, Felix A., author.
Survival by sea water / Dr. F. A. Ryan.
volumes cm
Includes bibliographical references.
ISBN 9788121213615 (HB) set
ISBN 9788121219754 (HB) vol. 1

1. Seawater—Physiological effect. 2. Seawater—Physiological effect—India. 3. Survival at sea. 4. Survival at sea—India. I. Title.

LCC G540.R93 2019 | DDC 910.54 23

Contents

Foreword

Recent floods in Chennai city, unprecedented by its magnitude and scale of damage, has once again brought into focus the issue of climate change. Very few are aware that soon after this disaster another flooding of a city, equally unprecedented, happened in northern England. These two disasters didn't happen by accident. On the contrary, they are part of a pattern that will keep unfolding fairly consistently in future. Climate change will bring in its wake severe floods as well as crushing droughts: in either case water will be at the centre of crisis. While severe floods might obliterate the sources of water forever, deadly droughts will dry up the last droplet. As Paul and Anna Ehrlich wroted: "Though it is irrational to state that people might actually kill each other for water, such a situation cannot be ignored not in the distant future." Future of mankind will witness a completely different kind of war; it will be a "water war".

One of the foundations of managing climate change is "adaptability." Viewed from this perspective the book, "Survival by Sea Water" by Dr Felix Ryan comes across as a harbinger of hope and growth for people all over the globe. The author asserts that we must turn to the sea positively in all possible ways. Ryan has come up with this path breaking observation after an extensive research of existing data across regions. He emphatically says that climate change will reduce renewable surface water and ground water resources significantly in most dry sub-tropical regions, exacerbating competition for water among sectors. He says in Thailand around Bangkok the water table fell by 80 feet and in peninsular India in many areas it fell by 200 feet. Texas and Oklahoma lost about 30 percent of their irrigated area due to falling water table. What is worrying is the water table steadily goes down and down.

There are primarily two issues relating to availability of water. First, the cost of it. In order to surmount the problem of adequacy of water technology of high cost is inescapably employed. While such projects require heavy investment thereby rendering provision of water expensive which the rich can always afford, the poor are completely bypassed in the process. In order to overcome scarcity the USA is buying water from Canada. Singapore is buying water from Malaysia and Indonesia. Second, even the expensive Reverse Osmosis (RO) method does not ensure quality of water fit for drinking. Many impurities including dissolved solids are not eliminated through this process. These twin issues are further compounded by the fact that in many countries, mostly developing, water distribution system is woefully inadequate. Around 60% to 70% of the people in these countries do not have access to safe drinking water. The situation is going to grow worse in not distant future. Robert Engelman, Director of the Population and Environment Programme observes that the ratio between population and renewable water is dropping below the minimum needed for health and economic development in more and more countries.

Against this context the alternative suggested by Dr Ryan brings hope for survival of the humanity. After sustained research and rigorous tests he has come out with about 40 methods of distillation of sea water into potable water. His research says that most of the communities live within a radius of 100 kilometers from the sea. It will be possible, he says, that the local governments can involve the target communities in creating small canals taking off from the sea shore into their neighbourhoods and provide connecting inlets into each of the households. By using reflectors, says Dr Ryan, the solar thermal temperature can be increased to 50 to 60 degrees Celsius and the sea water collected in household pots can be exposed to the sun. Water in the form of steam will evaporate fast in the cool breeze of the night filling the receivers. By this domestic distillation sea water is converted into potable water at least cost. What is remarkable is that this arrangement is scalable and the local communities are fully involved. Hence there is the most significant factor of community ownership coming in. Dr Ryan has innovated 14 such basic models to convert sea water into drinking water. He recommends that all public institutions like schools, colleges, hospitals etc should adopt these methods apart from households.

Survival by Sea Water is undoubtedly the most revolutionary concept advanced by Dr. Felix Ryan, an awardee of the United Nations Global 500 Honour in 1989. His observations and solutions are solidly based on extensive field research in some Third World countries. His call can be ignored at our own peril. His low-cost, people-friendly and environmentally sound methods of providing drinking water are here to stay. Only communities can decide.

Debendranath Sarangi, IAS (Retd)
Former Chief Secretary,
Government of Tamil Nadu.

Preface

Global water use has increased four times since 1940 and the scarcity of freshwater is felt everywhere. The less than 0.4% of the water available in our planet is not really available in uniform density everywhere.

Only 25 liters of water is available for an average Indian everyday estimated on a per capita basis while millions go without water to quench their thirst adequately. Over 90% of the water available in India is used for irrigation while globally three quarter of the water is taken away by agriculture.

The quality of water that is available is deteriorating day by day. The city of London observed quality deterioration in the quality of water as early as in 1845 when the foundation Trafalgar Square was built and the builders had to go as deep as 112 feet to find quality water. By the beginning of the 20th century England drew so much underground water for industrial, agricultural and domestic use that the water table fell to 236 feet below the surface level. The deep well which supplied water to the fountain in Trafalgar Square was abandoned.

In Thailand, around Bangkok, the water table fell by 80 feet and in peninsular India in many areas it fell by 100 ft. Texas and Oklahoma lost about 20% of their irrigated area due to water scarcity during the 20th century.

The problem of quality of water is getting worse everyday. Contamination by industrial waste, chemicals, artificial fertilizers and pesticides has effected quality and in several places the water is unfit for human consumption. In Florida over 1000 wells have been closed because of contamination and in California 20% of the wells have pollution levels above the official safety limits. In Britain 10% of the aquifers are contaminated and in Hungary 773 wells have been closed due to contamination.

THE INDIAN STANDARD 10500 of 1991 CLEARLY STATES THAT DRINKING WATER SHOULD BE FREE OF PESTICIDES.

In most of the Third World Countries 60 to 70% of the people do not get safe water to drink. Over 70% of India's water is polluted and 41 out of the 44 large cities of China have polluted ground water. As the book "Healing the Planet" written by Paul Ehrlich and Anna Herlich says, "fresh water may well turn out to be that requisite for many human positions and water security could be the factor that ultimately limits the scale of human enterprise". They point out that the scarcity is already a limiting factor of human progress in some parts of the globe. "Though it is irrational to state that people might actually kill each other for water, such a situation cannot be ignored not in the distant future".

All these facts, figures and projections are contained in the book "Water for Sustainable Development in the 21st Century" edited by Asit K. Biswas, Mohammad Jellali and Glenn Stout and published by Oxford University Press, Channai 600 006 (274 pages).

A study published by the Population Action International (PAI) in USA warns that by the year 2025 every third person in the World will suffer acute water shortage.

More than 335 million people lived in countries defined as "water stressed" in 1990. A water stressed country is defined as one that has less than about 1700 cubic meters of renewable water per year per person. Water scarcity is defined as availability of less than 1,000 cubic meters per person per year.

The study report called "Sustaining Water: Population and the Future Renewable Water Supplies" says, "present figures are projected to increase eight fold by 2025 to between 2.8 billion and 3.3 billion people by 2025". The difference between the two figures reflects how different rates of population growth could affect water availability, even over a period of just 35 years.

DEVELOPMENT IS ABOUT BUILDING HUMAN CAPABILITIES.

In 1990 out of the 28 countries that experienced water "stress" or "scarcity", all but 5 were in Africa or the Middle East. Djibouti had only 23 cubic meters per head per year and Saudi Arabia had 306. Kuwait had about 75 cubic meters. Tunisia had 540, Jordan 327, Kenya 636, Israel 461, Somalia 981 and Algeria 689. These drought-ridden countries will be joined at least by 18 other countries by 2025 and these will include Haiti, Syria, Peru, Iran and Nigeria.

Robert Engelman, Director of the Population and Environment Programme of PAI, who is a co-author of the report, warns, "Fresh Water that is renewed by rain and snowfall is a finite resource in a World with growing demand". He further warns, "There are no substitutes - as human populations grow, there is less renewable water for each person. In more and more countries, this ratio is dropping below the minimum needed for health and economic development".

The report has been hitting the headlines ever since it was published because of the following reasons:

1. It projects dramatic water shortages based on three population projections of the United Nations linked to the 149-country index that apportions known quantities of renewable fresh water on a per capita.
2. This is the first time the freshwater issue was raised by an NGO of International repute.
3. Scarce water can involve several countries in a major conflict and even in wars.

More than 200 rivers and large lakes boarder two or more countries and at least 10 long rivers flow through a dozen or more countries. Countries like the Netherlands, Egypt, Syria, Iraq, Cambodia and Sudan are the most vulnerable. All these countries rely on water resources that are out of their national boundaries, for a large amount of their renewable water. Their problem is most critical.

PUT ENVIRONMENT FIRST FOR DEVELOPMENT TO LAST.

Distilling sea water is probably the only known solution. Distillation yield/safe water; desalination does not. It removes only the salt molecules basically.

The concept of a "Water Stress Index" was pioneered by the Swedish hydrologist Malin Falkenmark. He identified thresholds for water scarcity and stress.

More hard facts can be found in the book "Sustaining Water: Population and the Future of Renewable Water Supplies", by Robert Engelman and Pamela Le Roy. It is available for US$ 8 from Population Action International, 1120, 19th Street, N. W. Suite 550, Washington, DC - 20036, USA.

United Nations hydrologists and metrologists met at the UK Institute of Hydrology during early December 1998 and agreed that urgent action is needed to prevent most of the plants /from dying of thirst and starvation.

Water scarcity is now slowly spreading to capital-rich USA, which tries to solve the water problem in other countries. Even the Big Brother is in trouble. The USA is now trying to buy water from Canada. A business company in Newfoundland called Mccurdy Enterprises is trying to export water from Springfield Gisborne Lake in the Canadian province, but the government of Canada and environmentalists are against making water an interstate trading commodity on environmental grounds. The environmentalists are absolutely right in their stand. What alternative does America have?

The model country, Singapore, is now buying water from Malaysia and Indonesia. It has a contract with Indonesia for the supply of water for several years, perhaps until Indonesia also runs short of water.

Introduction

Rains are scanty and the ground water level is falling year after year the world over. Hand pumps and motorised pumps, which were drawing adequate water from the ground last year are sucking air this year and frequently get air-locked. The expensive reverse osmosis (R. O.) method which requires costly diaphragms and motive power cannot reach the poor especially those living in utter poverty. R. O. plastic containers have to be sealed and transported enroute by stockists and retail distributors put up the cost of water to the end user.

Moreover, R. O. does not eliminate cancer-causing arsenic, bone-weakening fluoride and many other impurities. It cannot remove heavy metals, total dissolved salts, and toxic and human waste to conform to the quality standards prescribed by the World Health Organisation (WHO) or the Bureau of Indian Standards (BIS). All other hi-tech machines or methods are equally expensive, produce water that is not up to the mark and do not reach the poor while a adulteration of the water is alarming.

The crisis of water management in urban areas is even graver than in the rural areas. In our water supply system, there are enormous losses in the distribution system because of leakage and bad management.

As per the National Sample Survey (NSS) - 65th round, only 47 percent of urban household have individual water connections. It is estimated that as much as 40 to 50 percent of the water is lost in the distribution system. As the distance between the source and the delivery point increases, costs of power.

> God's time must come and what stands the test of time becomes classical. Once, water managers criticised the idea of exploiting the sea. Many thought it to be a quixotic idea but now they say only distillation and desalination of sea water can solve the problem of water and save the people from water-borne diseases.

Pipelines and transportation also increase. Governments find it impossible to subsidise or reduce the cost of water and therefore do not deliver water as needed. The poor are the worst affected as they have to spend much money and time to obtain water for want of individual house connections.

After researching for 50 years to reach safe and adequate water to the people, the contention is that every home, rich or poor, should distil its family needs of safe water on the terrace or backyard or on the sunshades over doors or windows by the steam distillation method using the freely available solar energy to get pure and safe distilled water. Nutrients part in distilled water are easily substituted by organic matter as explained in the following pages.

At 50 to 60 degrees Celsius, water in the form of steam evaporates fast in the cool breeze of the night and fills the receiver. To take solar heat to the boiling point, simple, small and inexpensive RYFO reflectors are used. These reflectors are scientifically explained with diagrams in the following pages.

There are several models (systems) for domestic distillation and the Ryan Foundation, founded by the author, has published 14 basic models in the booklet "How to Convert Sea Water into Drinking Water", which has gone into third reprint,

It is desired that homes and institutions, schools and colleges, hospitals, banks, etc. should follow these methods and distil sea water for their home consumption.

> Coming together is just beginning Keeping together is progress Working together is success Succeeding together is achievement.

The only pre-requisite is that sea water must be made available in close proximity to people's settlements by the government for the people to help themselves. How to make copious sea water available at different locations is explained in this book. We must turn to the sea positively in all possible ways, as recommended by the RIO Summit. Water supply must be decentralised and should not be traded with. There is no other way to solve our almost universal water problem, which is getting worse day by day.

> Greenhouse gas emissions will dry up water resources warns the UN Scientific Panel on Existing and future risks to Asia from Climatic Change; The report of the panel 2013 further warns that there is robust evidence of water availability being hit significantly with increasing emissions. "Climatic change will reduce renewable surface water and ground water resources significantly in most dry sub-tropical regions, exacerbating competition for water among sectors.

> The conventional wisdom of what is now taught as economics bypasses the poor, the very people who really need development.

Poverty threatens the environment but wealth is by no means guarantee against the menace. In fact wealth threatens the environment more than poverty does. the North is more responsible for global environmental destruction than the South.

The common man in the North depletes the world's natural resources base as much as 100 times more than the common man from the developing world does. Yet the North keeps blaming the South for environmental destruction. One wonders if this is due to ignorance, hypocrisy or arrogance of the North.

The North forgets the commandment of God, "Thou shalt not bear false witness against your neighbours".

Streams of water will flow through the desert, the burning sand will become a lake, and dry land will be filled with springs,where jackals used to live marsh grass and reeds will grow.

The Bible: Isaiah Chap 35 Verse 6-7

Chapter - 1

Second Green Revolution

The First Green Revolution focused attention on plant genes and soil chemicals. The Second Green Revolution must focus attention on women and water. Women and water can produce more and healthier food than cloned genes and deadly chemicals.

Good water management leads to good land management. Women for aeons have been producers and processors of food and guardians of family food security.

Nearly 70 percent of the worlds population lives in or within 80 kilometers of the sea coast.* It follows that sea water and saline agriculture must be exploited for human welfare.

Through various low-cost or no-cost desalination methods, women can be producers of potable water, besides being caretakers of water sources and supply systems.**

Genetic engineering and chemical technology, on which the first Green Revolution was based, are environmentally unfriendly, while women and water are environmentally friendly and are the most effective agents to bring about sustainable development for a better world order.

Thus, the Second Green Revolution must be a sea water revolution and women should be the main actors and activists in this revolution to bring about social change and economic growth in the field of agriculture, industry and common services.

---

* World Bank report - World Bank and Environment. Fiscal year 1992.

** Ryan Foundations domestic desalination methods are explained in the book of the Foundation "Water Management in Homes and Villages" and in the paper presented to the World Bank / UNICEF / WEDC Water Conference in Sept. 1996, now available as a brief pamphlet.

More on Rising Scarcity

According to the official statistics of the International Drinking Water Supply and Sanitation Decade the population of developing countries has been growing at an average of about 20,000 people per day, as a result of which more than a billion people were still without access to safe water in 1990 and some 1.7 billion had inadequate sanitation facilities. And the figures take no account of the depressing number of Water and Sanitation Schemes which are in disarray or disues.

Source: Water and Sanitation for All - A World Priority (recording of the Ministerial Conference on Drinking Water and Environmental Sanitation, Noordwijik, The Netherlands, 1990.

A report of the UN Commission on Sustainable Development adds,

> "By the year 2010, 26 countries, with a total pupulation of 416 million will suffer water scarcity, and 12, with a total population of 407 million, will be under scarcity stress. In 1987, some 80 countries, with 40 percent of the world population were already suffering from serious water shortages".

Water and sanitation problems are not confined to developing countries. In the newly independent countries of Eastern and Central Europe, "Water sources are heavily contaminated and infrastructure is breaking down. Industrialised; countries are also experiencing serious pollution of water resources, water scarcity and wasteful use. The volume and toxicity of industrial waste, intensive farming and excessive consumption all contribute to an environmental crisis requiring urgent attention.

Drylands cover approximately one-third of the land area of the earth. It is unrealistic to expect that water conservation and supply techniques could support more than a minor part of these drylands. Taking sea water to the drylands is the only way to develop them. This is simple common sense which hydrologists must accept.

In Thailand, around Bangkok, the water table fell by 80 feet and in peninsular India in, many areas it fell by 200 ft. Texas and Oklahoma lost about 30 percent of their irrigated area due to water scarcity during the 20th century.

Quote: "If the South stops pending just one percent of their defence budget which goes to the coffers of the arms suppliers in the North, then, they can get rid of the sad and unexcusable deaths of nearly 30,000 children every day, due to lack of safe water and sanitation" Gourisankar Ghosh, Chief of Water and Sanitation and Environment Cluster, UNICEF, New York.

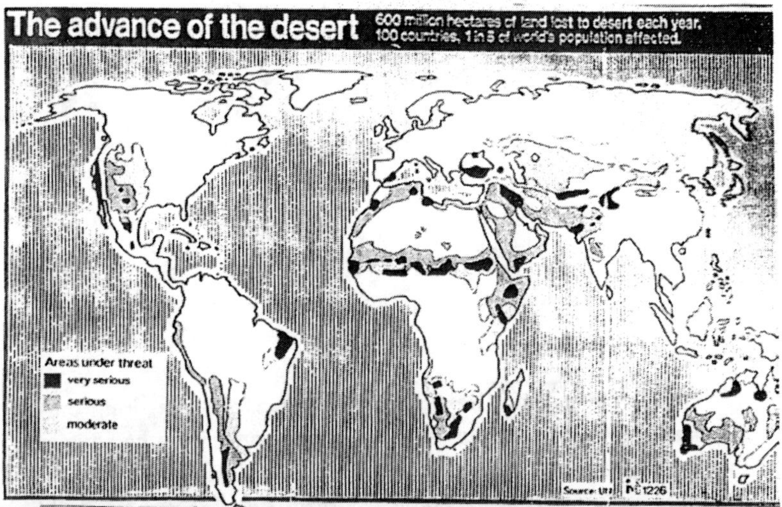

will not a 10 metre broad and 3 meter deep sea water canal bring prosperity to this dead desert land? Certainly, saline agriculture, domestic desalination and sea food can be promoted in this area and motor, wind and pedal boat navigation can be introduced. Why are national and global governments unimaginative and keeping their people in povety and Always crying about advancing deserts. God-given abundant resources are not explorted and that is the crux of the global problem.

Chapter - 2

Only Minor Dams are Feasible

The 30-year old Hirakan dam, which is 4.8 Km long on the river Mahanadi in Odisha, India, is now causing concern to the governments of India and Andhra Pradesh. In 1957, the dam was constructed for ₹ 100 crores to serve for 100 years but has already developed major cracks below the water level and experts are not too sure if the cracks can ever be blocked. Added to this trouble comes the realisation that the dam has not done for Sambalpur, rice bowl of Orissa, as was expected. The irrigation potential was estimated to be 2.5 lakh hectares but the coverage is only 1.58 lakh hectares which is a shortfall of about 37 percent. Moreover, the benefits do not offset the loss of about 45,000 hectares of fertile land which went under the waters of the reservoir.

Building, of minor irrigation dams has suffered almost total neglect in the hands of the government of India, perhaps because of its lack of glamour and small capital requirement, giving little scope for politicians, contractors and engineers to siphon off money allotted for their construction.

A minor irrigation scheme, according to the norms prescribed by the government of India, is one that has a culturable command area of upto 2000 hectares, which may be based on either on surface water or groundwater sources. Official documents put the potential for irrigation of this category of dams at 55 million hectares (out of a total feasible potential of 113 mha for all sectors).

Many tanks and small lakes that once dotted the Indian landscape have gone out of existence either because their anacuts were lost to residential colonies or their catchments were encroached upon by other ways. Even the ones that survive are poorly maintained and silted up. One reason for this panlous state may be that the time-honoured practice of the local population being involved actively in their upkeep has been given up.

In a corrupt country like India, where engineers, contractors and workmen steal cement and blue jelly stones and substitute them with sand (often sea sand) and brick jelly, big dams will not stand even for half a century. Most of the dams in India, according to the 1995 report of a World Bank team of dam experts, will break down at any time during heavy rains and floods and thousands of lives will be lost. If India cannot stop corruption and adulteration in the field of dam construction big dams should be given up in preference to small dams and sea water canals and minor barrages. The country should put down corruption before it dams rivers.

> In the 155 m high Miyagase dam which was completed in Japan in 1999, the cost of storing one cubic metre of water was 200 yens or $1.62 at the July 1999 exchange rate. This is estimated to be three times more than the cost of desalinating sea water and ten times more than desalinating sea water using electricity and heavy machines. In such mega projects the social cost is not taken into account. Bringing thousands of hectares of cultivable land under water and uprooting thousands of rural families from their settlements is not taken into account. Why then do countries still keep building high-rise dams ?

Chapter - 3

Economics of Reverse Osmosis

It is sheer madness to convert sea water into potable water by the reverse osmosis (RO) process, using costly Japanese machines, Gulf oil, and the Haban diaphragm, then filling the water in plastic bottles, each costing about ₹ 2, made in large-scale factories, Using heavy machines, scarce and costly electricity, inorganic chemicals and toxic raw materials, which cannot be easily disposed of, and on top of it all spending ₹ 5 or more to transport the bottles with the so-called potable water to a distance of 5 kms or less in road-polluting vehicles and selling the water to the urban rich at ₹ 15 or more per bottle, which includes publicity, promotional expenses, service tax and profit.

R O is an inefficient process as 25 to 45 percent of the total water drawn is rejected in the form of concentrated salt solution and several impurities. Moreover, pumping costs are high and does not always yield safe drinking water, without metal and chemical impurities within the limits prescribed by the World Health Organisation (WHO).

One cannot understand why governments are not converting sea water for farm, factory and domestic use by the RYFO methods using, wind,wave and solar energy explained here by the RYAN FOUNDATION and why they are not making sea water available to people in man-made canals and streams, especially in remote drought areas for domestic distillation in RYFO do-it-yourself stills which are assembled with mud or metal pots. From time immemorial, villagers have been using crude, unproductive stills for distilling country-made illicit alcohol and they certainly know the principles of steam distillation. RYFO stills are/very productive and convenient.

A down-to-earth water revolution is the need of the day and it has to be a sea water revolution of the people, by the people, for the people. Governments cannot usher in a peaceful revolution and they have never done it. They can only keep revolving around problems the way they have been revolving over the years and keep saying that the water table is falling year after year and the water quality is getting bad to worse. They will spend a billion dollars to instal bore: pumps to suck air, or saline water but will not spend ₹ 100 to get down to brass tacks. Third World governments, like to promote scarcities and problems to stay in power and to make their money. They derive their strength as rulers from poverty, problems and scarcity.

Chapter - 4

Qualities of Ozone

1. Ozone is an allotrope of oxygen with three atoms of oxygen.
2. Ozone has a very short half-life and reverts back to oxygen.
3. Ozone exists as a stable molecule in the stratosphere known as ozonosphere.
4. Ozone is formed from oxygen by corona discharge at higher voltage.
5. An Ozone generator has to be used at the site requiring ozone as it cannot be stored.
6. Ozone is one of the most powerful disinfectants and oxidizing agents known.
7. Ozone destroys all micro-organisms including bacteria, virus, fungi, spores, mold, etc.
8. Ozone also oxidises most organics making them more biodegradable.
9. The best advantage of using ozone is that it leaves no residual ozone in the medium being treated unlike all other disinfectants. This makes it the most environment friendly and preferred choice worldwide.

BUT

Ozone does not eliminate several harmful heavy metals contained in groundwater or reduce the percentage or density of the heavy metals, which are very harmful to health. Thus, ozone machines will not deliver safe water. Most of the bacteria, virus, fungi, spores, etc., can be destroyed by boiling the water to a high degree.

Distillation is the only way to obtain safe drinking water. After the sea water in the boiler is steam distilled, about half a litre of re-sidual boiled sea water from the boiler is mixed with about ten litres of distilled water to restore the lost nutrients and taste. Now and again one can drink pure distilled water but not every day and regularly as it will take away the essential vitamins and other body chemicals. Less than 2 percent of disolved salts and heavy metals is required by the body metabolism not only of human beings, but also of animals, birds, fish and plants and trees, weeds and reeds.

DEVELOPMENT, IF DONE WITHOUT CHANGING THE CORRUPT SOCIAL STRUCTURES WILL ONLY STRENGTHEN THE HANDS OF THE EXPLOITERS.

Chapter - 5

Water Only for the Rich

In the report "Desalination plants a Durable option to meet water Shortage", in the HINDU dated 1st July 2004 the water experts brought together by the CIT recommended the RO or the reverse osmosis process, even though RO does not eliminate dissolved salts, heavy metals and toxic wastes. It is because of RO water that Coca-Cola and Pepsi-Cola got into trouble.

One expert stated that it will cost only 5 paise to produce a litre of water. Apparently he has not taken into account the high cost of RO machines, the interest paid on the heavy investment, the high cost of the machine membrane, which is to be replaced about once in six months and of course the cost of electric power. He has not taken into account also the cost of bottles, bottling, labelling, packing storing, transporting, distributing, paying duty and 'Mamool'. All this will add up to ₹ 15 per litre.

Probably, the expert had in mind only 10 percent of the city's population within a radius of 10 kilometres. What about the millions of poor people who do not get even an ounce of safe water ? Gandhiji said that India lives in her villages but today India is dying in her villages for want of water.

Agreed, wo live in a rich man's world. But water is a poor man's necessity as well. The constitution of South Africa has it that access to safe water is a human right, while the Supreme Court of India recorded in its significant judgement on the 6th Dec. 2000 that there is a duty of the State under Article 21 to provide clean drinking water, access to which is a fundamental right of the citizens.

Only solar-cum-thermal distillation of sea water, for which the technology is available, will reach the unreached poor for 5 paisa per litre. Heating, evaporating and condensing (distillation) were known to the stone-age man. It is said jokingly that the second sin committed by Adam and

Eve was teaching the world how to brew illicit alcohol by the distillation method.

If the CII gets the government to supply sea water to the citizens, as done by the Government of Hong-Kong, it will solve the water problem.

Chapter - 6

Chennai! City Water

In February 1997 the Department of Water and Sewage Examination under the Directorate of Public Health showed that 16 percent of the representative samples collected from different parts of the city from hand pumps and lakes were bacteriologically poor and hence unfit for drinking. The analysis revealed that almost all points of distribution contained algae and micro-organisms like Escherichia coli - the bacteria which causes enteric infections. The study reported that practically water in no part of the city was really safe. That 28 out of the 173 samples contained micro-organisms, as admitted by the concerned government agency itself, could be an understatement of the ground reality according to some NGOs and public health experts. The available quantity and quality of water is much worse now in 2014.

The report mentioned that the description of the quality of water at some points was alarming and specifically pointed out that the hand pumps at the Thiayagaraya Street junction in Pudupet and at Langs Gorden Road were "turbid and Whitish" and had a bad smell. The water contained micro organisms and 8 milligrams of iron per litre which quite excessive by any standard. The METRO Water Board does not seem to be bothered about the poor quality.

Taps at Virugambakkam, Kamaraj Road, and AVM Colony also had an unacceptable colour and smell. The samples contained 1.5 milligrams of iron per litre. Samples from Saligramam were equally bad and contained 3 miligrams of iron. Samples collected from the Ashok Nagar overhead tank, the K.K. Nagar head works and from Pallipet, Thiruvanmiyur, Besant Nagar, and Kilpauk and Porur were "whitish and hazy and bacteriologically poor and contained unusual figures of hardness and chlorides".

At some locations in Basin Bridge, Kasimedum Korukkupet, Saidapet, Waltax Road, Choolai, T. Nagar, Virugambakkam, Saligramam. Pulianthope,

Kottur, R.A. Puram, Adyar, Mylapore and Chepauk the samples were all sub-standard and unfit for human consumption. The Metro water Board is morally bound to keep the public informed every year about the quality of water in different locations. Does it do so?

The Department of Public Health, which received the report on 30th December 1996, exposed the filthy conditions surrounding the water supply lines. At K.K. Nagar, out of the four mechanical filters, only one was working at the time of the study and all the valve chambers were found under dirty stagnant water.

The report notified that at 17 places there were sewage blocks and in Saligramam, people were using "pit taps" by making holes in the supply lines and collecting water and in other places cycle tubes were used for siphoning off water.

"The Hindu" reported in February 1997 that during 1996, the incidence of diarrhea, typhoid and cholera was much higher than in the previous year. This is the true picture of the drinking water supplied by the Metro Water to the citizens of Channai, on the whole the situation is much worse today than what it was two decades ago.

What Man Has Upset Man Can Reset

A problem-solving project is proposed in a study prepared by a team of experts from STEAG, an industrial firm in Essen, at the request of the Bonn government. According to the German view, the rising demand for water and other domestic purposes can only be met in future through the desalination of sea water. Even today, about 10 to 12 million cubic meters are manufactured in this way daily. By the year 2020, the worldwide demand could be around 4000 million cubic meters.

If governments cannot give the poor, healthy potable water, let them give at-least sea water. There is a world of wealth in sea water and thousands of livelihood activities based on marine resources and saline agriculture. Sea water, sea fish, desert sands and saline agriculture can rehabilitate humanity and feed 10 times the present population of our planet and give them the purchasing power to purchase the food that is available, thereby wiping out poverty in the coming years.

Chapter - 7

Plight of Perambalur

(A case study)

In the months of February, March, April, May, June and July, in several villages of the Perambalur district of Tamil Nadu State in peninsular India, people can't get a pitcher of water to drink or cook. They walk several miles and spend half a day to get a bottle of water.

In particular, Alathur, Ariyalur, Kolakkalratham, Anaipadi, Sendurai, Palur and Perambalur are almost totally dry. The district did not get during the best monsoon season even half the usual rainfall due to the changes in climatic conditions and falling of the water table.

Of the total population of 12 lakhs more than half depend on farming for their livelihood. Three lakh people are landless farmers. For want of rain, over 2 lakh hectares are left fallow adding to the problems of unemployment, food scarcity and malnutrition.

All the 400 tanks in the district are almost dry. The British rulers had interconnected many of these tanks to store water throughout the year but the ancient tanks and their inter-connections have all been broken and neglected now and the authorities say there is no money to make them functional.

Out of the 40 overhead tanks, 25 have been out of use for the past several years. As a friend remarked, they stand as silent monuments of the district, as a sort of testimony the vision of the British rulers and the indifference of the "couldn't-careless" policy of the governments that followed after independance.

Responsible people report that most of the deep borewells of the area did not yield water from the time they were installed or failed within a short time as the water table fall. This, however is the common story from many parts of the world.

In recent years, some money, though inadequate, was provided by the Government of Tamil Nadu to construct more overhead tanks which were not really necessary, as the ones already existing are not being used. Money was also provided for installing heavy horsepower motorised pumps for deep borewells. This was done knowing fully well that the subsoil water was too low and too saline and unfit for human consumption. The local authorities who are always attracted by money, constructed several overhead tanks to leave the people without water. The hopes of the people have been belied but the contractors and engineers made their money.

Money or overhead tanks and jet pumps will not and cannot solve the water problem of these areas. Sea water canals and domestic desalination is the only solution. Where governments cannot provide even water to drink people should help themselves. Under the order of the day people organisations can provide water for the people. Moreover many Third World governments have proved their incapability. Governments which stay in power because of the poverty and illiteracy of their people will certainly not cut the roots of their tree that is the roots of their poverty.

> India now has over 17 million energised wells and an esteemed 10 million non-energised wells. While most of the open wells are dry, borewells are sucking air.
>
> The World Bank has played a major role in promoting non-sustainable groundwater exploitation, which is the most significant indication of the water crisis in India, Gujarat in particular.

Let There be no land without Peasants and Peasant without Lands. Land and water belong to the human family.

A Common Scene in Somala with a long coastal Belt.

Chapter - 8

Sea Water Distillation - The Reality

Sponsored by the IDRC of Canada, three experts, Yatcs, Woto and Tihage did R & D on Solar power desalination in Botswana in 1990. They report in their published case study that the deep boreholes turned out to be blank or saline and desalination seems to be the only solution to the water problem in Botswana. They state that salt water is struck more often than potable water and even with sophisticated prospecting techniques, it is difficult to tell whether the under ground water will be sweet and whether the yield will be significant. They further record that at least 55 percent of boreholes drilled in Kogulagadi District are salty and dry. In many areas, they fear, it is unlikely that any potable water will ever be found.

They are also of the opinion that the reverse Osmosis process is not feasible in rural areas.

After designing and setting up several stills in Botswana, they advise that sea water in the evaporation tank should not exceed half a centimetre high in the floor of the tank for the evaporation to be quick and the rate of evaporation to be fast. They conclude that the traditional solar desalination system with slanting glass roof does not give adequate water and it is not the answer for providing drinking water for the rural poor.

Ryan Foundation R & D and observations in several African countries, including Somalia, and in India regarding traditional glass roof or plastic roof solar stills are similar for discouraging.

The solar cum thermal stills and all clay thermal stills innovated by the Foundation seems to be the only answer so far to providing potable water for the rural poor at almost no cost.

To construct convensional solar stills glass sheets or transparent plastic sheets are required. Both are made by capital-intensive factories in far away places and do not reach the poor at a price they can afford. Moreover, stills, being common Village or Community property, lead to family quarrels and group clashes and their maintenance becomes nobody's responsibiliy. Taking all those factors into consideration. Ryan Foundation recommends thermal stills or thermal cum solar stills on a family ownership basis. Some of the best models of the Foundation where renewable energy does not pose a problem are explained in this book.

The very scientifically built glass roof solar still at the million dollar shrimp farm of the ITC at Tuticorn in peninsular India, personally supervised by this author and assisted by two highly qualified engineers (financed by the ITC), yielded good but inadequate water. There is no point in investing ₹ 10,000 to collect 10 litres of potable water. No doubt, a row of such stills constructed by the Central Salt and Marine Research Institute of the Government of India at Jamnagar yields more water and more salt, but does such a large investment justify the quantity of potable water produced? Moreover, can so many sheets of glass of such large size be carried to remote villages ? What will be the loss on breakage and cost of packing and transportation ?

According to Ryan Foundation research and analysis, thermal-cum-solar mud stills hold out better chances of success. Some experiments are under way in different countries. During peak hours, the heater attins a temperature as high as 160 degrees Celcius, but if the thermal still is made of sheet glasses, it cannot reach the rural masses; moreover, it will cost about ₹ 500 which the rural poor cannot afford. Transportation costs will be in addition.

The Energy Environment Group (EEG), based in New Delhi, has now come out with a feasible model for rural areas. This is a mud cooker (not a still) but uses polystyrene in the place of glass. The cooker is like a cardsboard box that can be placed on any surface. It does not use glass covers end reflectors and still attains a temperature of 120 degrees Celcius during peak.

> Sea water can be desalinated and distilled for human consumption and used for flushing toilets and cleaning roads, for saline agriculture, cultivating halophytes (salt loving trees) for promoting sea fish, sea weeds and reeds and also house boats and floating markets. Further, to make dry areas humid, to make trees and plants flourish, birds to sing and bees to buzz hours. Presently, it costs about ₹ 250 per box but the price can be brought down considerably.

There are possibilities of combining or altering this model with the no cost models field tested and promoted by the leaders in the field,

Solar Cooker International of California. Ryan Foundation is now making a study of such combinations. Possibilities are bright for attaching the EEG mud still combined with a Solar Cooker International model and attaching it to a RYFO 4 mud pot thermal still using only a cheap transparent plastic sheet as an imported component. Hopefully, Ryan Foundation should arrive at the final model of solar cum thermal still. In the meantime what is explained in this booklet seems the best solution.

On a commercial scale by far the best method known to Ryan Foundation for giving drinking water to the rural poor without asking them to pay for at is the Saudi method which has been explained and published by the Foundation in several books and journals and discussed in UN seminars and made part of UNDP Secretariat document for the Rio Summit. This American technology uses renewable wind, wave and solar power on a commercial scale in stainless steel stills mounted on the sea shore 22 miles from Jadah in Saudi Arabia. By this method, fields can be irrigated by the desalinated water and, perhaps, organic fertilisers can be mixed in the tanks meant only for cultivation. R & D on this project can lead to several improvements.

These projects, however, can be undertaken only by corporate bodies or companies and registered NGOs, as to put up one unit to provide water to about 1000 families or to cultivate 1000 hectares the cost will be about ₹ 10 lakhs or US $ 33,000. The technology is so simple that unqualified engineers can commission these projects.

Ryan Foundation expects/requests multinationals, international NGOs such as Oxfem, Action Aid, Christian Aid, Misereor, Aga Khan Foundation, Ford Foundation, Rockfellow Foundation, Tata Foundation and church agencies to promote these giant stills. Only NGOs can promote and administer them successfully and not government agencies. They should be owned by NGOs or not-for-profit companies answerable to shareholders, who may be consumers, well wishers, multinationals, churh agencies and international foundations such as Tata Foundation, Aga Khan Foundation or Ford Foundation. "Equality Unions" may be formed to promote and manage these giant stills. To know more about "Equality Unions" read the book captioned "Finding New Routes in old paths; Linking Cultural Needs to Technical Knowledge; Appropriate Technology Inspires Developing Societies : Concept, Controversy & Clarification By Felix RYAN, Published in USA by Bethany Books, 10391, Stockholm, SWEDEN.

The RYFO 4 pots still can also be made on a giant scale and solar rays with magnifying glass can be used to heat the sea water in the bottom container and this system will be much cheaper, quicker and more efficient and trouble free than the other systems. Money making multinationals should sponsor this technology to meet their social obligations. We consider the RYFO 4 pots method even better than the Saudi method though it has not been tried so far. Resolutions to promote the 4 pots still on a giant scale may perhaps be passed by women activities in the forthcoming UN conferences at Denmark and Beijing.

The World Bank, UNICEF and the GEF can perhaps promote and fund such projects instead of sinking tonnes of money into boreholes and agrevating the problem of water world over by reducing water tables. New grounds have to be broken to bring about a new world order.

"We need new and meaningful decisions, not repetitions of what we have already agreed and promises of good intensions". - Gro Harlem Brukdtland, Prime Minister of Norway, Chairman of the World Commission on Environment and Development, (in her foreword to "Our Common Future Reconvened" April 1992.

> Development of water resources is the surest and shortest path towards poverty alleviation and improvement of the quality of life of the people.

Chapter - 9

Groundwater Fast Vanishing in North India

A news release from NASA's JET Propulsion Laboratory (JPL) said on Thursday Thirteenth August 2009 that using NASA Satellite data scientists have found that ground water levels in north India have been declining by as much as 33 cm that is much more than one foot every year over the past decade. The conclusion of the researchers is that the loss is entirely due to human activity. The release says that 108 cubic kilometres of ground water disappeared from the aquifers in Punjab, Rajasthan, Haryana and Delhi and adds. This is enough water to fill Lake Mead three times and lake Mead is the largest man-made reservoir in the US of the size 28.5 acres between the states of Arizona and Naveda.

Co-author of the investigation, Isabella Velicogna of JPL and Irvin of the University of California said, "Using Grace Satellite observations, we can observe and monitor water levels and variations in storage in critical areas of the world without leaving our desk."

This is the true and scientific picture of the ground water position in north India. The picture is similar if not worse in east, west and south India and in almost all countries in Africa.

Of the 445 million hectares of uncultivated land in the world, 201 million ha. (over 45 percent is in Sub-Saharan Africa.— Down to Earth, Oct 2012.

Hydrologists and water experts know that on an average cultivable soil that goes dry during summer one centimeter of rainfall will go down the soil by about 1 foot and 11 centimetres of rain will go down by about 11, 12, 15 or at best 20 feet, if the soil is loose and the water is flowing well. It is therefore amazing to read that 11 centimetres of rainfall revived hand pumps in the region which are usually bored to an average depth of 150 to 200 feet in Gujarat.

Therefore, the most practicable and viable policy will be to draw water from the sea in open unpaved canals, as now being done, in the Gulf and desert countries.

If governments, are incapable of providing treated clean water to the masses they should atleast make available to them sea water for their survival. Certainly, sea water is better than no water and can be easily and inexpensively desalinated by the rural poor using - RYFO methods.

Chapter - 10

Packed Water Usage

On the 22nd of January 2013, the Times of India reported that unsafe packaged water continues to flood Chennai city. The Bureau of Indian Standards (BIS) seized a large quantities of packaged water in Chengalpattu with false ISI marks.

In Thalambur several 20-litre bottles with the brand name VALLABH AQUA was found by the BIS officials and the marketing of these bottles had no permit from the BIS. Action has been taken against the firm under the Indian Standards Act 1986. The offence is punishable with a fine of ₹ 50,000 and/or imprisonment up to a year.

Such illegal units of packing water continue to come up in the suburbs of the city and in the neighbouring districts of Kancheepuram and Tiruvallur.

About 60 lakhs of packaged water is consumed in Chennai everyday which is said to be the highest for any Indian city. Every house seems to be using packaged water for drinking and cooking. Around the city there are about 450 brands of drinking water sold in sachets and plastic bottles. The Times of India report revealed that more than 15 lakhs of 250 millilitre sachets and 3.5 lakhs of 20 litre cans are sold in the city every day.

Health experts say that this accounts for the sharp increase in water-related diseases.

> Interest in desclimation still high, but people are realising that desalination is n't a magic fix to the state's water issues, says Tom Luster, Coastal Commission Water expert in San Diego.

Chapter - 11

Reports of Bron Smith

The Well-known Health Capsules, published by Bron Smith, report that a study of the 100 most popular brands of bottled water showed that one-third of the bottles contained cancer-causing arsenic, trihalomethanes bacteria or other contaminants and every fifth bottle contained man-made chemicals and only one bottle contained Phthalate at an acceptable.

Bron Smith further published for the benefit of the public that people now consume twice as much bottled water as they did years ago and the demand for bottled water continues to increase. He adds that it is surprising that bottled water is less regulated than tap water and that it can be just as toxic. The Environment Protection Agency (EPA) requires cities to test drinking water supplied to the public daily but to test bottled water only yearly. (In India the testing is done at random, occasionally or never).

In a speech before his death Dr. Schumacher author small is Beautiful of (1974) said, " Development is a process where you collect money from the poor people in rich countries to give it to the rich people in the poor countries". Development he said, could not centre around bringing the technologies of the highly industrialised countries to the developing countries. "This kind of industry has no future. Nature cannot stand it, the resource endowment of the world cannot stand it and the human being cannot stand it." quoted from the book Global Mind Changes by Willis Harman by the Development Forum journal of the UN Secretariat- March-April, 1989.

All people have a right to have access to safe drinking water - Mardal Plata Declaration-1977

Fedrico Mayor, Director General, UNESCO said at the International Conference on World Water Resources in June 1998 that in 20 years, per capita annual water availability diminished by about one-third. It is still diminishing year after year.

Chapter - 12

Bottled Water Unsafe (2007 Report)

Reuters cites a talkly Cynthia Sass to the Fitaesa Summit in Dallas, held in March 2007 and underscores her message that bottled water is not necessarily healthier or safer than tap water. Twenty five percent of all, bottled water is actually repacked tap water, according to Sass.

In a recent survey of consumers most of the people interviewed said that they drink bottled water because they believe it to be purer and safer than tap water or well water and also for taste and convenience.

Sass claims that nearly 70 percent of all bottled water in the US is sold within the same state to escape FDA regulation and one out of five states do not regulate the bottled water industry.

She notified that tests on 1000 bottles of 103 different brands of bottled water found man-made chemicals; bacteria and arsenic in 22 percent of the bottles.

Contamination is found also in tap water, Tap water tested was found to contain lead, pesticides and arsenic Sass points out.

While several people can tolerate exposure to trace elements, some people, such as those with HIV, cancer or asthma as well as pregnant women and those recovering from major surgery, cannot stand this water.

(Cynthia Sass is a sports nutritionist in Tampa, Florida, USA; American college of sports Medicine).

Water in Sachets are unfit (2007 Report)

The Corporation of Chennai found water sold in sachets unfit for human consumption in July 2007. The public analyst at the Food Analysis Lab found a rise in acute diarrhoea, in the city. According to official sources a total of 120 sachets and 33 PET bottles were picked up at random in the second week of July, out of which more than were found to be bacteriologically of poor quality and unfit for drinking.

Brands like Aqua Magic, Wave, Spice, Prince, Yoga, Ripple, Hi-Tech and Dolphin were among the poor-quality ones.

The Health Officer of the Corporation of Channai said that legal action will be taken against the manufacturing units under the Prevention of Food Adulteration Rules, 1954.

Chapter - 13

Access to Safe Water - A Human Right

Housing and water are both emphasised in almost all international documents on human rights. The documents referred to are:

1. The Universal Declaration of Human Rights.
2. The International Covenant of Economic, Social and Cultural Rights.
3. The Convention for the Elimination of all forms of Racial Discrimination.
4. The Convention of the Status of Refugees.
5. The International Convention on the Rights of the Child.
6. The International Convention on the Elimination of all forms of Discrimination Against Women.
7. The Vancouver Declaration on Human Settlements.
8. The Global Shelter Strategy.
9. The Agenda 21 of the Earth Summit.
10. The UN Committee on Economic, Social and Cultural Rights (general comment).
11. The Centre of Human Settlement in it's Housing Rights Strategy.
12. The UNHCHR in it's position paper for Habitat II.

And yet, the UNDP (UN Secretariat) has not declared easily accessibility and availability of adequate healthy water at no cost as a basic human right.. On the contrary, UN agencies recommend the pricing and sale of water without taking into account the vast oceans of water which Nature (God) has provided for free use by humanity, beasts and birds, plants, trees and all living beings.

The talk about stopping the spread of deserts without spreading (taking) sea water into the deserts and dry lands for saline agriculture and desalination, promoting sea food and canal transportation makes no sense.

UNDP will do well to take the slogan SEA WATER IS BETTER THAN NO WATER. The motto of the SURVIVAL BY SEA WATER GLOBAL MOVEMENT monitored by Ryan Foundation International is "THE SEA AND THE TREE HOLD THE KEY" (for poverty alleviation).

> If we don't take water and jobs to where people live people, will move to places where they can find water and they almost never find healthy water. Therefore in their new environment they consume dirty water and fall sick and water-related deaths are on the increase the world over.
>
> Land degradation and desertification are common in all the five countries of central Asia which proclaimed their independence from the former Soviet Union.

Chapter - 14

Sustainable Water Supply

Sustainable Water Supply - Sensible Innovations

People of our planet are aware of the growing global water scarcity, desiccation (decrease of groundwater) and related problems. Several ways are being tried out to identify new sources and to make the water supply service sustainable. Some successful ways of achieving sustainability are explained here for others to follow.

1. Several years ago Hong Kong which is very short of freshwater, introduced the dual pipeline system - one to supply healthy drinking water and the other to take sea water for flushing toilets and sewerages and for washing roads, pavements, etc. The system is very successful and can be profitably copied by many countries. With the availability of reinforced concrete pipes, fibre glass pipes and rigid PVC pipes, this low-technology, low-investment method is very easy to install.

2. In Sharjah, next to Dubai in the Gulf, an excellent swimming pool has been built using sea water. The water flows into the pool directly from the sea through filters on one side and returns to the sea from the other, thus maintaining a steady inflow and outflow of clean water into the pool. This is a model which can be copied in several countries, star hotels, university campuses, residential colonies and tourist centres to make scarce potable water supply sustainable. More and more sea water swimming pools should be encouraged all over the world. Some of these pools may be used for rearing tasty fish, such as Chanos Chanos (milk fish) or pearls, spot, which multiply very fast.

3. In Dubai, domestic sewage water, which contains negligible chemicals filtered and used to water highway and roadside gardens, plants, parks and avenue trees.

This system is also now successfully followed in water-scarce Coimbatore district in peninsular India but on a small scale.

4. In Ede, a city in eastern Netherlands, about 3500 houses are being constructed with dual water supply system as in Hong Kong but with a difference. Here, one line is used to deliver treated drinking water and the other to supply untreated filtered rain water to be used in toilets, for washing animals, for washing cars and for watering gardens. Sea water is not used as in Hong Kong.

5. The latest innovation to convert sea water into potable water using solar energy with magnifying glasses, advocated by the India-based Ryan Foundation (RYFO) has still not taken off even though the UNEP, the World Bank and several water experts see tremendous possibilities in this method ushering in a new-generation water supply system.

6. Tapping water flowing in small streams and rivulets and storing them for domestic supply and for cultivation is now being done in several hill stations in India and elsewhere. Nepal is a striking example of this method.

However, much of such hill slope water can still be blocked and used for sustainable water supply.

7. There are several household low-investments, low-technology methods to distil saline or sea water that have been innovated and published widely by the Ryan Foundation in its book "Water Management in Homes and Villages" and reproduced in many journals in India and abroad. These methods must be demonstrated and popularised in water-scarce areas to make water supply sustainable.

Availability of sea water in man-made canals is a pre-requisite to follow these methods.

8. From another angle, a shift has to be made from chemical agriculture which consumes plenty of water, to saline agriculture which requires only sea water. About 100 plants, trees, weeds and reeds (halophytes) have already been identified by various world organisations that can be cultivated with sea water (such as coconut, palmyrah, Acacia holosericea, Salicornia SOS, etc.). Silt grass (Paspatum Viginatum), Buffalo grass and Bermuda grass are all halophytes. Spirulina, a blue-green algae, is yet another. Ryan Foundation has a list of most of these halophytes and has been conducting R & D on them through NGOs networking with them in different countries.

A recent finding of the Foundation is that in the family of coconuts, the East Coast Tall of India exhibit higher drought and salinity tolerance, and then comes the Nigerian Tall.

The role of the government should shift from that of a provider to that of a promoter and facilitator.

Coconut water can be a very healthy and partial substitute for drinking water and therefore coconuts should be intensively cultivated in the 83 coconut growing countries of the world, says the Foundation.

9. When the cold sea currents meet the warm land mass, fog is formed. In many parts of the world, particularly on the south-west coast of South America and more particularly in Chile, where the conditions that produce fog are almost constant year around, this method is feasible. The method of tapping the fog is explained in the book "Water Management in Homes and Villages" (RYFO).

The poor and dirty Chilean village Chungungo, with 300 inhabitants, all fisher folk, is the only place in the world so far where the supply of drinking water is harvested over the top of El Tofo at an altitude of 400 to 1200 metres. Countries which have dense fog during season can tap the fog and store the water to meet the summer scarcity.

World Bank Newsletter Findings No. 74 of November 1996 which reports about sustainable development of Africa's water resources, particularly in Aub-Saharan Africa (ASA), states; "The cost of exploiting new water supplies is rising sharply. In addition, the investment for new infrastructure-irrigation systems, domestic water supply systems and hydro-power will be enormous. US $847 billion will be needed to provide clean water and sanitation services to all households in SSA over the next ten years. And if African countries add another 8 million hectares to irrigation over the next 10—15 years, the additional capital costs will amount to nearly US $30 million."

The Newsletter adds that an "estimated 320 million hectares of vegetated lands have been degraded over the past several decades on SSA."

Isn't it then sensible to introduce and promote saline agriculture and desalinate sea water for human consumption and irrigation? Israel, Arizona and some Gulf countries have already given a lead in this field of science.

10. Several of the other methods to conserve and manage water, such as molehill catchment ponds, sea water lagoons, anti-evaporation floats,

rain water soak to conserve and manage water pits to store water, etc. are explained in detail in the book "Water Management in Homes and Villages" by this author.

> Sea water for eradicating poverty, unemployment and rural migration
>
> "The Copenhagen commitments - (at the World Summit for Social Development) marked a significant advance. They speak of the 'eradication' and not just the alleviation of poverty said the UN Secretary General Boutros Boutros Ghali at a ceremony to launch the International Year for the Eradication of Poverty (1996).
>
> While globalisation has benefited societies, extreme poverty, homelessness and the creation of a burgeoning "underclass", unemployed youth, with no prospect or hope for the future, stand out glaringly as one of the greatest challenges of our time.
>
> However, all these problems can be solved, by digging sea water canals, not only for transporting goods and people but also for promoting fish and saline agriculture and for producing innumerable varieties of raw materials needed for village and cottage industries.
>
> This will open also enormous opportunities for domestic desalination and saline agriculture programme stated above. Hundreds and thousands of sea-water plants, weeds, trees and reeds have now been identified and widely published and they are ONLY to be launched and harvested for human welfare.

For the foreseeable future, sea water desalination is likely to continue to be a lifeline technology to obtain potable water for the water-scarce, oil-rich countries as well as for island nations without any alternative.

Converting oil into water is an unsustainable, short-sighted policy but the RYFO six tank commercial stills which distill sea water using reflectors and magnifying glasses are the best to desalinate water and reach water to the unreached, especially to the rural poor.

> Digging and sucking the ground deeper and deeper for water is like killing the goose for the golden egg." At some point this 'house of cards' will collapse. When it does, India's grain harvest could fall by as much as 25 percent. A wild wave of famine is round the corner.
>
> According to world watch, the situation in Pakistan is even grimmer, and grain harvest area per person in Pakistan was expected to fall from 0.31 hectares in 1950 to 0.80 hectares in 2000 and 0.30 hectares in 2050.

Water is unevenly distributed. Some countries such as Canada, Norway, Scotland have surplus water, but other regions such as the Middle East, Africa, central Asia and Singapore are desperately short of water. Ironically, these latter regions will have a leaping population growth during the next 50 years and their children will go thirsty, hungry and under-nourished. Egypt has a severe problem since 96.5 percent of the country is arid desert and water supply from the Aswan Dam is limited by the 1959 Nile agreement with Sudan. To meet the needs of the growing population, Egypt plans to reclaim land in the eastern and western deserts, but after the year 2000, the major constrain to development has been lack of water, not land.

The Saudi project is a deviation from the project concept presented by the author to the World Bank/UNICEF/ WEDC GLOBAL Water Conference at Delhi. His concept of boiling sea water in four large containers by solar energy by reflecting, enhancing and concentrating the heat on the boilers is followed in Saudi with a difference. The author's concept was only to cool the steam into condensed water (continued other side).

Saudi Method—a Deviation

In the fall edition of IETC's INSIGHT, Mr. Toyaro Goto's article "Freshwater from the Sea" raised a timely and important issue of growing urgency. INSIGHT readers may find this further discussion of desalination both useful and informative.

Water problems in the Middle East are getting more and more acute every year. International conferences on water have been noting the problems and possible solutions over and over without bringing relief to those who suffer from water shortages. The Ryan Foundation, based in India, which advocates simple and inexpensive desalination, has taken a positive approach to this problem and encourages the exploitation of sea water for survival. Below is a brief explanation of several of the methods the Foundation promotes.

North of Jeddah, Saudi Arabia, sea water is being converted into fresh water on a large scale using solar power. Powerful lenses magnify sunlight by 40 times to produce steam and the steam is used to run coolers, which freeze the sea water. When sea water freezes, the salt collects on the outside of ice crystals and can be separated by washing the crystals in fresh water. The crystals arc then into freshwater. This method uses wave, wind and solar energy, all abundant renewable energies, to convert sea water into potable water. The Jeddah plant is producing millions of gallons of water annually, enough to meet the domestic needs of 12,600 people daily.

Further research is bound to simplify this technology to make it even more economical than it is today. Windmills can be used to lift sea water into stainless steel boilers and freezers, and a row of these units could bring large tracts of land under freshwater irrigation, promoting food production, employment and income, in addition to giving people water for drinking and bathing.

Direct applications are also available. Man-made sea water canals, such as the canal cutting across Dubai or the Bucking Canal running parallel to the east coast of peninsular India, can be commissioned for direct use. Canals can be used for navigation, desalination and saline agriculture and for promoting sea food production.

When sea water is made Sea and brackish water can be available in open-cut canals, desalinated by simple evaporation and condensation methods (photo by local residents can easily Damrong Juntawonsup entry to UNEP perform desalination using photographic competition 1991/92). simple evaporation and condensation methods. In addition, canals offer water for bathing cattle and for keeping toilets clean. Today, 40 percent of the potable water used worldwide is used in the toilets of urban centers. With earthenware and rigid PVC pipes, sea water can be used for flushing toilets.

Scientists are also finding new crops that can grow on salty soil irrigated with sea water. Hundreds of plants and trees have been identified in a USAID book, "Saline Agriculture," and in 1990, The US National Research Council published the results of a four-year study of hundreds of plants that can tolerate salt and sea water. Ben Gurion University is demonstrating the use of halopytes (salt-loving plants and trees) in Askelon, Israel. Dr. Pasternak at the university believes that growing field crops from sea water is the key to the future of desert agriculture and points out that cotton has so taken to salt water that yields have been increased by 20 percent. Oil and medicinal plants are already being grown near the ultra-salty Dead Sea and in Eilat on the Red Sea.

Sea water is part of the solution to the global water problem, and these are some simple methods to harness sea water for human progress. What is often needed for us is the political will to provide water to facilitate sustainable development. Political will still seems to be lacking because many of these do-it-yourself methods do not require capital intensive, high technology, which is often favored by governments.

Felix A. Ryan, UN Global 500 Environmentalist

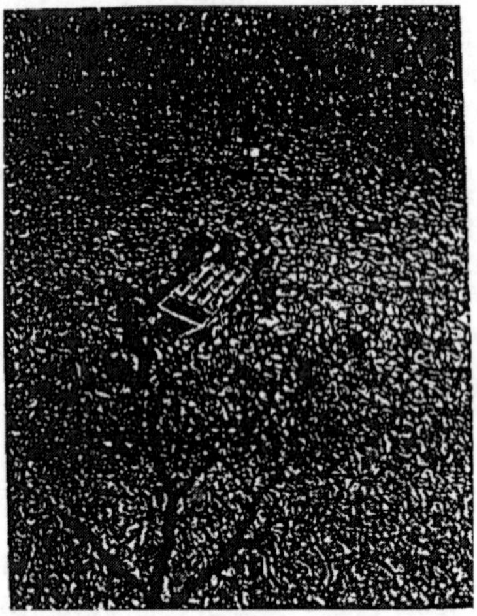

sea and brackish water can be desalinated by simple evaporation and condensation methods (photo by Damrong Juntawonsup. anty to UNEP photographic competition 1991/92).

But the American company which put up the Saudi plant used the steam to run a generator, use the power produced to freeze sea water remove the salt crystals floating on the top surface of the ice blocks, and then melt the ice with solar heat. This method gives salt free water but does not remove the impurities present in sea water. The method is as good (or as bad) as the RO Process, The RYFO method yields absolutely pure distilled water, to which when common kitchen vegetables are added safe water with necessary nutrients is obtained. The RYFO method is simpler, cheaper and safer.

Water must contain some protein, carbohydrate, calcium, iron, vitamins, enzymes, minerals, 1 sodium, potassium and so on...To iodise water add about 50 milligrams of potassium iodate to 10 litres of distilled water.

High Levels of Fluoride and Nitrate

A Study of the Water Supply and Drainage Board of the Government of Tamil Nadu During August 2001 Reported that Underground Water in Over 44 Percent of the Sources in the State is not Potable. About 9 percent of these water sources contain excessive fluoride, over 10 percent have more than normal levels of Nitrate and 18 percent have excessive disolved solids (Tds) and 7 percent contains excessive iron.

There is highly excessive fluoride in water sources in Salem, Erode and Dharmapuri Districts in Salem, Dharmapuri and Tiruvannamalai nitrate concentration is unacceptably high. All this is due to use of excessive fertilisers.

Moreover, an analysis of protected water sources in 23 districts, of Tamil Nadu during 1994-95—1998-99 revealed that the quality of water in about 65 percent of the samples was far below drinkable standard during 98-99, samples were collected from 1002 different locations from rivers, lakes reservoirs, openwells, bore wells and other protected supply sources and analysis proved that over 60 percent of the samples were unfit for human consumption.

A report released in 2001 by the Comptroller and Auditor General of India notes that since 1994 the quality of water in Tamil Nadu has not improved in anyway.

God gave the Israelites water from the rock (Exadus. 17:16) or they would have died without water now God gives the Israelite water from the sea and water in pipes laid parallel to treated water pipes. By and by many countries are turning to the sea.

Chapter - 15

TDS in Water

TDS stands for total dissolved salts in river, lake, well or tap waters not in sea water. Water flowing from a river or rain water on the ground will dissolve minerals from the soil over which it flows. Therefore, the water running in canals, streams and channels can be contaminated by pesticide run-off from black and grey water discharged by residential buildings, farms and factories. When this water flows through metallic or plastic pipes, it picks up more solids, which further dissolve into the water. Ground water which passes through several layers of different types of soils can also dissolve and absorb minerals on its way. All such dissolved salts in is referred to as TDS (total dissolved salts) which is expressed in parts per million (ppm). These salts can also contain other particles or pollutants such as clay, silt, debris, insecticides, etc., which do not dissolve. The Bureau of Indian Standards (BIS) has fixed the upper limit of TDS as 500 ppm. BIS also states that where water is scarce the TDS limit may be anywhere between 500 and 2000 ppm. TDS does not measure contaminants like fluoride or arsenic and BIS has prescribed other standards for fluoride and arsenic. The human metabolism requires some minerals dissolved in water. So zero TDS is not desirable. It washes away vital minerals, enzymes and vitamins from the body. Moreover, a very low TDS affects the pH level of water which should be 6.5 to 8.5 according to BIS specification. So 7 PM of TDS is an accepted minimum level, though BIS does not specify the minimum level.

> Hard water contains too much of magnesium and calcium. Between 20 and 120 milligrams per litre is moderately hard. Above 120 mg per litre it is hard water. Only below 20 miligram per litre the water said to be soft.

TDS in excess, especially with salts of calcium and magnesium, leads to hardness of water; which causes scaling in household utensils.

Among the water, purifying systems the RO process removes large molecules like salt molecules but does not remove small molecules and impurities. If the RO membrane is changed frequently and the process repeated several times almost all TDS can be removed by the RO process in which case some minerals and vitamins should be added to the RO water to restore the lost nutrients needed by the human body.

The RO process rejects about half the water that is fed into the machine and this highly concentrated water with salts and minerals is unfit for any other use.

Rain water is water distilled by nature. If properly harvested it will gather small amounts of minerals from the ground and make the water absolutely fit for drinking with the desired amount of TDS.

Any water analyst can test water for TDS level. TDS meters are easily available today and costs less than ₹ 1000.

> There are good models of sea water canals in drought areas, which also provide quick water transportation. These waterways can mostly be seen in central Asia, Gulf regions and Middle East countries. Israel is a forerunner in exploiting the sea and she is the largest exporter to Europe of fruits and vegetables grown in the NAGVI desert sands.

Chapter - 16

The Arsenic Problem

In the past three decades or so, there is an underground water problem known as arsenic poisoning in West Bengal and Bangladesh and now in some parts of Uttar Pradesh as well. Arsenic is a semi-metal and a large amount of it can be found in sea food too.

Arsenic is a slow but, sure killer. It has become a global menace of the 21st century and poses a threat to health, progress and sustainable development. It causes skin cancer and other diseases.

Arsenic is found in open as well as closed wells. It penetrates the sub-surface aquifers (large amounts of stored underground water) and cause an incurable disease called 'arsenicosis'. Reports reveal that arsenic has affected millions of people in Argentina, Mexico, Chile, Taiwan, Thailand and eastern parts of India, in particular West Bengal. It is observed that starting from Bangladesh it is slowly spreading towards the south of India.

When we go deeper into the earth and suck water with energized pumps and motors, the chances of drawing water laced with arsenic becomes greater and greater.

The numbers of affected people are increasing every year. Fr. Julian S. Doss, S. J., who did a research programme in West Bengal, reports that 2,700 villages of 74 blocks in 9 out of 18 districts are affected by arsenicosis. Six million, he says are drinking arsenic-contaminated underground water and about 3 lakh people are suffering from various skin diseases by being exposed to chronid inorganic arsenic toxicity. The poison has made many people crippled and children orphaned. It has also resulted in young couples.

Many people believe that arsenic develops a disease akin to leprosy or Parkinson's disease, but ignorantly they continue to imbibe the poisoned water that they get. No scientist seems certain about the action of arsenic on the human system. What is clear, however, is four distinct stages in the evolution of the deleterious effects but with the patient manifesting no symptoms - (1) arsenic in the urine and body tissues, (2) the clinical or dermatological stage with pigmentation (darkening of skin melanosis) on the back, chest and limbs and swelling of hands and feet (oedema) and also the hardening of the skin into modules, often on the palms and the soles (keratosis), (3) the complicated stage with enlargement of internal organs such as the liver (hepatomegaly), kidney and the spleen and (4) the malignant stage with the onset of gangrene (skin, liver or bladder) and cancer (carcinoma).

To eliminate arsenic from water, read the booklet "Contaminated water" and the pamphled 'distil sea or saline water for Drinking both by the same author and published by the Ryan Foundation International.

Desalination by the reverse osmosis process or by any other known method of treating water cannot eliminate arsenic. A known and safe solution is distillation by the stills of different models, small and large, invented by the Ryan Foundation (RYFO) to meet the needs of different families and institutions, write-ups on which have been published by BOOKS FOR CHANGE, 139 Richmond Road, Bengaluru 560025, India (e-mail: bfc@actionaidlndia.org).

Peteris Vittata Sucks Arsenic

The American scientist, Mark Elles of the Edenspace Systems Corporation in Dallas, Virginia, says the species of fern called peteris vittata suck arsenic out of tainted water and in less than a day reduces the concentration level below the safety limit set by the U.S. Environmental Protection. Agency. The U.S. has set the limit of arsenic in drinking water at 1.0 milligrams per litre. Elles says the ferns are grown directly in the water, similar to the reed-bed systems currently used to remove organic waste, and they grow and multiplies easily in warm countries.

The fern was identified three years ago as a hyper-accumulator of arsenic and holds as much as 2 grams of arsenic per kilogram plant matter. Moreover, the fern is hardy and fast-growing and can be used repeatedly for successive tank loads of tainted water.

Phytofilteration does not produce an arsenic-rich chemical sludge and so the sludge can be disposed off without harming the environment.

Researchers in the Jadavpur University employed some filtered beads for removal of arsenic from arsenic-contaminated water. The National Test House, Kolkota, Asian Arsenic Network, Japan and Thailand reported the usefulness of these filtered beads. This is more or less genuine. Much better is the application of a new filtering medium made up of oxides of aluminium and other metals. It is manufactured by a US organisation. Its superiority has been verified in India and it works well in the deep bore wells of Shantipur in Nadia District. If these methods are further studied and improved, they will save millions of lives.

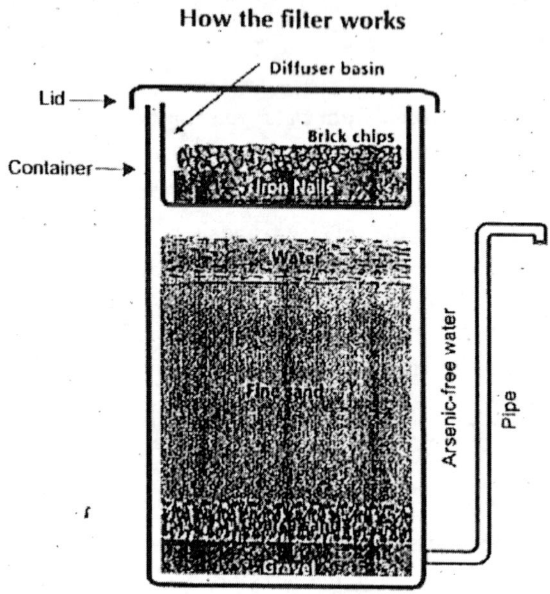

Kanch Arsenic Remover

The Massachusetts Institute of Technology, USA along with an NGO in Nepal called ENPHO have developed a low-cost (pro-poor) household arsenic filter which removes the carcinogenic chemicals, pathogens, turbidity, iron and odour. The filter is called KANCH and comprises of sand and iron nails. Iron nails exposed to air and water rust fast and produce hydroxide particles. It is common scientific knowledge that ferric hydroxide (rust particles of iron) is an excellent absorbant of arsenic. When the water contaminated with arsenic is poured into the filter, the toxic element is absorbed by the surface of the ferric hydroxide particles. After this, the arsenic-loaded iron particles are flushed into the layer of the sand bed below. The fine layer of sand traps the arsenic-loaded iron particles in the upper portion of the sand layer and thereby removes more than 90 percent of the arsenic in the water. The contraption can be made in homes by family members and will cost only about ₹ 200 in India.

Occurre nce of Arsenic in Water

Arsenic is locked-up iron oxyhydroxides. These minerals are stable when there is plenty of dissolved oxygen in the water. But when bacteria digests organic matter and uses up the oxygen, the minerals dissolve and release the arsenic they hold. This explanation has been accepted by most scientists as the means by which arsenic is released into aquifers in India, Bangladesh and many other countries according to the book "Venomous Earth" by Andrew Meharg (Macmillan).

He points out that one out of every 60 people is living in an area where they may be exposed to drinking water with dangerous levels of arsenic. Two to three times as many may be getting water with lower, but still undesirable levels of arsenic.

Water Hyacinth Removes Arsenic

(A recent finding)

The De Montfort University of the UK set up a research team led by Parvez Haris. The team has found a powerful tool made of dried roots of water hyacinth (Eichhorbua crassipes) removed from a given quantity of tap water all the arsenic within an hour of mixing the powder. The levels of arsenic dropped from 200 parts per billion (ppb) to below the safety limit fixed by the World Health Organisation of 10 ppb (reported in the Journal of Environmental Monitoring, April 2005, Vol. 7, No. 4). A biogeochemist, Andrew Meharg of the Aberdeen University says the method is an advance over chemical-based methods, since several of the affected households can find water hyacinth not far from their houses.

Sono Arsenic Filter Kitchen Model

An associate professor at the George Mason University, Fairfax, Virginia, USA has developed a filter which removes arsenic from drinking water. SONO is the name of the innovation and it is meant to be used in water treatment in the kitchen.

The speciality of the SONO filter is that it does not require the use of any chemicals in the process of filtration. The filter is made by using two metal buckets placed one on top of the other with a composite iron sheet in between, which sandwiches a layer of iron plate between two layers of sand. Iron is a natural absorbent of arsenic and effectively removes arsenic in the water passing through it.

The ex-factory price of the filter is about US$40 and the product use is guaranteed for 5 years.

The big question is, how about removing other harmful metals, semi-metals, chemicals and toxic waste to make the drinking water absolutely pure and safe?

The poor will remain poor if they depend on the technology of the rich. And people will suffer ill health and short lives if they depend on the drinking water supplied to them by their governments.

Developing countries have been made to depend totally on development aid from developed countries; digging wells, planting trees, taking care of babies, building schools - practically nothing happens without development aid. Is it not aberration to use foreign money for laying one's roads, collecting one's garbage, building toilets? Hundreds of these activities are being done in many countries with foreign money and a large size of that public money is being set apart by politicians and bureacrats for private gains. Problems of poor countries are never going to be solved by rich countries. It is not correct, happy or healthy for the people of a nation to live for more than half a century on massive and repeated foreign aid.

Negligence towards approaching environmental problems with new imagination and vigour will amount to our failing to face all responsibilities to future generations, to those yet to be born, to those who have no say whatso ever in the decisions we are taking or not taking, which will seriously improve or impair life on this planet, which is the common property of all human beings.

A study conducted by the Indian Market Research Bureau for the UNICEP and published by the National Drinking Water Mission of the Government of India in 1990 indicates that faced with a choice between a hand pump and an open well people often opt for open well water for drinking and cooking.

Working of Hand Pumps

On the spot investigation by UNIDO Expert Dr. Felix Ryan, along with his German colleague, in different parts of Sierra Leone in December 1992 Dr. Ryan checked 100 pumps with his own hands.

Picture - 1

Takes photo as members of his team from the Ministry of Industry, Sierra Leone, make an in-depth investigation .. the failure of so many hand-pumps. Problems were discussed with the local people of the villages.

Conclusion

The conclusions of the investigations were recorded and published by Dr. Ryan in his eye-opening booklet. Pumps without water followed by the booklet' "Survival by Sea Water" which was part of UNDP secretariat official document for the R.

The Summit passed a resolution to promote sea water desalination by all feasible methods and to persuade governments to exploit the sea to provide potable water for their people but no visible action has been taken so far either by the governments or by the UN which organised the summit. Clearly, another case of "BUSINESS AS USUAL".

Picture - 2

Chapter - 17

Drink Healthy Water

The poor will remain poor if they depend on the technology of the rich. Gandhiji said India lives in her villages. Today, India is dying in her villages. Water is the main problem. The RCO process is a technology of the rich and does not remove heavy metals, dissolved salts and toxic waste to the level prescribed by the World Health Organisation. In other words, Reverse osmosis does desalination but does not purify water and make it safe for drinking.

It is not possible to separate heavy metals, dissolved salts, and toxic and human waste from water. It is like trying to separate ink and milk mixed with water.

Bottled RO water, which we purchase and drink is not at all safe or healthy for human consumption. Bottled water has made India the cancer capital of the world. Like in India water in Bangladesh, Pakistan and many other countries of the world carry cancer-causing arsenic and fluoride, which attacks the bones.

Therefore, every home must distil the available water or sea water in the four pots contraption used by tribals and others from time immemorial. Please refer to the book "How To Convert Sea Water Into Drinking Water" published by Books For Change. No. 139, Richmond Road, Bangalore- 56 00 25. email -bfc @ books for change, info authored by Dr. Felix Ryan, whose latest innovation is as simple and easy to use as a pressure cooker. When four bottles of sea water is boiled for about 20 or 30 in a domestic stove absolutely pure and healthy distilled water is obtained. see chapter 161.

One cannot survive by drinking only distilled water, from which vitamins and enzymes etc. are removed totally. Some common kitchen culinary items listed by the Ryan Foundation must be added to distilled water to restore the lost nutrients.

Water scarcity and bad quality of water are getting out of control mainly because the demand for water in our planet doubles once in 21 years, according to the World Bank and the population of the world doubles once in 30 years, according to the United Nations Fund for Population (UNFP). Nature does not replenish even 50% of the ground water we take out every year. The situation is precarious and calls for policies and programmes with foresight to avoid the impending disaster.

Sea water can be distilled easily by the very many methods explained in this book with diagrams.

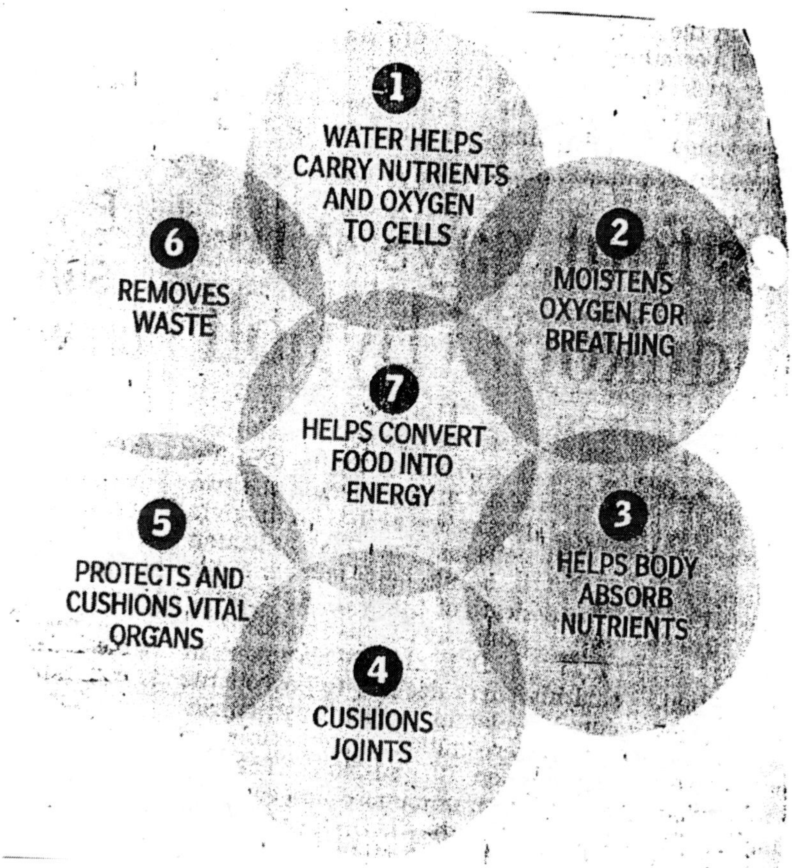

The entire village system of the globe is heading for a disaster and by the end of this century water will be an extremely hot issue.

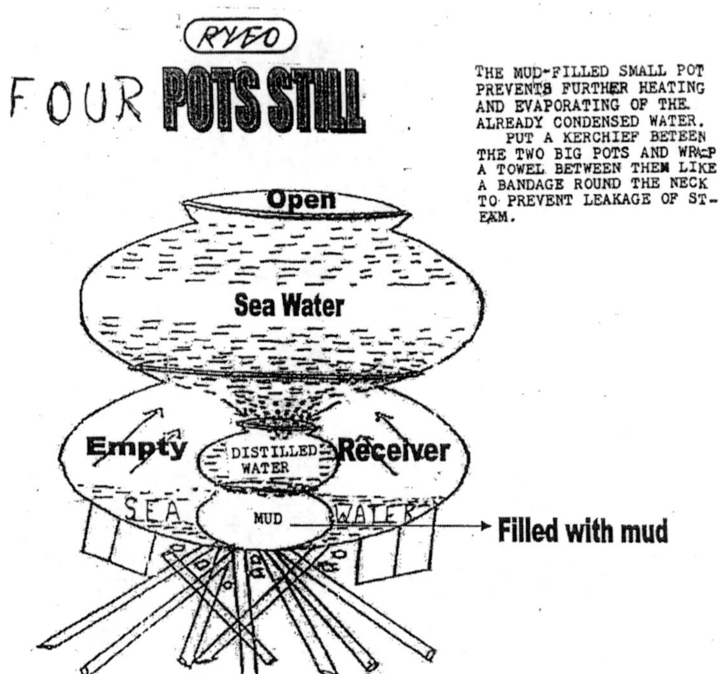

The Mud-filled Small Pot Prevents Further Heating and Evaporating of the Already Condensed Water. Put a Kerchief Between the Two Big Pots and Wrap A Towel. Between them like a Bandage Round The Neck to Prevent Leakage of Steam.

Chapter - 18

Ryfo Pot and Bucket Still

Instructions For Use

This is a very simple and very easy to assemble and dismantle still innovated by the Ryan Foundation.

1. Take a big bucket of about 40 centimetre opening diameter.
2. Take a small bucket of about 30 centimetres opening diameter.
3. Take a large pot mud or metal, of about 20 centimetre in diameter
4. Wrap an old towel around the pot and place it inside the big bucket as shown.
5. Before placing it, pour sea water into the big bucket up to the brim of the vessel inside the bucket.
6. Place the setting on your terace or in the garden in such a way that the rays of the sun fall on it.
7. The sea water in it will evaporate; the steam will hit the bottom of the pot which contains cold sea water, condense and drip into the container kept in the middle of the bucket.
8. You may use a reflector made of aluminium ar stainless steel to reflect the rays of the sun. To begin with the metal bond commonly used by construction workers may be used to reflect the rays of the sun. If necessary three or four bonds may be used.
9. For more heat, set a 1 foot diameter magnifying glass which will quickly take the water to boiling point.
10. If a magnifying glass is not available put the whole setting on a three stone campers stove and apply fire.
11. Now and again remove the warm water from the pot and introduce fresh cold water. Colder the water, quicker the condensation.

12. A saucer-like lid with holes is necessary to prevent the already condensed water in the receiver from evaporating.
13. Pack the cavity between the big bucket and the big pot with an old towel to prevent the steam from leaking out. The packing must be compressed hard all around the pot.

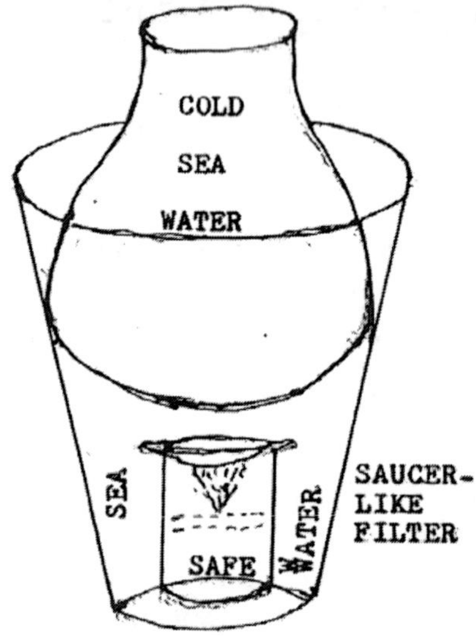

An Old Bath towel is conveniently folded is put round the pot before the pot is put into the Bucket.

Taken from 'TAKING ACTION, (an environmental guide) published by the United Nations Environmental Programme (UNEP) in 1995.

Freshwater Resources Reproduced By: Courtesy of the UNEP

Taking Fresh Water From The Sea

Since up to 80 percent of humanity lives near coastal areas, many communities are turning to the sea as a source of freshwater. The methods of converting salt or brackish water into fresh water vary from large urban systems to simple home-made devices, and include both evaporation and filtration techniques.

Since the early 1980s, Professor Felix A. Ryan of Madras, India, has researched ways to bring simple desalination technology to the rural poor. His organisation, the Ryan Foundation, has promoted many innovative methods for communities to gain access to clean water, some of which can be easily constructed but of such easily-obtainable materials as cloth) plastic tubs and clay. Ryan, who has received a UNEP Global 500 award for his environmental achievements in developing countries, believes it is better for villages and households to provide their own source of drinking water, rather than relying on central governments, hundreds of miles away to bring it to them in long systems of unreliable pipes.

In one of his techniques, called the "domestic still," Ryan places a small tub inside a larger tub. In the larger tub are placed two 750 ml. bottles of salt or brackish water. A plastic sheet is then placed over the top of the larger tub, with the centre depressed down into the small empty tub (but without touching it) so that it looks like an inverse cone. The contraption is left all day in the hot sun. The heat causes the salt or brackish water to evaporate. The steam rises up onto the inside of the inverted cone, and runs into the smaller tub as freshwater. This system can yield 750 ml. of freshwater on a hot day. If heat is applied below the larger tub, more fresh water can be obtained. For best results, Ryan says the system should be left overnight, so the water can condense.

Ryan continues to work indefatigably on behalf of all aspects of sustainability in the developing world. From his base in Madras, India, the professor produces numerous books to help the poor in the South to gain access to innovative technologies, that can help improve their standards of living in a way that is both sustainable and affordable.

Sea water is better than no water. Saline agriculture is better than no agriculture. When sea water goes more than 200 feet into the ground it does not salinate the top soil.

From somewhere between the city of Broom and port Headland or from Derby in the North Eastern coastline of Australian open sea water canal may be cut and water taken across the Great Sandy Desert and from there to the Gibson Desert.

Chapter - 19

The Cooling System

When a hot dish is kept on the dining table with a lid, a good amount of water drips from the lid when it is lifted. The science behind this is simple to understand.

When hot vapour comes into contact with atmospheric temperature, the vapour condenses and drips down as beads of water. Colder the temperature that the hot vapour meets, quicker the condensation.

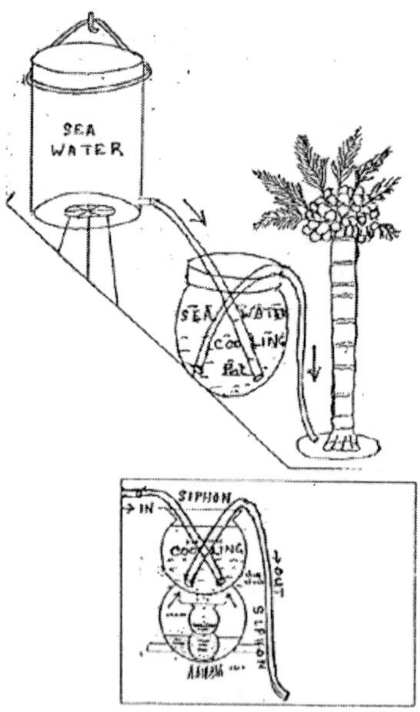

The mud filled small pot prevents further heating and evaporating of the already condensed water. Put a kerchief between the 2 big pots and wrap a towel between them like a bandage round the neck to prevent leakage of steam

In the RYFO domestics till the sea water in the funnel-shaped lid must be constantly kept cold as it gets heated by the hot vapour rising from the still. The best way to keep the water in the funnel-shaped lid cool is to have a big drum, as big as a standard diesel oil drum at a height of about six feet from ground level and the sea water in it allowed to flow into the funnel-shaped lid of the still through a rubber or plastic tube and flow out of the lid through an outlet as shown in the diagram.

The sea water that flows out of the pot must be of room temperature as it is meant to cool the hot steam. This water may be led through a garden hose to a coconut tree or a fish pond or to water any halophyte (salt-loving plant or tree) to promote saline agriculture.

The sea water that flows out of the still is safe drinking water, when this cooling system is adopted will produce more water in a short time.

To distill any liquid there must be a quick heating system and a fast cooling system.

Governments of the world must make sea water easily accessible to people through man-made sea water canals, lagoons, ponds or swimming pools.

Design by the Author

Abundant water must be made available to people for them to progress. 'Sea Water Tanks' recommended by the Ryan Foundation is an ideal method to make sea water accessible to people.

Water scarcity is increasing all over the world day by day and the demand for water is increasing more than what nature can replenish. Sea water is the only cost-effective way of meeting the global demand for water.

Global demand for water doubles once in 21 years and the global population doubles once in 30 years. Nature does not replenish even half the groundwater we take out every year. The next global war will be for water.

Jordan finds it very difficult to supply water, for her people, due to a general lack of water resources, in the country. It is basically an arid country, with very few aquifers and very little groundwater.

Due to this, drilling deep wells has adverse effects on the quality of water and duration of its yield, and on the environment. In spite of this acute problem, Jordan is supplying healthy potable water for over 80 percent of the people for domestic use, agriculture and industry by exploiting the natural God-given sea, and, remarkably, at a much lower price than many other countries in the world that sell water to their people. Access to safe water is a human right; Why should one pay for one's human right?

"It is no longer a question of just paying more money to make water available", observed Biksham Guja, Manager of WWF International's freshwater programme "There is no simple techno-economic fix either".

To comfortably ensure the availability of freshwater on a progressively sustainable basis, the whole question of the water cycle, the quality and quantity of water, has to be looked at, from an ecosystems and cross-sectoral approach".

We may add that, as we cannot create or manufacture water on our planet the only other alternative is to exploit to advantage the God-given sea.

"Water, water everywhere drop to drink so scarce and rare why not then you Vaporise Saline sea and waves that rise water from sea has no price. Easily made sweet and nice turn to sea if you are wise".

Chapter - 20

Is Boiled Water Safe to Drink?

No doubt, boiling water for 15 to 20 minutes kills harmful bacteria and their spores, fungi, protozoon, cysts, viruses and eggs of helminthes. When we drink boiled water, we take into our body the soup of all these dead bodies and also swallow the dead bodies. So it is wise to filter the water through a cloth before it is boiled and filter it again after it is boiled.

All over the world, groundwater contains many heavy metals, such as cadmium, lead, mercury, silver, aluminium, calcium, copper, iron, magnesium, potassium, sodium, zinc and many other metals in various proportions in different locations. Boiling does not remove or reduce the percentage of these heavy metals. On the contrary, boiling concentrates them as the water evaporates and reduces in volume when boiled.

In the same way, groundwater contains several dissolved salts, such as calcium chloride, magnesium chloride, nitrate, ammonium cations and many others in different levels in different locations and they vary in proportion in different seasons of the year. Frequently, the percentage of these salts and heavy metals exceeds the level permitted by the World Health Organisation (WHO). These chemicals and metals cannot be eliminated or reduced by boiling but will only be concentrated in the boiled water.

The reverse osmosis (RO) process separates the salt molecules and one or two other large-size molecules when the water is pushed through a costly membrane in the RO machine.

Instead of boiling on a stove, as an alternative, water may be filled in glass or plastic bottles with half the bottle painted black lengthwise and left lying on the rooftop or on a corrugated metal sheet under bright sun for a whole day or two. When the temperature reaches 50 to 55 degrees Centigrade, the thermal effect of the sunlight and the UV radiation kill the water-borne microbes. The black portion of the bottle should touch the roof. (See picture).

A chlorine tablet made out of chlorine (ALAZONE) may be added to two litres of filtered water to make thė final yield a little more harmless. But these tablets are available only in cities and towns.

Excessive lead in drinking water leads to vomiting, anaemia, loss of appetite, convulsion, and damage liver, brain and kidneys.

Excessive arsenic leads to lung cancer gastrointestinal ulcer in gastro intestines and damage to the kidneys. Starting from Bangladesh, arsenic has now spread all over India in groundwater.

"The pesticide, herbicides and fertiliser, toxic lead, in our body is now so heavy that if we had our urine checked, we would see that we are excreting about 5 -7 different chemical compounds", writes Mrs. Menaka Gandhi, former Minister, Government of India.

> Once the ten most important lakes of Madras were Vysarpadi, Perumbur, Paravallur, Madavaram, Chetpet, Spur Tank, Nungambakkam, Mylapore, Mambalam, Kottur and Kalikundram. None of these exist in today's concrete jungle. Where have they gone?

Environmental News Service states in Volume 2 of its 2003, report that Los Angeles has significant levels of chlorine by-product, as well as a substantial concentration of arsenic, high enough to pose cancer risk. The San Diego water supply has high levels of trihalomethanes along with ethylene dibromite, lead and cancer causing radioactive elements.

"Heavy metals, chemicals, dissolved salts and toxic waste cannot be eliminated in the water supplied to the people. Separating them is like separating ink, milk and water mixed in a bottle. Only distillation can give pure water, purer than rain water. Distillation is the answer for our drinking water problem and good health and not desalination".

The centralised vacuum flash chamber system for giving safe water is very cost intensive, requires non-renewable energy to operate and will not reach the unreached poor in remote villages. The Chennai-based National Institute of Ocean Technology has commissioned a vacuum flash chamber for distilling (not desalinating) sea water and providing safe drinking water.

Lakshadeep Island (India) to supply 100,000 litres per day for 60,000 families. They have not revealed to the public the total cost of the plant and the working expenses. Does end justifying the means?

The best methods to get safe and pure water are the Ryan foundation methods of steam distillation or organic treatment on a domestic scale.

Today, at least a million die from drinking polluted water every year and possibly a lakh from air pollution. At this rate in ten more years 10 million will be killed because of bad governance.

Chapter - 21

Facts about Water Purifiers

The Indian Standards 10500 of 1991 clearly states that drinking water should be free of pesticides, while a hallmark judgment of the Supreme Court of India pronounces on December 6th, 2000 that drinking water is a fundamental human right and there is a duty of the state under Article 21 to provide clean drinking water to its citizens. Alas! These directives are never followed. India merely specifies standards and does not enforce them and adulterators are not punished.

Any machine or device that claims to purify water has to remove chemical physical biological adulterants including heavy metals, pesticides and dissolved solids. Contaminants have different characteristics, such as size, water solubility, molecular weight, volatility, size and ability to react with oxygen. Thus, treatment of water is a multi-pronged process and selecting appropriate technology to change these characteristics is very important, difficult and expensive.

Most water treatment systems combine two, three or more treatment processes. They are ultra-violet (uv) treatment, reverse osmosis method, disinfection and filtration technologies and so on. Experts say that 51 percent of purifiers in use are uv based, while 42 percent are reverse osmosis based and the others use resin and candle filters. While this is the method followed in India, only Percent of the RO systems are used outside India; excepting in the Gulf countries, where mostly the RO machines are used. The drawback of the RO system is that about 70 percent of the water fed into the machine and thrown out as waste is highly polluted.

Healthy Water Must Contain, Within The Limits Prescribed By The World Health Organisation Some Protein, Carbohydrates, Calcium, Iron, Vitamins, Enzimes, Minerals, Sodium, Potasium and so on.

Activated Carbon filters use blocks of powdered carbon or carbon blocks. Powdered carbon removes more contaminants, including volatile organic chemicals, herbicides, pesticides, chlorine and many other chemicals commonly found in the tap water supplied to the public. It also removes bad taste and smell. However many heavy materials and bacteria enter the filter cartridge especially if the carbon is not densely packed. The filters, which are costly and imported into India, have to be changed frequently to let the water that is pushed through the membrane pass through it.

Ion exchange is used chiefly to remove hardness in water caused by carbonates and bicarbonates of magnesium and calcium. These filters, which have a resin base, draw out magnesium and calcium ions in water and replace them with sodium ions. Ion exchange filters can remove copper, iron, lead, radium, barium and nitrates. They can also remove, to some extent, selenium and arsenic. As of now most manufacturers of filters integrate the ion exchange system into other water purifying systems.

The RO method is quite popular now, especially in India and the Gulf countries, and the manufacturers of the system claim that the system can eliminate almost every contaminant. This, however, is a total myth.

In the UV light disinfection systems water is passed through a chamber and exposed to UV light, which destroys viruses and bacteria. The life of the UV lamps is limited and they must be replaced periodically. UV alone cannot be used for water treatment because water has very many contaminants. However, whatever UV cleans or eliminates, it does in seconds, whereas chemical disinfectants take many minutes.

Cleaning the water with alum and disinfecting it with chlorine do not reduce pollution by pesticides or toxic metals. Possibly, Chlorination could aggravate the situation as it could oxidise the pesticides.

The impurities commonly rejected by the RO system are calcium, sodium, chloride, sulphate, total dissolved salts, bacteria, tri-halomethane, polycarbonate biphenyls' and some organics and heavy metals. There is a debate amongst reputed water experts about RO removing essential salts and nutrients. Many opine that it does.

It is estimated by RO machine manufacturers and dealers that 5 million households use water purifiers in India of which about 2 million households use the RO machines. Supposing there are five members in a family 15 million liters per day (mld) will be consumed at the rate of 3 liters per head for cooking and drinking.

If 50 percent of the water fed into the machine is rejected as waste about 100 million liters of water will have to be fed to get about 50 mld per person. The RO industry spokesman says that the operation and maintenance cost of RO membrane filters is ₹ 20 to 30 per kiloliter which amounts to

4 to 6 lakhs per day and on annual maintenance cost of ₹ 15 to 23 crores. This is in addition to the Corporation or municipality water bill.

It is because of this heavy cost that a bottle of RO water which costs ₹ 5 ex-factory, is sold to the consumer for ₹ 15. Is there no cheaper alternative for the RO- water? Yes, "distillation" on a domestic scale is the positive answer; the only answer known. (One liter of the water from the boiler is filtered and added to liters of the distilled water to restore the lost ingredients and taste). Cumin seeds (jeera), curry leaves, holy basil leaf (tulasi), coriander leaves may all be put into the jar collecting the distilled water to restore the lost nutrients and taste.

The following quote by N. Bala Ganesan appeared in the New Indian Express on August 27, 2006.

"Permit me some pedantry, regarding mg/liter and parts per million. Take three one liter bottle of distilled water and pour out fifteen drops of water from each. Now add fifteen drops of a liquid toxin, say Cobra Venom (which attacks humans much like some pesticides) to the first bottle. Shake it vigorously and add fifteen drops of that mixture to the second bottle and shake that vigorously. Finally, add fifteen drops from the second bottle to the third and shake again. The third bottle will now have 0.001 mg/liter or 1 ppb of Venom. A sip of the mixture will not hurt you. But would you let your children drink two liters a day of it, for decades to come, knowing the effect to be cumulative."

The effects of sustained long term exposure, even at "Sub-threshold levels," to pesticides aren't minor. They include birth defects, growth retardation, behavioural abnormalities, musculo-skeletal degradation, immunological deficiencies, liver and kidney damage, brain and central; nervous system impairment and cancer. So the water we drink must be pure and safe. Machine made water purifiers do not yield such safe water. Water steam distilled by solar energy must be promoted the world over for better health and longevity of people.

> The water table in parts of south Morocco, has dropped 50 meters over the last two decades, records UN DTCD news (Vol. 4 No. 2). In the Manila area it has dropped nearly 100 meters in the last 20 to 30 years. As a result of lower water tables due to over-exploitation, pumping costs have soared; springs, wells and streams have dried up. The report further records that over-exploitation also makes groundwater supplies more vulnerable to contamination. In Senegal, for instance, fluoride is spreading into the drinking water system.

Chapter - 22

UN Water Policy Must Change

from land water to sea water

Water, the very source of life from land water to sea water on earth, is in short supply. With the mind-boggling boost in population growth and the steadily rising poverty line, the paucity of water has developed menacing proportions. The number of poor people - defined as those living on less than a dollar a day - is also shooting up. In short, the very poor rising population has no water for daily existence or even for marginal cultivation.

The impending peril is reiterated by the renowned water scientist, Sandra Postal in her article in a Technology Review of the Massachusetts Institute of Technology; "Water may seem to be everywhere, but for the rising problem of the world's population, there may soon be hardly a drop to drink. Or to use for growing food, supporting industries and cities, and preserving life-giving eco-systems."

Top-ranking officials of national and international governments meet in global conferences and thoughtlessly decide that water must be treated as an essential commodity and the poor must be made to pay for it. There might be willingness to pay, but what about the capacity?

According to a recent IMF (International Monetary Fund) report, some 1168.55 million people are below the poverty line in India, much higher than the figures for the late nineties and early twenties.

The obvious conclusion is that if poor people in India and most of the developing countries have no capacity to pay and if payment is insisted

> Nothing is swindled and wasted in the world so drastically and at every stage of transmission as the money given to developing countries as aid to relieve poverty and hunger.

upon, the resulting scarcity of water for household purposes, is bound to continue, causing an increase in sickness, malnutrition and death.

As a way to avoid the impending catastrophe, the big corporations and multinationals should cut down their wasteful expenditure of hundreds of millions of dollars, and the money thus saved each year could help many a poor country to provide distilled and/or desalinated water from the sea and grow more food (as in Israel), and also increase export earnings.

Also, instead of allowing national and international governments to dig deeper and deeper, sucking the earth dry and then declaring that potable water is getting more and more scarce, these big corporations and multinationals should take the responsibility of taking sea water into the land through man-made canals, such as the Buckingham Canal in Chennai (India), which could be connected to dry rivers to help the rural poor, in particular, to help themselves to desalinate or distil the water by many methods published by the RYAN Foundation and obtain safe drinking water without depending on the government supply system, which is both expensive and undependable.

The UN Conference on Human Environment held at Stockholm declared that access to pure air, pure water and a clean environment is a fundamental human right. Indeed so. Land, air and sea belong to the global family, of which billions are dead, millions are living and trillions are yet to be born!

Natural justice demands that governments provide fundamental human rights, such as access to fresh air and safe drinking water. They should not be taxed, as taxes of every kind are already imposed on goods manufactured, sold, imported, and exported, on houses, land or personal and public vehicles of transport, etc. etc.

For the average citizen, existence is an endless process of paying direct or indirect taxes, of one kind or another, for every need or act of his. Essential services like telecom, electricity and postal services are heavily taxed and so are large amounts paid as registration fees, stamp duty, wealth tax, capital gains tax, etc. If properly (and honestly) utilised, this large slice of revenue should be more than enough to pay for fundamental rights like safe drinking water. If this money is not enough, there are avenues for increasing taxes on luxury items like tobacco, alcohol, cosmetics, expensive garments, etc.

Water is a fundamental right, and a basic necessity for survival; coercing people to pay for it in any way is a crime. It is harsh and mindless measures like this which lead to public outcry and revolt that make democracy give way to dictatorship or communism.

Paradoxically, we find even now in interior and God-forsaken villages, foreign and locally made liquor (whisky, brandy, rum, etc.), Pepsi, Coca-cola and other branded bottled drinks. In recent years, there have been attempts to push the sale of bottled mineral water, but the high cost puts it out of the reach of the poor villager and it is consumed only by the rich.

A World Bank Newsletter, FINDINGS, which reports about Sustainable development of Africa's Water resources, states:

"The cost of exploiting new water supplies is rising sharply. In addition to the investment for new infrastructure - irrigation systems, domestic water supply systems and hydro-power is enormous. US$ 947 billion will be needed to provide clean drinking water and sanitation services to all households in the sub-Saharan African households over the next 10 to 15 years, the capital costs amounting to nearly US$ 30 million, in addition."

Wouldn't it be then a wise policy to promote agriculture using sea water, and also desalinate and/or distil sea water for human needs? When access to food has been declared a basic human right by the UN, it goes without saying that safe drinking water is also a basic human right.

An estimated 320 million hectares have been degraded over the past several decades in sub-Saharan Africa, entirely for want of water for cultivation. It is sheer lack of foresight and negligence on the part of the United Nations system and wrong funding. Their short-sighted policy does not take sea-water into consideration even though the RIO summit emphasised that the sea must be exploited in all possible ways to solve the global water problem.

There are technologies and devices galore invented to harness and exploit the sea water. Western ideas, gimmicks, platitudes and gadgets have all proved to be useless in developing countries like India. There are many enterprising businessmen in the West, who make money by manufacturing and supplying sophisticated salt-removing RO machines, which are most unsuitable and unaffordable by the poor in the Global South.

More and more safe and cheap water alone can make development aid, however munificent, productive, purposeful and sustainable. The only economical (and practical) way to provide safe water for ALL the poor is to divert sea-water to dry and parched lands, and to desalinate and/or distil the water for drinking, using several available renewable sources of energy.

In particular, solar energy or energy (heat) produced by easily made foliage briquettes. Germany has shown the way for making briquettes and pellets.

> World Bank development aid has contributed to the increased occurence of drought and desertification, and a growing water crisis - Vandana Siva in the HINDU paper dated 26th May 2000.

Almost all water experts, technologists and environmentalists agree that the RYFO 4 tank still is probably the finest innovation so far to solve the water problem. This method, which has been published by the UNEP and many other methods suitable for different locations and situations, must be taken to water-starved people all over the world. As a pre-requisite to make the RYFO system work well and reach the remote areas, saline agriculture and sea water canal transportation must be promoted.

Saline agriculture is better than chemical agriculture. It is the sea which ultimately holds the key to our progress and prosperity the sea and the saline tree (halophytes) are resources of the poor and they are available in abundance the world over.

> "People who come to the cities come from rural places where the soil is exhausted. The desert has come to the city - and it is having a really big impact", said Sarah Muwonge, acting Mayor of Kampala when over 130 mayors from Africa, Asia, Latin America and Europe met in June in the German city of Bonn for a forum on "Cities and Desertification". Muwonge continued, "Already Kampala is exhausted; it has no more land. If nothing is done, unemployment is going to increase, and we shall have a lot of illegal settlements, and we just can't provide for them all".
>
> If water for cultivation is made available to rural people, they will protect their land, invest on it, work in it, enjoy secure tenure of their land and protect the land, leading to food security and self-sustenance.
>
> Let us look into our safe water problems with fresh eyes and in a different light.
>
> The statement of SHARED CONCERN presented in the Citizen's Fifth Report and signed by numerous environmentalist, scientists and activists across India is now available as a report with the Centre of Science and Environment, 41, Tughlakabad Institutional Area, New Delhi 10062. This report says that Jaisalmer, a tourist centre now has so much water, having moved from its historical use of rain water to expensive pumping of groundwater, leading to cracks in the fort.

Chapter - 23

Transporting Water: A Problem

(Supply point to delivery point)

The basic factors of production are not only land, labour, capital and organisation but under the order of the day, water, fuel, transport and communication must be added. In our planet, as of now, we have enough land and labour and thanks to cell phones transistors and televisions and computers we also have adequate communication facilities but not proper transport, especially to carry water from one village to another or from the supply point to the delivery point. Carrying water from one point to another is a difficult and time-consuming task for most villagers in most of the Third World Countries. The picture in the next page shows a common village scene.

In certain parts of India, Kutch, for instance, cows, baffalows and donkeys are also used to transport water. However, the most convenient way is to have well-dug canals or gutters closed with granite slabs Cuddapaha stones, bamboo mats and or reinforced concrete slabs, Plastic pipes are convenient but not advisable as bandicoots bite them and make holes as this author saw it happening in some African countries. Plastics are not degradable and those that are degradable are very expensive.

Bamboo pipes are cheap and convenient only if the joints are stuck and plastered well. The way of preparing bamboos as water pipes is shown in a following page. In the USA, the UK or mainland Europe we don't see women, often very poor and is sickly, carrying water in pots and buckets as shown in the picture. This time-worn practice can be put an end to only if a network of sea water canals, channels and gutters are provided by the people themselves, by NGOs or by village panchayats in India, Self-help is the best help.

There is No Use Conquering Space,The Moon and Orbits When Millions of People Don't Even Have Safe and Healthy Water To Drink, Made Available Close By, Let Women Come First And Then The Moon. Let The Water Business Be Taken Over By Wind, Wave, Solar and Garbage Energy For Women.

People must realise that the economic process is a physical process because it is always based on the transformation of materials with the aid of energy.

The first law of thermodynamics teaches us that energy and matter are not created but can only be - transformed from one state to another. The second law of thermodynamics states that every transformation results in losses.

Chapter - 24

Restore Animal Power

Diagram 1

The whole family has to go 20 miles away from their village to collect a drum of contaminated brackish water to be kept and used for a week. The rulers of the country have no sense or shame.

Let the load of water, stones, sand or grains shift from the head and shoulders of sickly women and undernourished children to the back of animals, which are made to carry weight from time immemorial.

Diagram-3. It is easier and more comfortable to carry water in automobile tubes than on pots over the head.

Diagram 2

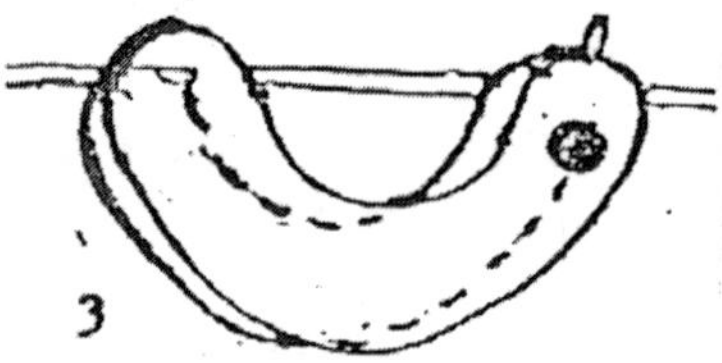

Diagram - 3

Diagram - 4

Reproduced courtesy of the UNV NEWS-Sept, 1990 (United National Volunteers).

Taking food and water to the un-reached. Picture shows donkey used as a load carrier.

Chapter - 25

Restores Animal Power - Explanation

(explains diagrams in the previous page)

Food-for-work projects in the province of Tigray in Ethiopia seen in diagram 2, Efforts are being made to complete 12 dams before the arrival of the next rains. Over 1000 men and women are working on each dam. Report from Africa Recovery, March 1988. Photo by Edith Simmons/ Unicef.

Employing people for manual labour and giving them food for their work is a good idea in poverty stricken and drought-ridden areas. But they should be taught to do the work without strain and suffering. The UNICEF photo marked No. 2 shows women, probably mothers with several children, sweating it out and breaking their backs.

Diagram No. 3 shows an appropriate method of carrying water to construction sites. The idea is simple and costs almost nothing. Two women walking one behind the other as seen in the picture should carry a pole on their shoulders - one end resting on the shoulder of each woman. A piece of cloth or towel may be folded and used as a shoulder pad.

Diagram 3 shows a discarded automobile tube of a lorry four wheelers that shift sand, stone and cement to the site which may be put on the pole, half of it hanging on either side of the pole as shown. The brass valve in the tube must be on top of the pole as shown. Two holes of l0 centimetres diameter may be cut out from the tube on either side of the valve (on one V wall only) and the two holes may each be at a distance of 10 centrimetres from the valve. The cut and removed circular piece of the tube may be stuck or stiched at one end of the hole to serve as flaps to close the two holes to prevent water from trickling out. Water should be filled through both the holes to balance the weight. One tube can hold more than four pots of the type shown in the picture.

Donkeys can be commissioned as water carriers, especially in villages. They are strong and steady beasts of burden. The advantage with donkeys is that one need not purchase feed for them as they eat grass, corn cobs, paddy straw, paper waste, cardboard boxes and several leaves in dry or wet form.

Spanish donkeys are the best in the world. They can carry considerable load, which now falls on the heap of women and children. Donkeys from the Rann of Kutch are the second best in the world and the one from Sikkim the wild ass known as kiang are also excellent load carriers. These three breeds must be crossed, intercrossed and genetically improved to serve humanity.

Chapter - 26

Water Lifting Made Easy

Usually women and children fetch water for the family from wells, poods, lakes or rivulets. Many women sell pots of water to families that are not able to fetch for themselves. Thus, they make a livelihood or some side income to supplement their family income.

Usually, women carry the water pot on the head and sometimes they carry two pots, one on top of the other. Sometimes they bend their body sideways and make the water pot sit on their hip.

The diagram shows an easy and convenient way of carrying water. Two pots of the sane size and weight are tied with ropes to a pole and hung like a village weigh balance. The pole is about two metres long and the middle of the pole is placed on the back of the neck. A small pillow or cushion may be kept on the neck to serve as a pad to take the weight of the pole and the pots without strain.

An easier and more comfortable way is to use two car or lorry tubes to carry water. After removing the metal valve, water is filled into the hole which is about 6 cm in diameter. While carrying the tube on the shoulders the hole remains right on top of the shoulders and so the water does not spill out of the tubes. Two tubes of equal size filled with an equal quantity of water will be well balanced and the rubber tubes will not hurt the body of the women in any way.

However, Ryan Foundation would prefer women and children to carry tube loads of water on donkey backs or on a bicycle. A bicycle can be loaded with tubes of water by a women or a teenage girl. They can ride the bicycle while carrying the empty tube to the water source and push the bicycle when the tubes are loaded with water.

Tubes will also occupy less place inside their shacks. Tubes loaded with water can be hung on the walls.

A warning must be given in this context that if sea water is not made accessible for the poor there will be more and more trading in unhygienic water and the poor, who have no money to buy food, will have to shell out for water.

There is no way forward until our policies are made innovative.

Chapter - 27

Sea-Water Tanks

Several projects and proposals of Ryan Foundation International require easy and quick access to sea water. The Suez Canal and the Panama Canal are man-made sea water canals for ships to pass. But the water of these canals is used in several other ways as well. The Panama Canal across the Isthmus of Panama, which links the Atlantic and Pacific oceans, is a huge undertaking which was completed in 1914 at a cost of about 340 million dollars, an enormous sum in those days. However, to meet the requirements of desalination plants and saline agriculture such colossal investment is not required. Then can be met with very small investment, which even a villages can afford for their own benefit.

All that is required to provide sea water for people living not far from the coastal belt (80% of the world's population lives within 100 km of the coastal belt) is to give them water in three cubic metre tanks built one km away from the coastline and connect the brick and mortar tank to the sea by brick and mortar pipes (without using cement, as even the big buildings of olden days were built).

Usually the level of land near the coastline is about 5 feet above the sea level. The connection may therefore be given 5 ½ to 6 feet below sea level with a filter and lock arrangement inside the tank. Such an arrangement will always keep water in the tank to the level of the sea. The rise and fall of the sea level does not matter as it will not badly overflow the tank. When the tides are very high the tank will of course overflow a little. The overflowing water may be chanelised for saline agriculture or into a man-made lagoon or led back into the sea.

The open tank may be fitted with cross beams or bars in a check pattern and closed with slabs so that dust and dirt do not fall into the tank.

A report published in 2014 reflecting the joint judgement of US intelligence agencies had warned that the use of water as a weapon of war or a tool of terrorism would become more likely in the next decade. The Inter-Action Council comprising more than 30 former heads of state of government, has called for urgent action.

From one tank the water may be taken to another similar tank in the next village by underground pipes or closed gutters and from the second village to the third and so on. If the inlet pipe or gutter from the sea is about 2 feet broad, about 10 to 15 tanks in about 15 villages can be connected to the chain and all of them will maintain the level of water to the level of the sea, even if 500 pipe lines draw the water from these tanks to different locations for desalination, saline agriculture and fish ponds. The supply will be perennial and the sea will not go empty.

Hundreds or thousands of such tanks may be built all along the coastline and connected and interconnected to make abundant sea water available to people everywhere. Sea water is better than no water and the sea has a world of wealth.

Where necessary, overhead water tanks may be built above or beside the ground-level tanks to lift the water up using windmills with canvas buckets, Persian wheels or animal-turned water lifters. The appropriate technology methods have been published widely by Ryan Foundation.

There is a common tendency among people, particularly rural people and agriculture specialists, to say that sea water will pollute the ground with salinity. But by the piped method suggested above by the Ryan Foundation, there is no question of salinity spreading into the soil and there is no room for complaint or conflict on this issue.

Some villages may have mortar-paved tanks in their villages 4 square metres in ares and one metre deep to rear sea fish. The urban rich may draw the sea water into their compounds and have their own swimming pools, fish and duck ponds, etc.

The Root Cause of Millions of People Being Unemployed, Underemployed, Under-paid or Living Below the Poverty Line is Scarcity of Water Even to Wash, Bathe and Meet Basic Needs and Cock of Proximity to the Scanty Water that is Available. In Many Rural and Urban Homes Family Members have to Hunt for Water and Stand in a Queue for Half A Day to Fetch a Pot of Water.

Newly built housing colonies can have the small RYFO desalinating system on top of the terrace and supply absolutely clean and healthy water for the flat owners free of charge after recovering the initial cost, which is minimal.

Large institutions such as hospitals, hostels, schools, colleges, churches, temples, mosques and residential settlements can have centralised large-scale RYFO plants-one, two or three to meet their in-house requirements and also to supply safe water to their neighbourhood.

Governments and local bodies should quickly swing into action to get these sea water tanks built and connected. Where they are unwilling or lethargic, the people can take the matter into their own hands and build tanks with contributed or donated money and connect the ground-level tanks with flexible hose pipe buried underground. A kerchief-like cloth may be tied at either end of the hose to serve as a filter. Even if these hose pipes are damaged by some construction workers, there is nothing much to lose or regret. The hose pipes may be repaired and restored or replaced and it will cost very little. Laying of such small hose pipes to draw water from the sea does not infringe any government law or rights of people of the area. However, it may be done with the consent and cooperation of the neighbourhood.

Hundreds of plants, trees, weeds and shrubs have been identified that grow well in sea water and these can be promoted extensively in rural areas for reforestation and restoration of the environment. Reforestation will provide fuel wood for cooking and rural women will be spared of their time consuming and backbreaking work of gathering fuel wood.

When plenty of fuel wood is made available close at hand, it can also be used to boil and desalinate sea water to get healthy potable water by the many cottage-scale methods published by the Ryan Foundation and thereby women will also be relieved of their burden of trekking long distances and carrying water on their heads and shoulders to meet the family needs.

> The seriously water stressed economies, stretching from South Korea and India to Iran and Egypt, are paying a high price for their water problem. The yearly global economic losses from water shortage are conservatively estimated at US $ 260 billion - Brahma Chellaney, geostrategist, in his book (2014).

By culturing sea fish in their family ponds, they will get plenty of animal protein for better nourishment and health. Fish oil, fish glue, fish feed and fertiliser may also be made use of in livelihood activities.

Villagers can also wash their cows and goats with sea water, keeping them clean and healthy and their surroundings cool and fine.

Public health has been defined as "ensuring the conditions in which people can be healthy". Thus, easily accessible, safe potable water is an imperative to ensure good public health, while accessibility to safe potable water is a basic human right. The WHO, UNICEF, World Bank and UNEP will be making significant contributions to public health as well as income generation and overall progress of the rural masses by providing potable water for them by the RYFO 6 Tank method, which has been declared feasible by water experts as well as NGOs, and national and international governments. It costs only half a million US dollars to desalinate water for 50,000 people. Villagers can maintain the system without any external aid or help.

> From the Great Australian Bight in the south of the country, water may be taken to the Great Victoria Desert. If all or most of these canals are dug and interconnected, much of the "desert Australia" can be made green to progress and prosper and to create a billion job opportunities inside and outside Australia and export trade and commerce and goodwill flourish.

> Potable water has a lower density 1,000 grams per centimetre than sea water 1.0255 grams per centimetre and tends to float over saline water.

Chapter - 28

Idea of Sea Water Channels

Bamboo poles may be cleaned and made into water pipes as explained in this book. Three pipes may be placed together parallel to one another and plastered either with tar (bitument) or with RYFO alternative cement as explained in this book. Through these pipes sea water can be taken fron one delivery point to another. The pipes may be buried one foot below the ground. Above where the pipes run the dry weather shrub VINCA ROSEA which gives red, pink and white flowers all through the year, may be planted to make the village look lively.

Craft cultivation is cultivation in a small piece of arable land close to one's home and such cultivation should be encouraged. It is akin to roof gardening and organic gardening.

From sea water cannals, lakes, ponds or lagoons water has to be led to the interior parts of the land and for this purpose 3' × 3' × 3.9 cubic feet channels will have to be built and there must be a regular flow of sea water through them falling into a deep well or a large tank in which the best edible fish may be introduced. At a distance of every 100 metres on the channels there must be cement tanks 2 feet high with a manhole cover. People who want water will lift this lid and collect the water. The lids must be fixed to hinges or else those who lift the lids may not put it back properly. In India and elsewhere people take away the manhole covers and sell them to scrap dealers or foundries. The manhole chambers will also serve as cleaning points. Hand pumps may be installed where necessary. The channels will be free of mosquitoes and cockroaches.

Chapter - 29

Tubes Reduce Evaporation

Picture is a common community tank in Rajasthan (India) and picture 2 is a traditional community well. Nepal, of which the country has hundreds, open community wells, tanks, and ponds are common in poor countries and are the basic source of drinking water, especially in the rural parts.

The major problem with these community wells and tanks is seepage and evaporation. By and large, in open storage, 25 percent of the stored water seeps into the ground and another 25 percent evaporates during a given period of time.

The RYFO solution is to inflate discarded automobile tubes of all sizes and float them in the community open water storages. Large-size truck and bus tubes are the best: as they spread out and cover the water sheet. As compressed air is not available in poor villages, they may be inflated with a hand pump or even blown in the mouth like a balloon. Tubes may be thrown one on top of another to cover the water if the well or pond is small. by completely protecting the water from direct sunlight and atmospheric air, about 20 percent of the total evaporation can be arrested.

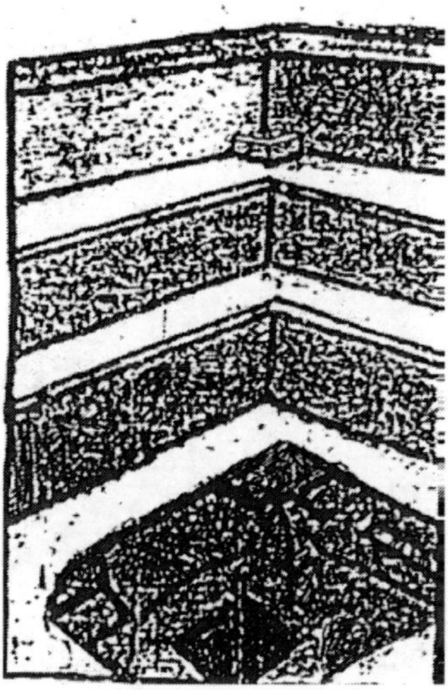

Moreover, in a well or pond so laden with tubes, the fish population multiplies fast. The smell fish hide under the tubes or inside or outside the circular edge and escape from the big fish.

Cleaning and reinflating the tubes may be done once a month.

The Creator knows that no life can exist in this planet without water, including plants and trees, according to His divine plan. Therefore, it is that He made the seas and the oceans and mighty rivers for humanity to exploit and utilise.

Chapter - 30

Micro Hydro Power

The people of Nepal have made a mark in the world for driving motive power for turning various machines to meet domestic needs. Most of the micro hydro power units of Nepal are owned and managed by individual villagers in their neighbourhood people use water-powered mill to hull rice, press even operate a band saw in some places. The journal hydronet from Germany, dated March 1989, from which the photo reproduced here is taken with reported that atleast 600 such hydro-powered water wheels were in existence in Nepal in 1989 and indicated that they were becoming more and more popular.

A typical water mill in the hills of Nepal Photo FAXT HYDRONET-3/89

Water to turn the turbine is diverted from a streamlet or rivulet by piling a row of rocks or poles to make a barrier. The diverted water is led to a pipe to flow on the turbine. A simple hut or shed is built around the turbine to call it a mill. The water filling the pit below the turbine inside the hut is led to a lower level through a gutter, as shown in the picture, and out into the open to join other rivulets or to form new streamlets, take its course and flow away.

The lost of a micro hydro power unit depends on the length of the lead pipe, the size of the turbine, the number and size of machines attached and the installation of additional attachments such as generators, bandsaws or grinding wheels.

Water wheels/generators can be erected on the slopes of mole hills on the slopes of lake or pond bunds or in catchment areas. A distance of three metres between the water level and the bottom level of the receiving slope or pit is adequare to make the system work efficiently. If the depth is not adequate, the length of the lead may be increased so that the water flowing through it gains force. Alternatively, water may be lifted into overhead tanks and allowed to fall on the turbines on the spot investigation by Dr. Felix Ryan. (Photo by Fakt).

Innovating for a sustainable planet is the best job on earth.

Chapter - 31

Ryan Foundation Wind Mills

The traditional windmills attached to bore pumps have failed all over the world. See RYFO Handout 337. Reasons for the failure have been explained in the Ryan Foundation booklets "Pumps without Water" and "Agrifilm for Better Storage and Use of Water". Nevertheless, wind power has to be harnessed for energy and Ryan Foundation recommands the trouble-free open-ground system where the maintenance can be done by villagers without training or expenditure.

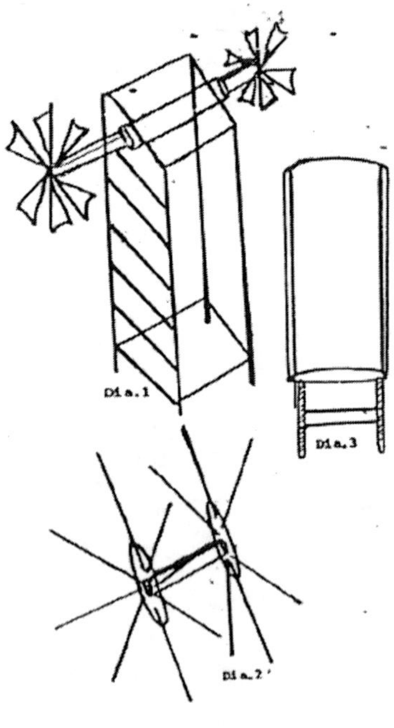

A rural wind-mill must be built of poles and canvas with a minimum of parts only for the axle and the ribs to fix the sails.

Design by the Author

Parts may be design made of galvanised iron [water pipes] of 5 cm diametre and the metal axle rod of 75 centimetres diameter. Two large metal spools with welded arms on either ends of the axle are also necessary. Diagram 1 shows a construction of metal parts [spool with arms] on one side of the axle rod. Two similar units are required like what is shown in the diagram 2. Alternatively, the axle may be mounted on two brick pillars.

Discs on either side of each spool should be as big as a long-playing gramophone record made of 12 gauge metal sheet. One centimetre diametre arms (ribs) may be welded to them in parallel position as shown.

HDPP woven cloth of the type used for packing cement, fertilizers or food grains may be stiched like pillow cases and put like a case between two parallel arms as shown in diagram 3 and tied or stitched to stay in position.

Water-lifting arrangements from open wells, tanks or sea water canals are explained in RYFO Handout 312 (No. 13). The x axle need not revolve to take the direction of the wind but the pole structure should be so built that one end of the axle points to the south and the other to the north.

We cannot direct the winds but we can adjust our sails.

> Land, sea and air belong to the global family of which billions are dead, millions are living and countless are yet to be born.

Chapter - 32

Play-swing Water Lifter

Errection cost about ₹ 50,000. The swing is fixed to a rotating metal pipe. To an axle fixed inside the pipe a bicycle free wheel and a large-size sprocket wheel is fixed. Both the wheels are connected with a cycle chain. When children play on the swing) the washer pumps go round and round inside the plastic pipes, gather water from the tank in the ground and fill the overhead tank. From the tanks water is taken to different locations by hose pipes or rigid PVC pipes. Details of construction and working are in the next page.

A RYFO INVENTION

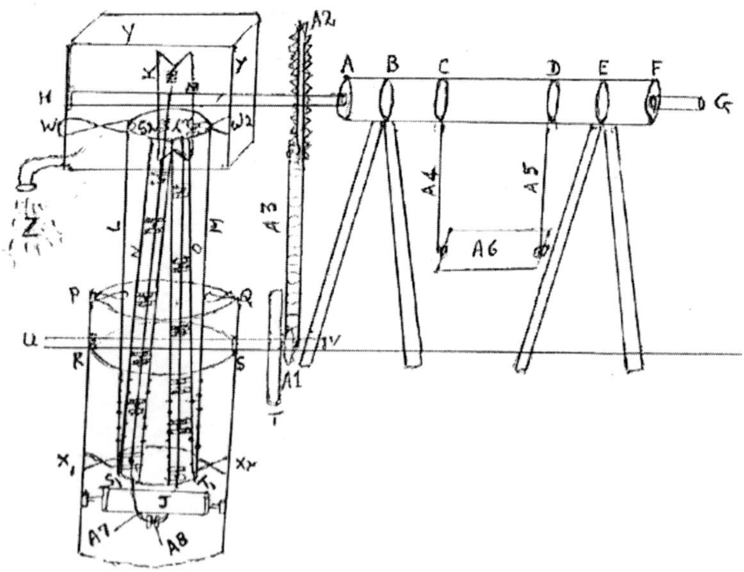

Design by the Author

Chapter - 33

Play-swing Water Lifter (picture)

Chapter - 34

Play-swing Water Lifter - explanation

1. AF is 3 inch diameter metal pipe. B and E are metal bushes welded to the pipe and the legs of the swing.
2. C and D are loose metal bushes through which the pipe passes. The pipe must rotate forward and backward inside these bushes.
3. C and D are fixed clamps with hooks for the swing rods to hang.
4. A4 and A5 are one inch diameter rigid rods hooked to the plank.
5. At AF the pipe is closed with thick metal discs and welded. Before fixing 2.75 millimetres holes are made in both the discs.
6. HG is a 2 inch pipe, quite long, and the end H sits on the metal bush fixed inside the water tank. HY is the tank.
7. K is a wide pulley of 12 inches diameter and through the middle hole in it GH passes.
8. Wl andW2 are clamps to hold the 4 inch plastic pipe.
9. X1 and X2 are clamps to hold the 4 inch plastic pipe at the bottom.
10. J is an aluminium roller fixed on metal bushes in the well.
11. LM is a plastic pipe 6 to 8 inches broad.
12. S1-S2 and T1-T2 are two plastic pipes each 2 inches in diameter.
13. A one centimeter thick nylon rope passes through the two rollers J and K.
14. Simple knots are made in the rope with a gap of 1 foot in between.
15. Washers cut to size out of a car tube or leather are fixed on either side of each knot and tied or stitched to stay in that position.
16. PQ and RS are metal rings like hula hoops fixed to the circular wall of the well and the long plastic pipes are fastened to them, preventing them from shaking.

17. OPERATION: When a child swings the cycle chain and turns the sprocket wheel, the cycle free wheel and the nylon rope goes round and round around the two rollers and inside the two plastic pipes.
18. One centimeter holes are provided in the three pipes that dip into the water.
19. Both the 2 diameter pipes pick up the water and empty the water into the overhead tank and the water is drawn out from the tank through tap 2.
20. All metal parts are painted with black anti-corrosive paint or bitumen (tar).
21. Pipe LM is meant to give stability and strength to the system.

The best known solution to the drought problem of Gujarat (in any state or country for that matter) is to drop sea water into several open wells along the coastline to recharge the depleted water table and lift it to a level reachable by open wells and bore wells.

Water is one of the core essential and basic necessities. Despite earth's majority of the cover is filled by water (97% by oceans), 2% of the water on earth is glacier ice at North and South poles and only 1% is available as freshwater suitable for humans. In recent years the water table is facing serious threat due to rapid population increase, industrial and urban development, over usage, climate change and global warming. With the increased exploitation of water, it has now become a serious issue. India is threatened with a difficult and tumultuous water future.

Good governance is fundamental for the success and sustainability of any national poverty alleviation strategy. Evidence shows that countries that are able to develop and maintain transparent government institutions are more effective both in combating poverty and improving the living standards of their citizens, reports IF AD/UN, Nov. 2000, No. 8.

Chapter - 35

Sea Water for Livelihood

In many Third World countries want of water is the main problem for initiating development. However, since 80 percent of the global population lives within 100 kilometres of the sea coast villages close to the sea may exploit the sea for their livelihood activities by taking sea water into their villages by flexible PVC pipes like the garden hose, by bamboo pole pipes or by open cut narrow channels and creating artificial ponds, lakes or lagoons in the village. If an open channel is cut, there must be lock and sluice arrangement to regulate the flow of water or else the village will be marooned during high tide. The flow through flexible tube will be easy, practical, economical and convenient to manage and maintain. By folding the outlet end of the tube or lifting the flowing end above the sea level, the flow can be stopped and even a child or illiterate woman can stop the flow. Water can be taken from one pool or pond to another.

When sea water ponds, lakes or lagoons are available in the midst of rural settlements, slums or squatter settlements, several activities become possible. Solar stills may be built to convert the water into potable water on a community scale and the hundreds of plants and trees that grow well in saline soil and sea water may be cultivated. Some such plants and trees are salt bushes such as Atripolex mummul aria, mangroves, casuarina, coconut, acacai, palmyra, prosopis Juliflora, Pongamia glabra (Karanj in Hindi), which gives rich oil used in soaps and medicines. All these and many more can be cultivated and exploited for human welfare.

When sea water ponds, lakes or lagoons with regular inflow and outflow back into the sea are available people will take to fishing, rearing crabs and turtles, growing sea weeds for making agar agar and preparing rabbit feed (rabbits and their by products can give millions of jobs for billions of people). (See the author's book on rabbits). All water based raw materials can be exploited in one way or another by processing and manufacturing industries and for generating employment. For instance, if plenty of fish is available, besides eating fresh fish people can dry, smoke and preserve fish and also make fish meal, fish glue, fish oil, chicken and cattle feed and fertilisers out of the incredible fish by the village method. All these methods are explained by the Ryan Foundation in several publications on appropriate technology.

Inexpensive village community-scale desalination plants may be built to provided drinking and water for drinking and use in kitchen gardens.

When sea water is allowed to flow into waste-lands in certain soils and over a period of time the salinity of the water is reduced and certain types of trees, plants and shrubs can be cultivated and mangroves promoted.

Water is the most basic necessity of life next to fresh air and it is getting scarce and costly in most parts of the world. Coastal villages must exploit this God given natural and renewable resource, sea water, for their livelihood and survival. For a detailed explanation of various ways of using sea water for survival see the booklets "Water Management in Homes and Villages" and the booklet "RYFO 6 Tanks Still" published by the Ryan Foundation.

Chapter - 36

Natural Harbour

A FJORD (otherwise called FIORD) is a narrow inlet of the sea and extends far into the land in mountainous terrain. It has steep, rocky walls on either side. Near its mouthy it is shallow, but deep inland. The water in a FJORD is calm and without fast and quick occuring waves. Therefore, fjords serve as natural harbour.

Norway has many fjords. New Zealand, Chile, Canada, Alaska and Greenland also have them.

The sognef ford is the largest fjord in Norway and the best known in the world. It is over 1219 metres) deep in parts and 180 metres long. Its rocky metres wall towers 1,524 metres above sea level.

Wherever fjords are found, they must be extended and taken longer into the land and deepened where necessary and sea water steam distillation plants must be set up. The same fjords may be used to trans-port large containers of water to provide safe drinking water to families living on either side and nearby.

We must make the best use of what God gives us on a silver plate.

> "Human development is a process of enlarging people's choices. In principle, these choices can be infinite and can change over time. The most critical are to lead a long and healthy life, to be educated, and to have access to resources needed for a decent standard of living. Additional choices include political freedom, guaranteed human rights, and self respect" - Human Development Report 1999 (UNDP).

Chapter - 37

Illustration - 1

A

B

C

The usual lock-and-sluice arrangement to let sea water into the land. It is inexpensive and safe and can be operated by one man. The number of locks depends on the breadth of the canal (outlet) and the amount of water to be drawn from the sea. The picture shows irrigation system in the Mescene area in Syria. International water agencies should install this system in water scarce countries to improve the environment, food production and employment.

The Buckingham Canal runs almost parallel to the world-renowned Marina Beach beside the seat of the government. It is just a kilometre from the sea near the Presidency College and the Government Hospital for Women and Children. After the British rulers left the canal has become a breeding ground for mosquitoes and a source of contagious diseases, in the city. Picture shows the canal lined by huts and filth near Chepauk. Note the TV tower at the rear end.

Mankind Must Understand at this Point in Its Development That Its Future Relies as Much on an Intelligent Use of Ocean Water as It Does Today on Rainwater-Unesco 1998.

Chapter - 38

Illustration - 2

Given the sluice to control the flow village people will organise themselves to construct big channels to take sea water into their village. Photo taken from Rural Development No. 11, produced by FAO for the UN ACC Task Force on Rural Developments, December 1989.

Even today the Buckingham Canal, which holds sea water, is navigated in certain parts. The Picture shows a fishing boat between the two tourist spots Kovalam and Mahapalipuram some 30 kilometers from the city of Madras (photo 1991). Note the heaps of crude salt made out of the sea water at the rear of the picture.

The idea of digging sea water canals parallel to rivers with about a 5 kilo metre gap between the rivers and the canals and promoting estuaries and mangroves in between them should be implemented wherever possible. When the rivers go completely dry, the sea water may be turned into the rivers to keep the rivers flowing and the ground wet for the mangroves and marshy land plants and trees to grow. Plants and trees that grow well in salt water and saline soil have been identified and widely published.

Eradicating poverty is not in giving doles but in building up self-reliance and a work culture for the underprivileged.

Chapter - 39

Drumstick Seeds Clarify Water

The seeds of Moringa Oleifera, the common drumstick/ which is a drought-resistant tree growing in almost all tropical countries is found to have excellent use in water treatment. In Sudan, people have been purifying water by using Moringa from time immemorial and scientists at the University of Leicester, UK have confirmed the science behind it, WHO recommends.

The light yellowish white kernel inside the winged seeds removed, crushed and mixed with a small quantity of pure water and stirred for about 5 minutes. This suspension is then strained and mixed in a pot of turbid water and stirred again for 10 minutes with a wooden stick. To clarify 40 litres of water, some 40 seeds are required.

When mixed with river water, the pulped seeds, according to scientists, give water-soluble proteins that bind the riner particles of sediments in the water and form what is known as floe. Viruses and bacteria get enmeshed in the floe and after some 15 minutes the collodial particles settle down at the bottom of the pot. The action of the seed pulp is similar to the result produced when alum (aluminium sulphate) is mixed with water to purify water.

There are several leaf, root, bark and seed (organic) methods of purifying water practiced by tribals in India and Africa. In India, people use the wiry roots of Vetciver (Vetiveria zizanjoides) for filtering muddy water. When water is filtered through Vetsiver in a pot, the water not only gets clarified but also gains a pleasant smell.

Another seed used is Kataka (strychnos Potatorura) and it is used in a similar way. Ryan Foundation experiments show that when crushed Moringa (drumstick) seeds and (Vet siver) are both mixed and used, the filtered water is much purer and safer. An effort is being made by the Foundation to make available to the rural poor ready-mix packets of powder of Moringa seeds and Vet siver roots which are easily available. The collection, drying, powdering and packeting of these two raw materials is a very good income generating activity for the rural poor and there should be one unit (family) in every village to prepare this powder (mixture) hygienically. Done well, it can be sold also to urban communities. The powder has good shelf life up to six months.

Storing Water in Copper Pots Overnight Kills Several Pathogens and This Method Has Been Corroborated by Several Research Investigations.

Chapter - 40

Banana Peels Filter Heavy Metals

Metals such as lead and copper get introduced into water from sources like agricultural runoff and industrial wastes. Through water, heavy metals contaminate the soil and pose health risks. Lead is known to cause cancer.

Traditionally, silica, cellulose, and aluminium oxide are used to extract heavy metals from water. But these remediation strategies are expensive and have side effects.

Scientists from Brazil have found that the peel of the fruit can act as a filtration medium and pull heavy metals from water.

Compounds in banana peels contain atoms of nitrogen, sulphur and organic compounds such as carboxylic acids - the same as those found in traditionally used material. These acids have a negative charge that can bind with metals in the water, which usually have a positive charge.

Plants materials like sugarcane, bagasse, peanut shells and apple waste are already being used to extract metals from water. But they mostly require chemical modifications. Banana can act as an absorbent (adhesion of heavy metals on the surface) without any chemical modifications.

The team in Brazil used a simple method to establish the properties of banana peels. They mixed 10 milligrams of minced banana peels (particle size 35-45 micrometers) in 10 millilitres of a water based solution containing a known concentration of metals. The mixture was then filtered and found to have a lower concentration of the metals.

They also compared the filtration capacity of the banana peel with that of other natural materials like peanut husk and saw dust. It was a higher capacity. "This is because of the higher levels of protein and reactive groups in the peels", says Gustavo R.de Castro, the lead researcher.

He notes that the extraction capacity depends on the particle size. The bigger the particle size the more the surface area, hence the greater is the filtration capacity.

They found that a purifier made of layers of minced banana peels could be used up to 11 times without losing its metal-snagging properties. Synthetic material can be reused more times, but natural material is cheaper (and easily available).

In the study published in the February 16 issue of "Industrial and Engineering Chemistry Research" the scientists tested only two metals-lead and copper for filtration.

Castro is confident that metals such as cadmium, nickel and zinc will also be removed when they come in contact with the peels. "In fact the material can be used in industry for the treatment of effluent and will be a low-cost alternative", he adds.

Source Down To Earth. April 16-3-2011.

> Banana skin, fresh lime and drumstick seeds act as natural flocculator and coagulator.

Chapter - 41

Duckweed - A Water Purifier

Duckweeds are tiny little floating aquatic plants which grow in abundance all over the world in fresh or brackish waters. They are Monocotyledons of the botanical family Lemnaceae, classified as higher plants. There are four broad genera in the family and they are:

(1) Wolffia (2) Wolffiella
(3) Spirodela (4) Lemna

The plant produces seeds and flowers which cannot be noticed by the naked eye. They grow very well in all climates, except in deserts and ice lands. Native species are commonly available even in saline water.

The weeds are not fixed to the soil but keep floating. Their reproduction in nature is vegetative and water birds and run-away rain water spread them. Daughter fronds (shoots) bud from reproductive pockets on the sides of the mature mother frond. A single mother frond can produce about 10 generations of progeny in about 10 days and continue the reproduction for several weeks before dying.

Under ideal conditions duckweed fronds double their mass in two days provided proper sunlight and nutrient is available. This growth is faster than for almost any other higher plant.

Fresh fronds contain 92 to 94 percent water. Grown under ideal conditions and harvested regularly, the fronds have 5 to 15 percent fibre content and 35 to 45 percent protein content.

A monoculture of Nile -tilapia and a poly culture of Chinese and Indian carp feed A readily on fresh duckweed. The weeds are also used as poultry feed and in the wastewater treatment process.

Almost any land is suitable to cultivate duckweeds but water should be available year round. Alkaline soil with very high pH should be avoided.

Some states in the USA and, some countries in Europe have introduced this floating aquatic weed to purify drinking water in storage tanks. Nearer home; Bangalore City and Guntur District in Andhra Pradesh have introduced the weed.

It cannot be said that the weed makes the groundwater absolutely safe to drink but it certainly eats away many of the dissolved salts and some heavy metals as scientists have found. Some water experts say that it is better to use this weed than the RO machine.

> The idea of self-reliance is the central feature of Appropriate Technology.
>
> Conservation means utilisation of natural resources in the most efficient way to benefit the greatest number of people for the longest period of time.
>
> The main engine of development is, and has always been, self-help and hard work and not aid. No amount of Government statutes or legal measures can substitute for the people themselves taking up the responsibility, especially when it comes to finding basic needs of life such as water, fuel and shelter, for survival.

Chapter - 42

Ryfo Four Post Still

1. Ryan Foundation has innovated several stills to give drinking water to the rural poor but some of them require the use of metal containers and plastic sheets, which are not easily available in remote villages. Even if available, the poor cannot afford them.
2. The four Pots Still does not require metal or plastic. It is easily assembled with earthern pots readily available with the rural poor. Two large-size pots and two small ones are taken. The small pots should go into at least one of the large pots.
3. They are set preferably in the open, on a three stone stove (campers stove). Sea water is poured into the big bottom pot upto the brim of the small bottom pot, which is filled with mud as insulation and to prevent it from rising due to boyancy. Sea or saline water is filled into the big pot on top and the stove is heated from below on low fire. Water evaporates even in room temperature but with some heating, evaporation is quicker.
4. Since fuel wood is scarce, costly briquettes made of foliage (dried leaf sweepings), crushed like tea leaves and made into lumps with locally available tree gum to bind are used for heating. (Tamarind seed, roasted, crushed and boiled yields good gum (glue). For details on alternative stove fuel, see pages 144 to 148 and page 105, 256 to 271.
5. HOW IT WORKS: Sea or saline water is poured into the bottom pot upto the brim of the small pot, as already stated. The big pot on top is filled with sea or saline water. When heated, the water in the big bottom pot evaporates, the vapours hit the cold bottom of the big pot above, condense, drip and fill the small pot. It is pure, healthy water and need not be boiled. Some salt, sugar, curry leaves or cummin seeds may be added for taste and smell.

6. As the water in the big pot on top gets warm, it is removed with a cup in small quantities and fresh cold water filled in the colder the water in the top big pot the quicker the condensation.
7. In between the two big pots there may be pinholes through which the vapour can escape. To prevent this, a thin cloth as big as a towel is twisted and wrapped around, like a bandage, between the two big pots and tied.
8. The two big pots may have a holding capacity of about 20 litres each and the two small pots, about 5 litres each small variations in size will not matter. The small pot filled with sand prevents the condensed water from getting heated and evaporating again. Under a low fire it takes about 30 minutes to yield 5 litres of potable water. For more, the process may be repeated or more stills may be set.
9. The most backward and poor people in almost all third world countries follow this principle in different methods and use different contraptions to distill country liquor (alcohol). If sea water is made available, they were know how to distil and will desalinate it.
10. Innovation: R & D and field tests were done by Dr. Ryan and his project associates, Mrs. Girija Sinnas, social worker and Dr. A.D. Monteiro, Chief Technical Advisor. UNIDO (United Nations).

Chapter - 43

Distillation for Safe Water-
not Desalination

Almost every year, after the monsoon showers, it is reported that the tanks and catchment areas are full and the water position is comfortable but two or three months later the public is informed that the storage has fast depleted and water supply will have to be restricted and regulated.

Metrowater looks at the storage position in relation to the usual annual demand but fails to take into account the nearly 50% of water that is lost through the corroded and perforated pipelines. Water is not lost so much between the pumping station and the water mains as between the water mains and the end user that is between an entry point to the residence and the taps in the residence.

In several places in the city, we see water leaking or even flowing from underground water pipes. If the Metrowater does not stop supply during specified hours of day and night, the leakage and wastage will be much more. The pity is that this large amount of water neither reaches the water table nor wets the top soil.

Think of the packed rows of street houses on either side of small roads, lanes and by lanes in Pudupet, Mylapore, Royapettah, Triplicane, Saidapet and other similary over-crowded places. These homes took their domestic connections from the road mostly during British rule or about 100 years ago using Birmingham pipes. All these pipes are rusted and with perforations small and big holes all along.

The homeowners are not willing to replace them because costs have gone up over a hundred times and they have no money to spend on repairs and replacements. So they simply ask Metrowater to lay new pipes or provide water through street pumps or tankers.

> The Next War is Not For Land, Not For Power, Not For Oil, But For Water -
>
> People will suffer ill health and short life if they depend on the drinking water supplied to them by their governments.
>
> The poor will remain poor if they depend on the technology of the rich.
>
> Safe water is the entry point for sustainable development.

It is just not possible to replace all the worn-out and perforated pipes because almost the entire old network of the city will have to be changed and this step will cost several thousand crores of rupees and take several years. We can only abandon the existing old network and lay new networks wherever possible at the ground level.

The second problem is the centralisation of storage tanks and catchment areas for a city of 5 million population, spread over extensive areas, in just three or four large lakes. Such centralisation is always troublesome. People near the pumping stations or water mains get more water and regular supply; others far out get little or no water or irregular supply. This is the common complaint today from the citizens of Chennai as one can read in the many letters to the editor published in news papers.

Regarding the supply of underground water, for the past three decades or so, the groundwater table in and around Chennai has been steadily falling by one metre or more every year due to lakhs of energized pipes and over exploitation and due to insufficient rains and a rising population. Bore pipes, which were giving water last year, are not giving water this year. The rapid drop of the water table is taking place the world over but in peninsular India, the fall is much more than the global average.

Many water managers think that the only alternative for us is to beg, borrow or buy water from Andhra Pradesh or Karnataka. This is totally a mistaken notion. The water situation in our neighbouring states is nearly as bad as ours. They too have the same problem of leakage and wastage. In their own states water supply is restricted and regulated. Many parts of Karnataka and Andhra Pradesh don't get adequate water except during heavy rains. Water agreements of the past between these states do not hold good today. Neighbouring states cannot give what they don't have for their own people.

This is the main reason why the Veeranam project failed. Veeranam Tank is fed by a tributary of the river Cauvery, which rises in Coorg. Owing to deforestation, climatic changes and increasing demand for water within Karnataka, no water is let out to fill Veeranam.

For the past 20 years, water experts have been repeatedly saying to National and International Governments that we have to exploit the sea in all possible ways as resolved in Rio de Janeiro. The booklet, "Survival by Sea Water", was declared part of the UNDP Secretariat official document for the Rio Summit. And yet, no noteworthy action has been taken in this direction.

The reverse osmosis (RO) process of desalination is not the solution. We cannot dismiss the RO process as totally unfit. RO machines are very expensive and the membrane used in them to separate salt molecules is also expensive and they have to be imported and replaced frequently. Indian-made membranes are of poor quality. Moreover, motion power (energy) is required to operate an RO machine. The fairly potable water it gives has to be bottled, sealed and transported which costs much money. A bottle of water taken out of the machine, which costs 5 paise is sold to consumers 5 miles away for ₹ 20. Thus, it is meant for the rich and not for the poor. This bottled water is no longer safe as several bogus companies are bottling filtered tap water without the RO Process Laboratory analysis has busted this racket through the press. Nevertheless, it is still a flourishing business in Gujarat and many other states of India.

RO machines are good for star hotels, schools and colleges, hostels and hospitals and multinationals for their house consumption and not to meet public demand. Governments should insist that large institutions have their own RO water management.

The performance of almost all RO machines installed by the Metrowater Board within the city and out in the districts is very unsatisfactory. Perhaps if left to private undertakings, the performance will be satisfactory.

Almost 70% of usable water in the Gulf region is produced by R.O. machines but we cannot compare oil-rich Gulf countries with water-starved countries without oil. Gulf countries convert their cheap oil into costly water. We cannot do that here in India.

Moreover, RO machines eliminate salt, suspended particles and some evil bacteria but not the many harmful heavy metals which sea water and groundwater in many locations contain. People must be given safe water without heavy metals such as arsenic, mercury, aluminium, Iron particles and many other metals. Banana skin peels granuled consume several heavy metals.

Recently the Government of West Bengal sealed bore pipes in eight districts because they contained excessive arsenic. In many locations, groundwater is not safe and has to be treated. Boiling alone will not do.

That is why water experts advocate sea water distillation and not desalination. Distillation gives pure and safe water. The basic principle of the RYFO method is evaporation and condensation as far as possible using solar energy, reflectors and magnifying glasses to boil and vaporise the water in large-size steel containers, each unit to provide water for about 10,000 people. Some kitchen ingredients are added to the lifeless distilled water to restore lost ingredients.

Excluding the land, a community project will cost about 22 lakhs in India say about 50,000 US dollars, Each unit will supply water continuously without recurring expenditure and with negligible maintenance cost. This system can be extensively decentralised with people's participation and thus the cost can be reduced.

The six-tank still (Ryan Foundation innovation) can supply abundant safe water to people even in remote villages and all round the year.

The innovation is basically meant to reach the unreached rural poor. According to the World Bank, 80 percent of the world's population lives within 100 kilometres from the coastal belt. The remaining 20 percent lives not far away from other sources of water.

> It is sad that Water is Becoming as Expensive as Milk and It Is Easier To Get Indian Made Foreign Whisky Or Rum In Remote Villages Than A Glass of Safe Drinking Water.

It is possible to draw or lift water from the sea by windmills on the coastline into large cement tanks and then transport water into stainless steel containers. A detailed project proposal is now readily available.

Such a project is working very successfully in Saudi Arabia as reported in several UN journals. In Saudi they use not only reflectors but also magnifying glasses which enhance the heat of the sun by forty times. The project provides safe water for more than 10,000 people.

The Ryan Foundation method is much simpler and cheaper than the Saudi method. It does not require magnifying glasses but only powerful reflectors (a German innovation), which are now made in India, all for ₹ 45,000.

Efforts are shaping in the direction of installing one 6-tank RYFO still near Chennai. Our Metrowater Board does not seem to be interested. Our governments have no time or patience for small people's projects. They consider the investment of ₹ 2 lakhs peanuts. However, with some support from abroad we can easily commission a unit. There is full backing in writing from the UNEP, the World Bank, The Third World Academy of Sciences and many other global agencies interested in water.

To conclude with an important note, access to safe water is a basic human right and God has made our planet with abundant water. Governments have therefore no right to tax water. If citizens of the world tolerate water tax, governments will soon tax the polluted air we breathe, God forbid.

> If the sea is not brought in before next summer, an expert warned acute drought in Gujarat will hit again next summer and every summer and the Government of India will have to shell out several lakhs of Rupees every year to save people in Gujarat. He also warned that such deep fall of the water table can cause tremors in Bhavnagar and surrounding areas.
>
> As warned, the area experienced 15 tremor shocks on 12 September and a couple of tremors on 15th September 2000.
>
> If the underground "hollow" is not quickly filled up with sea water, the expert fears, the land mass in the areas where recent tremors are occurring will cave in along with several buildings and the sea will move in.
>
> "Sea water is better than no water and saline agriculture is better than chemical agriculture", says the expert.

Chapter - 44

Slanting Glass Heating Method

The author demonstrates on the open terrace of his residence, his low-cost Cuddapah stone (granite) still, which yields 5-6 litres of distilled water by the use of direct solar heating and reflectors. If foliage briquettes are burnt under the slab, even 100 litres can be collected. The sea water feeder pipe and an opening to clean are behind the author. He will soon be mounting on a tubular stand a smaller and lighter model with glass sheets sloping on both sides which is twice, as productive. The distilled water collects in a 20 litre bottle on the floor as seen. The author's problem is how to fetch sea or unsafe water. Solutions to this problem and other details are provided in this booklet. A spoonful of some common kitchen ingredients and herbs are put into the bottle to restore lost ingredients and taste.

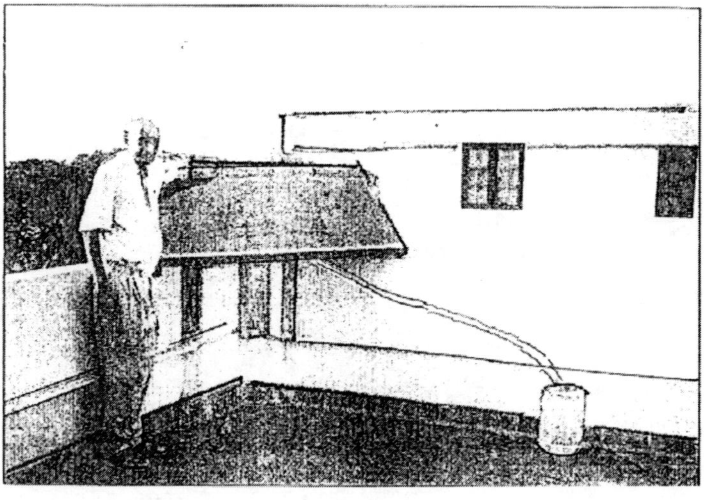

Nature, Women and Children Suffer Equally When Development Models Bring About Destruction of the Environment and Over-exploitation of Ground Water.

This Book on Water is in Spirit on Poverty Reduction and Growth Oriented Development.

Chapter - 45

Urgent Need for Domestic Distillation of Sea Water

Due to increase in population and climatic changes in many parts of the world, the water table has fallen below 150 feet and energized bore pumps are reducing the level further. In most places in the city of Chennai bore well water contains excessive selenium or arsenic and fluoride, making water unfit for human consumption as per WHO standards.

Centralised water storing and distribution in pipelines does not make available adequate water to the unreached. Scanty rains, deforestation, internal conflicts, inter-state politics, terrorism and sabotage are some of the reasons for this state of affairs. Therefore, every country, region, town, village and home must make its own arrangements. Domestic and community or family-scale distillation is the answer.

Government should provide a network of man-made sea canals and streams for people to distill sea water using mud pots. It has been demonstrated that steam from a rice pot, for instance, is easily tapped to convert three bottles of seawater into two bottles of potable water. This principle can be used on a commercial scale to cultivate fields and to supply hotels, hostels, farms and factories.

Wind, wave and solar energy must be combined to distill see water as done in Saudi Arabia. Transportation by lorries and pipelines must be avoided.

Conservation costs less than desalination

Saline agriculture with sea water should be encouraged, as in Israel or Arizona. Solar ponds and man-made lagoons must be promoted.

Sea water for eradicating poverty

The Copenhagen commitments (at the World Summit for Social Development) marked a significant advance. They speak of eradication and not just "alleviation of poverty".

While globalisation has benefited societies in many ways, extreme poverty, homelessness and the creation of a burgeoning" underclass" unemployed youth with no prospect or hope for the future, continue to stand out glaringly as some of the greatest challenges of our time.

However, all these problems can be solved by digging sea water canals not only for transporting goods and people, but also for promoting fishing and saline agriculture and for producing innumerable varieties of raw materials needed for village and cottage industries.

This will also open up enormous opportunities for domestic distillation and saline agriculture programmes. Hundreds and thousands of sea water plants, weeds, trees and reeds have now been identified and widely publicised. They need only to be launched and harvested for human welfare. This calls for a very large-scale mind shift of humanity from centuries old thinking and concepts.

> Protozoa - microscopic animal life found in water cannot be fully eliminated by the RO method.

A new set of tools is required now because the old ones have just failed to work anymore. The illusion that, sooner or later, the benefit of centralised development will flow down to the poor and the remote corners of the world should be abandoned forthwith. Instead, a model of development based on people's resourcefulness and ingenuity to achieve efficiency and social equity should be seriously followed and encouraged. Poverty cannot be eradicated without economic growth and economic growth cannot be brought if millions are still enslaved in poverty. We cannot afford to risk some more decades of stagnation through inaction.

The urban population of the world has tripled since 1950, according to the 1988 State of World Population Report. Big cities are almost criminally responsible for both liberal consumption of resources and production and accumulation of mountains of waste. Tragically, uncontrolled growth threatens to explode the present transport, health and sanitation systems in many countries.

Ten of the world's 12 largest cities identified will be in developing countries by the year 2020 A.D. (already several years have passed since this prediction). They are Mexico, Sao Paulo, Kolkata, Rio de Janeiro, Mumbai, Jakarta, Delhi, Bangkok, Dhaka and Buenos Aires.

But there is not enough water to go around, and the water shortage phenomenon is repeating itself in country after country, from the slums of Mexico to the overburdened farms of the China, from the Assent desert of the American South-West to the dead shores of the once vibrant Aral Sea. In many advanced countries, water shortage is crippling.

> Some Estimates Showed That Almost the Whole of India's Economic Growth Got Wiped Out By The Health Costs And Water Pollution Which Was Not Covered Economic Calculations.

The Aral Sea, once the fourth largest lake in the world, has shrunk in size by half, and its salinity has tripled since the 1960s - largely because the incoming water from the Feeder Rivers was diverted to irrigate cotton and rice crops of dubious benefit.

Thus, we are not only outstripping the precariously limited freshwater stock of our planet, but also poisoning the scarce and fast-depleting water resources that sustain all life.

It has been pointed out, "To comfortably ensure the availability of fresh water on a progressively sustainable basis, the whole question of water cycle and the quality and quantity of water has to be looked at from an ecosystems and cross-sectoral approach".

Consequences of pollution

The amount of polluted wastewater discharge in the world doubled between 1980 and 2000, whereas the global need for water doubles once in 21 years.

To add to the problem of shortage, there are alarming reports by the International Water and Sanitation Centre of Netherlands, that 2 million tons of human excreta pollute the rivers and ground water of our planet every day.

> Be it a revolution or just ripples of change it has to happen in the mind first.

For instance, in Latin America, about 98 percent of sewage water is discharged untreated. In China, two-thirds of the rivers are dangerously polluted. Three quarters of the rivers in Poland are so contaminated, that the water cannot be used for industry or agriculture, not to speak of human consumption.

Some rivers in Malaysia - 40 of them - are devoid of fish and other aquatic life due to high pollution. Sixty to seventy percent of the water that flows through the Passing River in Manila, Philippines, carries large amounts of untreated sewage.

Despite these known dreadful facts, not even half the number of projects financed and backed by the World Bank include, a sanitation component.

The potential for epidemics and water-borne disease is now dangerously high, just as it was in 19th century Europe, when measures to protect stored water from the polluting human population were taken only after cholera threatened to decimate the rich and poor alike.

> Greening the world and saving the environment and eradicating poverty caused by the polluted environment is the best way we can help show care and concern for others, for the have nots in particular. It is our religious duty and responsibility. Love your neighbour is contained in-providing abundant and healthy water for the poor and in greening our planet. All religions give importance to the provision of water. "And of water we made all living beings". (the Koran -Sura 33 -Verse 7). The Gulf countries provide plenty of potable water for the people by desalinating sea water by several methods. The Jews of Israel provide water by building sea water solar ponds in their desert country by novel innovations. Throughout India, Hindu families which have water will leave a mud pot of water on a stool on the roadside near their gate for passed to quench their thirst.

Chapter - 46

Minjur Plant - A Case Study

Requires huge investment on the machines and on maintenance. Needs motive power and maintenance engineers. Water has to be pipelined or transported from the production point to the consumer point. The RO process removes, besides salt molecules, only a small portion of heavy metals, dissolved salts and toxic waste.

Such units are good for meeting the home requirements of large industrial undertakings, universities, hostels, railway stations or airports and colonies of the rich. Such plants are totally unfit to provide absolutely safe and cheap drinking water free from arsenic, fluoride etc. for the rural and the slum poor, who do not have house to house pipe connections. Such projects can only be a drain on the government and will not relieve the thirst, poverty and hunger of the poor. What may be good for the advanced and rich North will not and cannot be good for the unadvanced and poor South. Small is beautiful and large is wasteful.

The much talked about alternative development has to be ushered not by undynamic and national and international governments that always adhere to old order and systems of doing things but by the people living at the grassroots as they perceive possibilities for action - Author.

If we want a common future we must work towards an uncommon future now.

"Man has lost the capacity to foresee and to foretell. He will end by destroying the earth." - Albert Schweitzer.

Chapter - 47

Gas For Distilling Sea Water - A Case Study

There is an interesting sewage treatment unit in Sangli District in Maharashtra which has been in use for about 40 years. Domestic sewage is treated by this unit by using the water hyacinth. The plant also produces biogas from water hyacinth weeds by anaerobic digestion and leaves left over after treating the water hyacinth are dried, crushed, granulated and sold as fertilisers.

Raw sewage contains several organic and inorganic chemicals that are dumped into lakes, rivers and water bodies. Water hyacinth can be used to remove these pollutants to some extent. In particular this plant absorbs metals and organic and inorganic matter. Further, absorbs ammonium compounds, nitrates, arsenic, carbon, phosphates and so on after all these are absorbed the weed residue can be crushed and granulated and used as a cheap and effective fertiliser.

The high moisture content in the weed makes it suitable for bio-gas production. It is found that it is better than cow dung since it decomposes faster than cow dung and produces a larger volume of gas. When cow dung and digested water hyacinth granules are mixed the volume of gas is much more and the quality of gas so obtained is much better.

This technology must be introduced and promoted all over the world as water hyacinth and cow dung are available in large quantities almost all over the world throughout the year.

The gas can be used to steam distilled sea water into safe drinking water.

Thailand produces a large quantity of soaps using the ash of water hyacinth, which contains Potassium chloride. Water hyacinth waste is added to all sorts of foliage, straw, groundnut shells, corn cobs,waste paper, etc. to make briquettes and pellets to steam distill sea water. There is no such thing as waste in this world there is a profitable use for everything.

Chapter - 48

Hospitals Spread Diseases

The water supplied by the Chennai Metrowater to the three and a half century old General Hospital, feared to be the premier institute for spreading diseases in the city of Chennai is of such poor quality, that when stored in a transparent bottle for few days, the sediments and floating matter are clearly seen. A senior surgeon of the hospital warns that anyone drinking it can be affected by water-borne diseases.

Many doctors of the hospital agree that the water supplied to the patients and visitors is pure only on paper. They report that at least 5 percent of the 10,000 of the ten or more patients that the hospital receives every day are at risk of contacting contaminated water-related diseases or of becoming more sicker than they were before presenting at the hospital. Doctors complain that they cannot rely on the reverse osmosis (RO) machines which treat impurities such as bacteria and not isotopes which are hazardous, like heavy metals such as lead or mercury.

They point out that when it rains, water enters the very old and improperly maintained sumps, laid flat on the ground where dirt and rust chocke the RO membrane of the machine which supplies water for the operation theatres and dialysis units. While the impure water problem is the same in other hospitals, government or private, THE PROBLEM IS APPALING in hospitals in many other developing countries. It kills thousands of people in the African continent, especially those in the northern part of the continent.

Only distilled water should be supplied to hospitals and operation theatres in sterilised containers and there should be a regular check and control on this.

Several people have observed that in several hospitals run by the government or municipal corporation or even private concerns, for want of water in the nearby taps the nurses fetch water from far-away taps in a stainless steel container, heat it over a spirit lamp and leaves it in the room. All nurses pg these hospitals use a new syringe for the first patients in the morning and for the patients who come later they dip the same needle of the same syringe in the same container of water that was heated in the morning for want of fresh water close by. This is a hard fact and a sad truth which veveals the problem of non-availability of safe water close at hand. There should be not only plenty of water for use close by but people should also have quick and easy access to water. This can be made possible only if the source of supply is the wide and deep sea.

> There should be a regular distilled water supply system to cater to hospitals, hotels and old age homes and the water may be called Metro Distilled.

Chapter - 49

Cry For Water

CRY FOR WATER: Water has gone scarce in the land of plenty too. Here, the U.S. President, Mr. Bill Clinton, hugs a weeping woman when he visited a water distribution area in Des Moines, Iowa, on Wednesday. Residents line up everyday for the precious commodity in the town.

Protozoa Microscopic Animal Life Found in Water

Water in sachets unfit for drinking

In what can be termed as a major crackdown on spurious packaged. drinking water business in the city the Chennai Corporation is all set to issue notices to manufacturing units selling unfit drinking water in sachets.

Brands like Wave, Yoga, Spice Aqua, Chiderry Palar, Hi-Tech, Magic Prince, Sri-Balaji, Coolbreeze, Ripple, Trust Aqua, Dolphin, VCR and Prince Life were among the poor quality ones. Interestingly samples of branded PET bottles have received a clean chit, sources said.

Zone/Dv	*Place of collection*	*Source of water (SACHETS)*	*Colilform*
Kilpauk/71	Gandhi Avenue	Wave	47
Kilpauk/71	Gandhi Avenue	Yoga	33
Ayanavaram/50	PM Road Spice	Aqua	47
Ayanavarara/57	Market Street Spice	Aqua	39
Ayanavaram/63	Sugalam Street	Chiclerry	39
Ayanavaram/63	SIDCO Nagar	Palar	140
Triplicane/86	Big Street	Hi-Tech	70
Triplicane/86	CNKRoad	Magic	49
Tondairpet/1	Thiruvalluvar Street	Prince	84
Fondairpet/2	Krishnamurthy Salai	Sri-Balaji	76
Tondairpet/2	GNT Road Cool	Breeze	95
Kilpauk/67	Anna Nagar (E)	Ripple	46
Kilpauk/68	EVR Salai Trust	Aqua	70
Kilpauk/68	Corporation complex	Dolphin	76
Kilpauk/67	West Club Road	Magic.	49
Ayanavaram/50	Paper Mills Road	Cool Breeze	95
Ayanavaram/61	Brickiln Road Trust	Aqua	70
Kilpauk/78	Valluvarkottam Hg Rd	Hi-Tech	49
Kilpauk/76	Sakthi Nagar	VCR	74
Nungam/104	Brindavan First St	Prince life	36

Source: Food Analysis Lab. Chennai Corporation, July 24, 2007.

Looking ahead, water shortages are not only going to intensify and spread, but users also will, have to pay increasingly more for their water supply. The world can ill afford to waste time. Make no mistake. Water poses a more intractable problem for the world than peak oil, economic showdown and other oft-cited challenges. Effective water management can help transform economies and power scarcity, writes Bralma Chellalaney in his book, WATER, PEACE AND WAR, Oxford University Press.

Chapter - 50

Unsafe Water in the City

The Bureau of Indian Standards frequently siezes thousands of bottles of untreated contaminated water in and around the large cities and towns of India. New illegal bottling plants continue to spring up in the sity.

In the city of Chennai alone 60 lakh litters of packed drinking water is sold every day. Almost every middle income home uses bottled water as the Metro water, which is not always treated, gets mixed with sewage water in criss-cross pipelines, which have cracks and leaks. Metrowater is also irregular in supply, as people often complain and the media keeps reporting.

There are about 450 brands of water sold in bottles and satchets as reported by the Times of India on 22nd Jan. 2013.

Due to the untreated water, there is a steep increase of waterborne diseases in the city and district headquarters.

What is more serious is the report of a leading , daily in Jan. 2013 that the Bureau of Indian Standards has been framed and has come under a scanner for taking bribes from the illegal and unauthorised units to refrain from even random checking for pernaitting the fraudulant use of ISI mark and for not questening labels on the bottles and satchets. It is reported that even some big bottles called "BUBBLE TOPS" had fake ISI marking.

It must be pointed out once again that in most parts of India water contains excessive flouride besides other harmful chemicals and a high level of total dissolved salts.

Most of the women in Third World countries assert that since the whole country drinks this water we will also drink it. If we break our heads about the quality of water, the quality won't improve.

Let the basic message of this book be repeated. Every family should steam distil the available water or sea water enough to meet the family requirement of the day. Water may be distilled on a kitchen scale as explained in this book and stored for 2 or 3 days. Some kitchen ingre-dients listed in this book should be added to the distilled water to restore the lost nutrients.

> By 2030, the global middle class could top 5 billion, up from 2 billion today. These people will inevitably demand higher living standards, as well as greater freedom, dignity, and justice. Why should they settle for less?

Chapter - 51

Metal Eating Algae

A unicellular algae called "chlamydomon as reinhardli", is prised for its metal eating properties in water. Bio-remediation researchers are further enhancing the value of this algae to maximize its ability to absorb trace metals found in water. The algae is easily available on a large scale. Easily and economically it can be genetically engineered and multiplied rapidly. Compared with the methods of chemical extraction of metals from water in use now, it is more environment friendly, less costly and more effective.

Researchers have found an effective method now to genetically enhance the algae, which involves attaching a protein known as metallothionein to the algae. The protein enhances the cells metal-picking capabilities by five times. When surrounded by high concentration of heavy metals, the cells grow three times faster.

It is a major step forward in reducing heavy metal pollution and remediation of contamination in and around great lakes, ponds, catchments areas and storage tanks.

This algae must be promoted and multiplied all over the world to reduce and even get rid of metals in Drinking water to the extent possible.

God has given us this algae and the duck weed to make contaminated drinking water potable. After treating with the algae and the duckweed if water is boiled or exposed to ultraviolet rays of the sun, as explained elsewhere, it will be safe for human consumption.

Chapter - 52

Let the Water Flow Again

The plight of the urban poor in Tamil Nadu and Andhra Pradesh is pitiable at the best of times. For millions of families, it becomes appallingly worse when the surface runoff from rainstorms enters their homes, huts and surroundings laden with decaying garbage and human waste left in the open as the result of a lack of waste disposal provisions and sanitation facilities. This aftermath of the storms also leaves behind new open ditches and surface drains in many urban settlements, which get clogged with all sorts of waste, filth and debris. This leads to flooding during rains and causes pollution of the water courses, or aquifers, which receive the filth-borne water. This will worsen in Chennai if the over 6.5 km stretch of the Buckingham Canal is not desilted.

The Buckingham Canal, the total length of which is 940 km, starts from Beddaganjam in East Godavari District in Andhra Pradesh and ends at Marakanam in Cuddalore District in Tamil Nadu. Within Tamil Nadu, its length is 165 km and it takes in a lot of flood-water during the Monsoon.

The Buckingham seawater canal in the city was dug during the great famine of 1878 to give work, food and water for the people. It is still a clear water-way from the Pulicat Lake on the Andhra Pradesh-Tamil Nadu border, but pollution is acute near the industrial suburbs and it becomes a constricted drain near Madras Central (railway) station. The Railways have taken over a stretch of the canal for platform extension, further reducing the width and now waste from Central Station is dumped under the old and new bridges.

The Ennore Thermal Plant discharges about 600 cusecs of water and all this is, at present, discharged into the sea. There was a proposal to let about 150 cusecs into the Buckingham Canal, the maximum the canal can take. If this is done, the Canal will become navigable upto the Cooum crossing and will also get flushed. As the Cooum and the Buckingham

Canal will then have about 5 feet depth of water, they can be used for mass transport by regular ferry service using pedal or power boats and country boats.

In 1993, the Municipal Corporation of Madras prepared a detailed and comprehensive plan to deal with floods and cyclones and identified several low-level areas within the city prone to water-logging. These included 58 in Kilpauk, 30 in Tondiarpet, 46 in Pulianthope, 20 in Basin Bridge, 26 in Nungambakkam, 10 in Ayanavaram, 15 in Kodambakkam, 12 in Triplicane, 41 in Mylapore and 21 in Saidapet. The easiest and cheapest way of raising these low-lying areas is to fill them with the sand removed from the Buckingham Canal while deepening and widening the Canal for inland water transportation.

When the canal is deepened, widened and, all obstructions on the way are removed or bypassed and when the low-lying areas, pits and potholes in the City are raised all the surplus and runoff water of the City will flow into the canal as in the past. Closing the Canal even in small stretches will only create additional problems, as even broad and long stretches of built-up rain- and drainwater systems cannot receive and conduct all the rainwater that now goes into the Canal, without flooding the surroundings and slums. If more seawater is let into the canal, the rain and seawater mix in the Canal can be let into lower lying areas outside the city for saline or semi-saline agriculture. It can be mixed with filtered city sewage and used for cultivation as is done in Dubai and in Coins-batore nearer home.

It will not be possible in our lifetime to remove completely and permanently the slums and squatter settlements on the banks of the Canal, as most of the dwellers there have no other source of water even for washing their feet or cattle. It is also not possible to stop them from throwing their garbage and debris into the Canal! Under such circumstances, the Corporation of Madras would be well-advised to consider maintaining a regular flow of water in the Canal and keeping its banks clean and healthy as suggested in reports of the two British teams sent to Madras to study the problem of the inland waterways of the City.

Nothing has an Uglier Look to is Than Reason,When It is not in our Favour.

Such cannals must be widened and deepentd for pedal boat navigation and for people to distill the water for drinking by the pro-poor distillation methods widely published by the Ryan Foundation International.

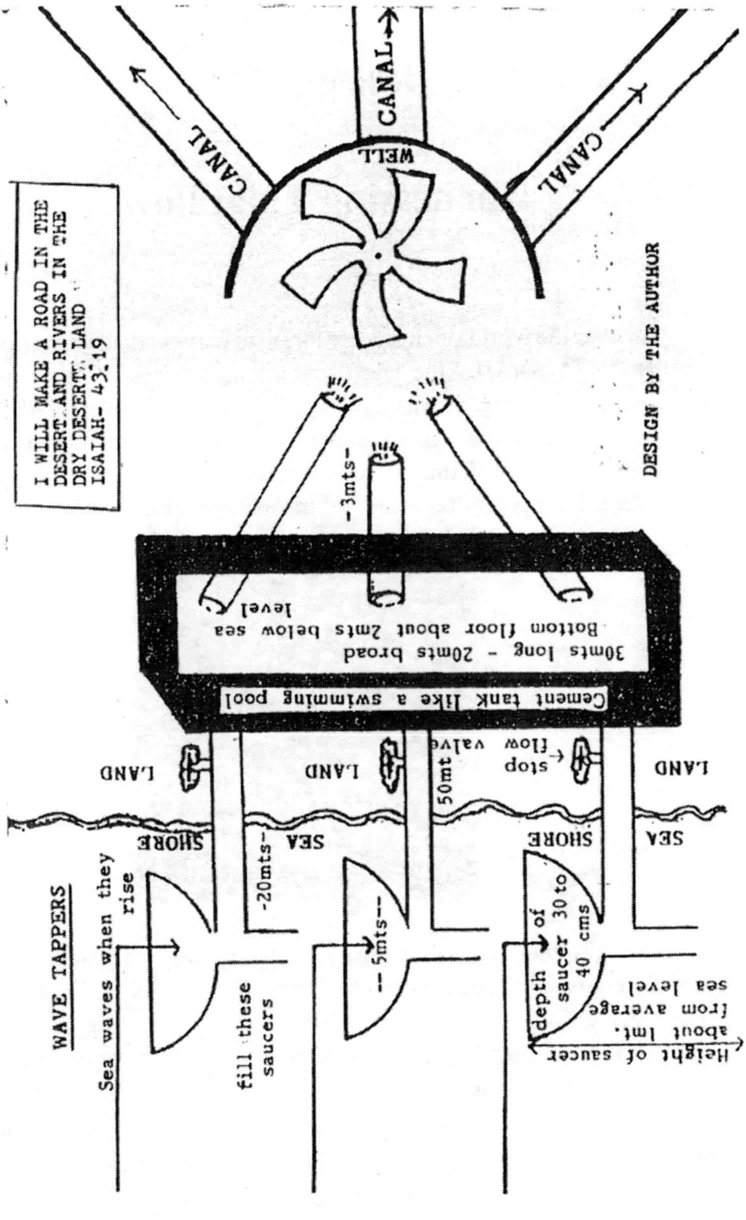
I WILL MAKE A ROAD IN THE DESERT AND RIVERS IN THE DRY DESERT LAND
ISAIAH- 43.19
CANAL
CANAL
CANAL
WELL
-3mts-
Cement tank like a swimming pool
30mts long - 20mts broad
Bottom floor about 2mts below sea level
LAND
LAND
LAND
stop flow valve
50mt
SHORE
SHORE
SEA
SEA
-20mts-
-5mts-
WAVE TAPPERS
Sea waves when they rise
fill these saucers
depth of saucer 30 to 40 cms
Height of saucer about 1mt. from average sea level
DESIGN BY THE AUTHOR

Chapter - 53

Harnessing Tidal Power

All over the world scientists are trying to harness tidal waves to produce energy, "SEA SNAIL" the latest invention, is a new type of turbine to generate tidal energy. It was developed at the University of Aberdeen in Scotland and hailed as a revolutionary technology to get more energy that is scarce world over. A miniature model of the sea snail was demonstrated at the ALL Energy Conference held in Aberdeen, Scotland.

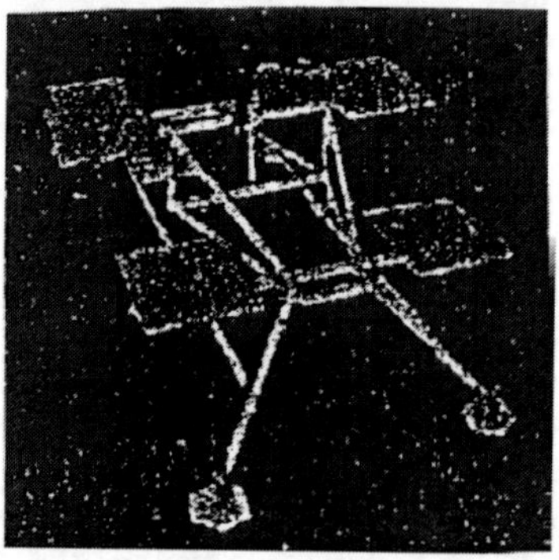

The equipment which is now a prototype measures 20 metres long and will undergo trials in the sea shortly. The full-size "SNAIL" is a fabricated tidal device that can be installed at a very low cost in shallow or deep waters. The adaptability of the machine will significantly increase the number of suitable sites for turbines and reduce the cost of instalation.

The invention is being tested in the Eynhallow Sound in ORKNEY and when commercially viable is expected to exploit the massive energy potential and distil sea water into safe drinking water.

All admire good development but only a few make that happen for posterity to prosper.

Chapter - 54

Ryfo Wind-Cum-Wave Generator

(Renewable Generator)

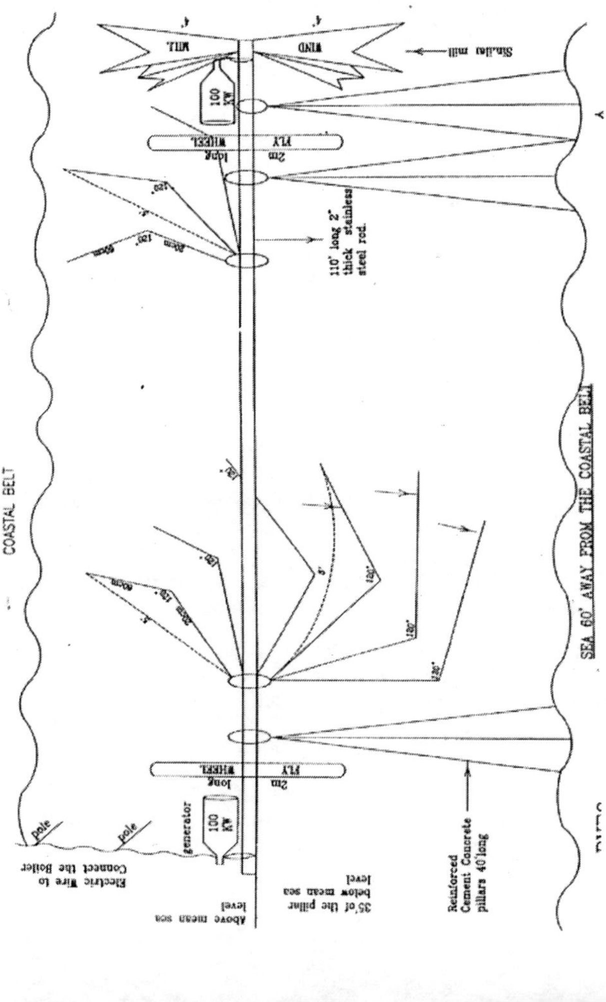

The two generators are mounted of the two ends of the axle.

There are spider turbine with a distance of 10, between them. In other words there will be 10 turbines in oil fixed of both ends of the axle.

WAVE LIFTERS - Diagram.
(explanation next page).

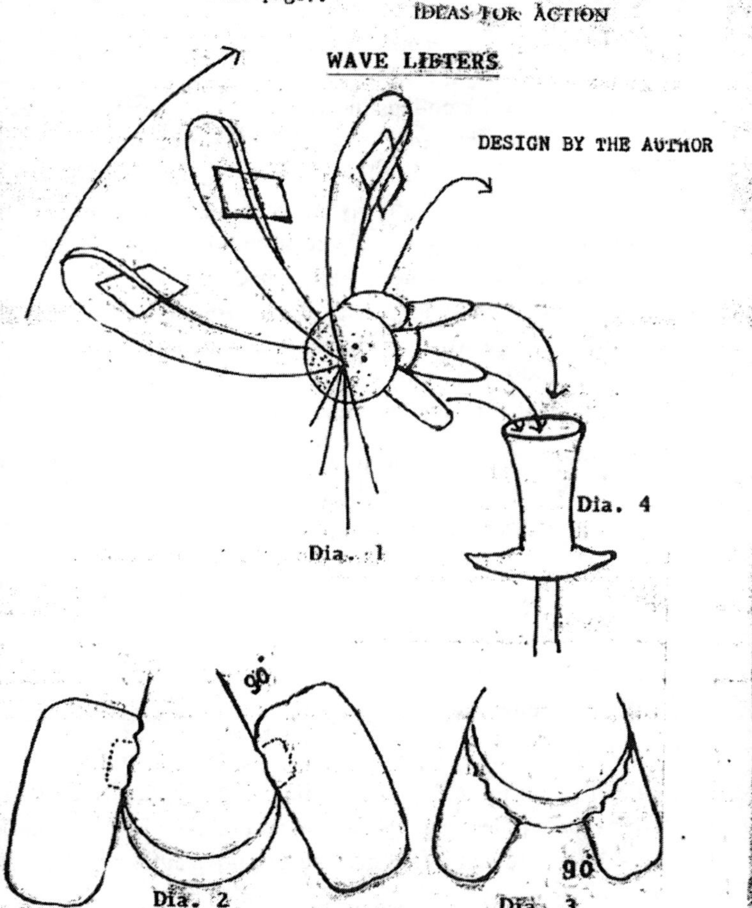

Chapter - 55

Wave Lifters Expanation

The buckets and the collapsible flaps are the special features of the windmill designed for the purpose. Each bucket wing seen in diagram is Two metres long and each wing has 30 cm by 15 cm flaps on either side fixed on hinges to work like butterfly wings. When folded towards the axle of the mill the flaps will fold only to a certain extent (to form about 90 degrees with the bucket wing, as shown in diagram 2) as there is a block (stop) in the rear of the flap to prevent it from folding further. Thus, when the wind blows the flaps will be in open position as shown in diagram 2. But this will happen only when the buckets with water move upwards and the wind will because of the wide open wings make the buckets move faster.

After the buckets empty the water in the receiving tank as shown in diagram 4 and move downward the empty flaps will collapse and hang downward as shown in diagram 3. In other words, when the buckets pick up the waves (water) and move upwards, the opened out flaps will face more of the blowing wind and move faster and when they return empty downwards, the collapsed flaps will not offer wind resistance. By this arrangement the mill will operate efficiently.

A hole is made on the existing wall of the well just above the tank as shown.

A pair of casuarina poles may be tied like a ladder 10 to 12 metres long. Two or three poles may be joined and tied together. One end of the ladder must be buried half a foot into the ground, slanting on the well wall as shown.

The other end must come almost in line with the middle of the well and on the very top crossbar (rung) a pulley may be tied or hooked.

Waterproof canvas cloth (raincoat cloth) may be cut and stitched in the form of a long tube of 15 centimetres. Out of the same material a bucket

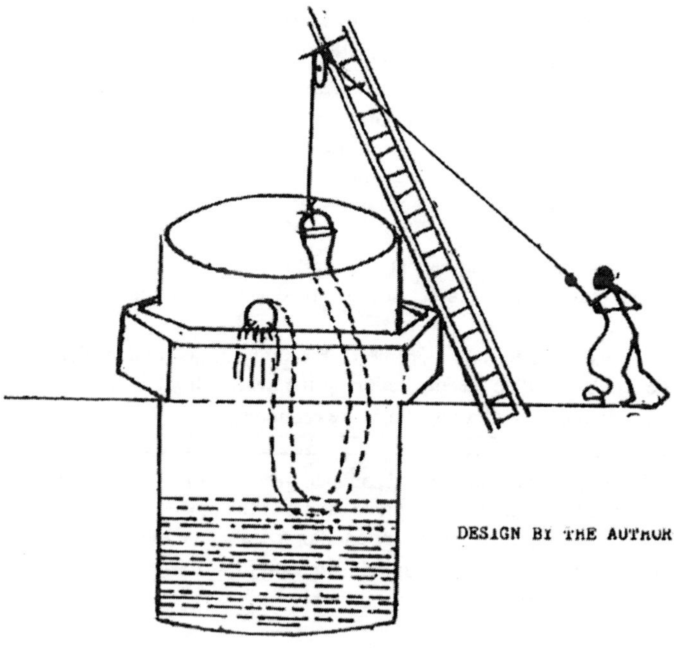

Lift Water from Sea Water Wells

may be stitched without the bottom. The bucket may be 40 centimetres long, 30 cm in top diameter and 15 centimetres in bottom diameter. The open bottom end may be attached to one end of the long canvas tube as shown and the other end pulled out from the inside of the well through the hole that was made and a metal rim of 18 centimetre diameter fixed to the delivery end of the canvas tube so that the delivery end does not fall back into the well.

The total length of the canvas tube should be equal to the length between the delivery hole in the wall and the water level in the well (plus 1.5 metres in case the water layer goes down).

A quarter centimetre thick metal ring is fixed to the opening end of the bucket and a semicircular ring attached to that ring as shown. When the two centimetres thick nylon rope or fibre rope tied to the bucket and taken round the pulley is let loose the bucket dips into the water and picks up water and when the rope is pulled back, the water in the tube gets emptied into the storage tank around the well.

Every day two families may fill the tank in turns and the others asked to take three buckets of water from the tank. This will save time, trouble and labour for the village and there will be happy community participation. The system can very well be introduced as a first step in community co-operation and people's participation in rural areas.

The all round plight of sub-Saharan Africa is extremely serious, all because of dreadful scarcity of water, besides other less important reasons. Civil wars and climatic disasters have taken their toll, swallowing funds, destroying infrastructure and preventing the extension of coverage.

The debt burden, and subsequent structural adjustment programmes, has drastically reduced budgets, for what are still regarded as social services.

The terms of trade have become increasingly unfavourable. Prices of export products have fallen steeply, reducing the availability of much needed foreign exchange and thus affecting the entire economy.

The UNESCO Director General sounded the alarm on World Water Day 2002 by describing the looming water crisis as one of the most critical challenges facing the world today, Two figures are enough to conceive the world population has tripled over the past century and the demand for water has increased sixfold.

Chapter - 56

Simple Sea Water Inlet

This is a model of a typical sea water inlet system leading into canals and inland waterways to flush the filth stagnating in the waterways, make the waterways navigable and keep the city clean and cool.

The pipe shown in the picture may be one or half a metre in diameter depending on the receiving in and flowing-out capacity of the canal. The pipe should be made of reinforced cement, fibreglass or high density solid PVC (polythene) pipes but not out of metal as metal will rust in sea water.

Instead of round precast pipes, square pipes the sides of which are built out of brick and mortar will also serve the purpose. Such square gutters can be built even by village masons. The inflow control system should be built with care and calculation.

Governments of Third World countries can construct this type of inlets with their own funds without asking the World Bank or the EEC for money. When constructed, several sea water based activities can be promoted and the water from the canal led into villages in small man-made open canals and the country will move towards progress and prosperity.

A simple, family scale method of desalinating sea water for drinking and cultivating plants and trees and promoting several sea based industries been explained by the Ryan Foundation (included in this book).

Stainless steel filter boxes to prevent sea sand from flowing into the waterways and lock and sluice arrangement to regulate or stop the flow of water are also explained. The stainless steel or fibreglass filter box about - cubic meter in size may cost about ₹ 2000 or US$ 200 and transporting these boxes from where they are fabricated to far away Third World countries will cost money. As an alternative to the filter box the mouth of the pipe into which sea water enters may be closed with a sheet of rubber (car tube opened out) or leather, and tied. The middle portion of the rubber or leather should be hammered with a screw driver and perforated with hundreds of slits. Sea water will enter these tiny perforations but the bulk of the sand carried by the water will not enter. However, by this method the water that comes out at the other end of the pipe will not have force and so mini generators cannot be worked to produce reasonable amount of electricity.

Humanity can survive in the 21st century only by harnessing sea water.

Israel a very big exporter of fruits and vegetables to the USA and Europe, grows them in the Negev desert cultivated by sea water. This is a well known fact.

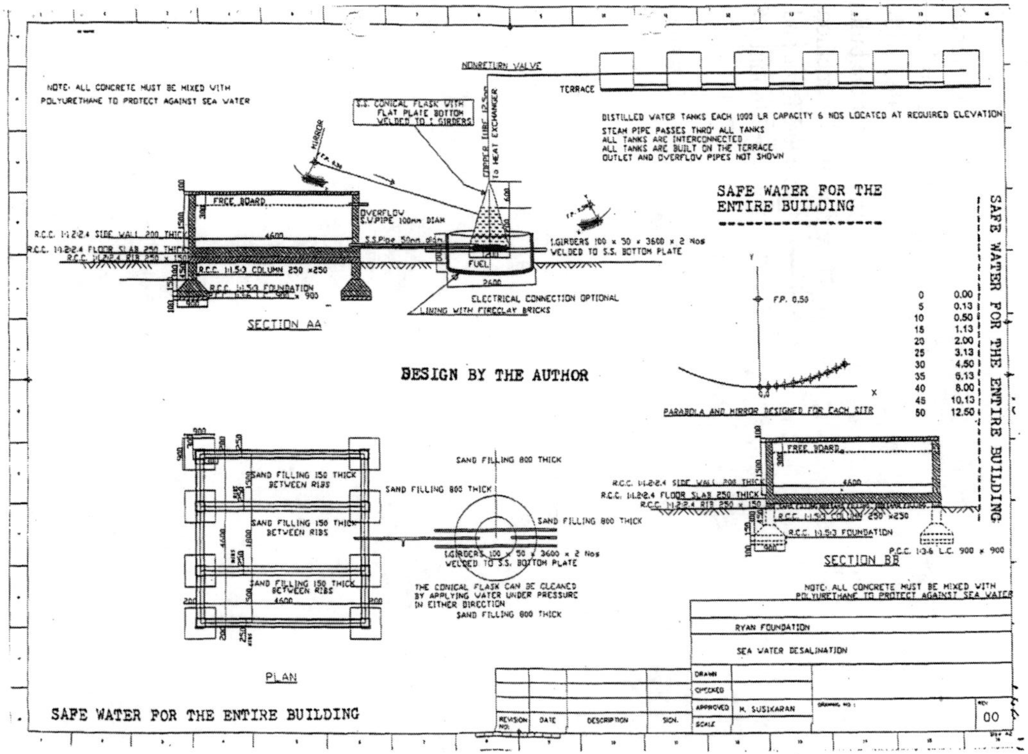

Safe water for the entire Building.

STILL FOR COLONY OF 1000 FAMILIES

DESIGN BY THE AUTHOR

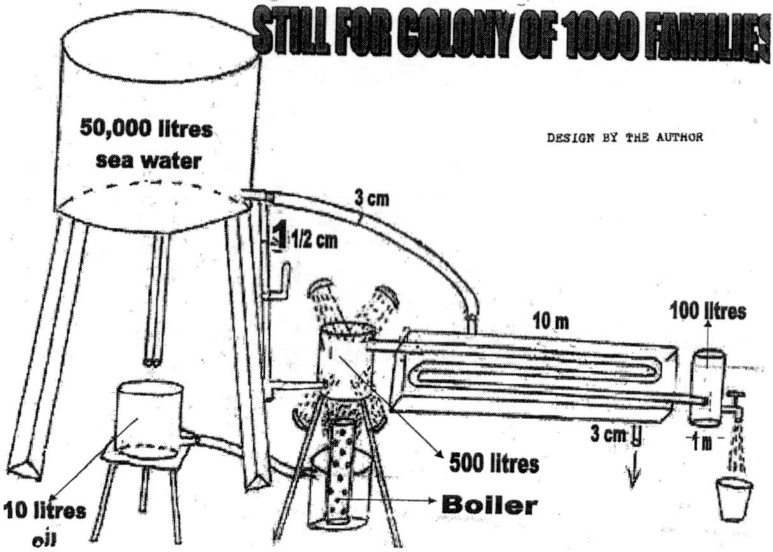

Chapter - 57

Fresh Water from the Sea

How it is converted is explained in this diagram:

1. An overhead water tank of 5000 litre capacity is built.
2. Below the tank an open drum of 10 litre capacity is placed on a stool or pillar.
3. This drum is filled with all sorts of waste oil, discarded car engine oil, fish oil, castor oil or any oil from pods and plants and trees. Methods of extracting the oil are explained in this book.
4. There is another stool beside the water tank of 500 litres sea water holding capacity.
5. Beneath this stool a small open drum with an empty beer can is placed and the beer can is perfurated on all sides.
6. Tailor's waste, paddy straw, corn cobs, waste paper, groundnut shells, tapioca peels) are stuffed into the beer can. (the and soon may be larger than a beer can) can container.
7. Oil from the small drum below the water tank is connected to the burner drum which holds the beer can with a rubber hose or cycle tube.
8. By this arrangement oil from below the water tank will keep flowing into the drum above the burner, which will have a regular flow of sea water from the big overhead tank.
9. Water in the drum above the burner must boil to produce steam. If the heat of the sun is inadequate, parabolic solar metal reflectors and parabolic magnifying glasses may be placed to enhance the heat to over 200 degrees Celcius. Different types of reflectors are shown in this book.
10. The steam from the boiler is led into a tub broader and longer than a bath tub.

11. The cold sea water from the overhead tank is led into this bath tub through metal pipes or cycle tubes.
12. There must be about a dozen metal pipes with hair-pin bends or joints on either end.
13. In this mini swimming pool the cold water cools and condenses the steam into distilled water which flows into the receiving tank which has a tap through which safe water is drawn out.
14. The bath tub and the several long pipes in it may be increased in size and number to get more water. The capacity of the tank may also be increased.
15. Throughout the day the water in the boiler must be left to steam and from nightfall to early morning it will cool and condense.
16. Different oil-yielding seeds and pods and easy methods of extracting the oil and also the way to extract oil from the unwanted parts of edible fish, from non-edible fish and from dead fish one also described.
17. Briquettes and pellets made out of paddy straw, the foliage of any tree, corn cobs a coconut and palmyrah waste should be used liberally as an alternative source of energy.
18. Burnt-out residue should be used as a fertiliser.
19. Some common kitchen ingredients mentioned in this book should be bundled in a small cloth into a fall as big as a table tennis or a regular tennis ball and soaked in the distilled water to restore the lost essential ingredients the water.
20. Sea water must be made available to people everywhere through canals, channels, closed gutters, ponds, lakes, lagoons or catchment areas.

Fluorosis present in water beyond permissible limit can cause discolouring and mottling of teeth and in advanced stages can even lead to damage of bones and joints.

Although successive governments over the last five decades have, in all, spent around ₹ 29,000 crores to make safe drinking water accessible to all, the problem of fluorosis contamination continues to stalk policy makers, administration and the scientific community.

A recent study has revealed that nearly 44,000 crores would be required to ensure supply of safe drinking water to the country's entire rural population; of this around 10,000 crores will be needed to simply improve water quality. -The Times of India, New Delhi, 23-1-2001.

Good governance is fundamental for the success and sustainability of any national poverty alleviation strategy. Evidence shows that countries that are able to develop and maintain transparent government institutions are more effective both in combating poverty and improving the living standards of their citizens, reports IFAD/UN, Nov. 2000, No. 8.

Water is one of the core essential and basic necessities. Despite earth's majority of the cover is filled by water (97% by oceans), 2% of the water on earth is glacier ice at North and South poles and only 1% is available as freshwater suitable for humans. In recent years the water table is facing serious threat due to rapid population increase, industrial and urban development, over usage, climate change and global warming.

Chapter - 58

Rough Cost Estimate

Ryan Foundation sea water distillation plant to provide about 50 liters of safe drinking water per family in a ready to occupy block of 8 to 10 flats each flat having a separate water tank but interconnected.

1.	Ground-level sea water tank of 30,000 liters capacity.	Rs	1, 50, 000
2.	1/8" stainless steel cone which has about 17% chromium mix (alloy) and is fixed in the middle of a 2 feet deep circular cement tank	Rs	6,000
3.	Cromium alloy stainless steel pipe and necessary fittings to connect the water tank and the cone	Rs	1,000
4.	Copper tube leading from the cone and passing through all the water tanks on top of the flats	Rs	3, 000
5.	Small accessories like couplings, bends, washers, taps, etc.	Rs	1,000
6.	Two solar reflectors as shown in the drawings in separate pages	Rs	30,000
7.	IIT Cam press to make briquettes out of foliage, paddy straw, Corn cobs, wood shavings, coconut pith, fibre, husk, etc	Rs	15, 000
8.	Two electric heating elements (optional)	Rs	6,000
9.	Labour charges	Rs	10,000
10.	Transportation charges (if made in Chennai, Coimbatore or Tuticorin)	Rs	10,000
11.	Miscellaneous	Rs	5,000
	TOTAL	Rs	2,36,000
	Note: In no case does the expenditure		
		Rs	2,50,000

Chapter - 59

Bamboo Pipes to Carry Water

Bamboo, of the grass family, grows fast and high under hazardous conditions of soil and climate. It is a highly drought resistant forest shrub-tree offering several livelihood possibilities and yet neglected. Bamboos make excellent sewerage pipes in remote rural areas where they can be grown and processed.

Ryan Foundation recommends standardisation of bamboo pipes to lengths of 3 mts and 8 cms. internal diametre. Mature bamboo may be cut at both ends uniformly with a saw and projections outside may be chopped with a chissel and hammer and then the inner cavity cleaned and broadened. To do this, a long iron rod long sharpened at one end like a may be used,

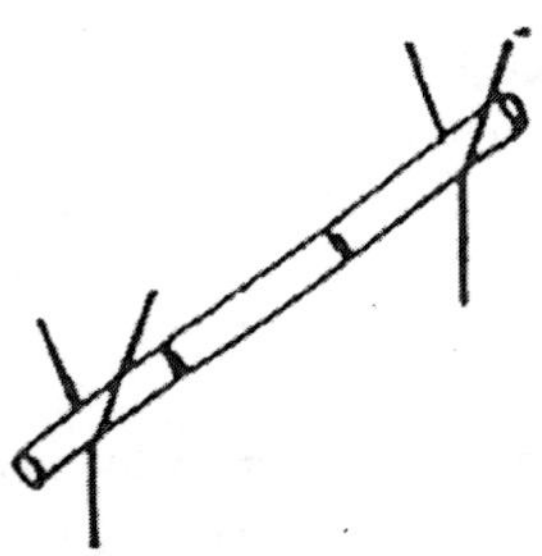

the bamboo may be placed on two 'Y' stands, as shown in the diagram and chiselled from either end. Only in this position we can see through easily and knock out internal projections and smoothed the ruggedness, whether used above ground or under all bamboos have to be treated. The best rural technology for treating is to boil Castor oil with neem, vasaka and datura leavee (3 parts) and turmeric, chillis powder and wood ash (one part) and paint the bamboos with the solution inside and outside. On a commercial scale the solution may be prepared in a long concrete pit-tub and the bamboos put into it and soaked for about a week. Pit-tub and the bamboo put into it and soaked for about a week This treatment prevents bamboos from cracking or being attacked by white ants. If bitumn (far) is available it may be melted and painted outside (not inside). The inside must be as smooth as possible if far is not available, used cement or fertiliser bags (plastic or fibre) or locally woven crude fibre cloth may be wrapped around the bamboo and tied.

Bamboo bits 30 centimetres long and to 10 centimetres in inner diameter may be similarly prepared and used as sleeves to join the pipes where required. Mortar mixture, explained in RF Handout 294, may be used as a binder to plaster the joints.

These bamboo pipes may also be used to take water to the fields and for domestic supply or drinking water.

In the Tanzania regions of Arusha, Rukwa, Morogoro and Dar-es-Salam several bamboo processing and installing units have been established financed by the Dutch government (US $ 140,000), Norad (US $ 196,000), SIDA (US $ 280,000), and the KEC. The processing method in different in Tanzania from what is recommended by the RYAN foundation. For RF processing only local materials available in the village need be used and nothing need be imported.

Raw materials being free, each 3 metre long pipe may be sold for ₹ 10 or US $ 1 (one) and the industry can provide full employment for atleast a dozen families of the village in the cultivation, processing, laying, repair, replacement and maintenance and it is an ideal industry for rural women.

Delivering safe and healthy water at a reasonable and affordable price, and providing sanitation for everyone, is technically and economically feasible. But do we have the universal political will to do so?

Chapter - 60

Tapping Sea Water by Utilising Bamboo Pipes

There are several ways of drawing water from the sea into land to fill dry ponds and lakes. The appropriate technology shown in the Ryan Foundation booklet "Pumps Without Water" quite inexpensive and feasible. Yet, all rural communities may not be able to instal them with their own initiative and financial resources.

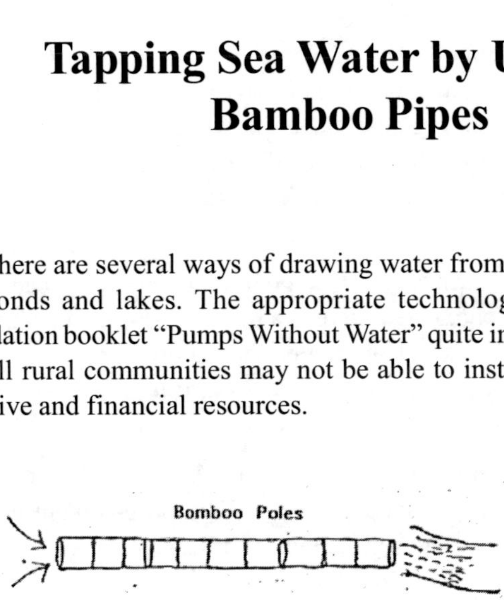

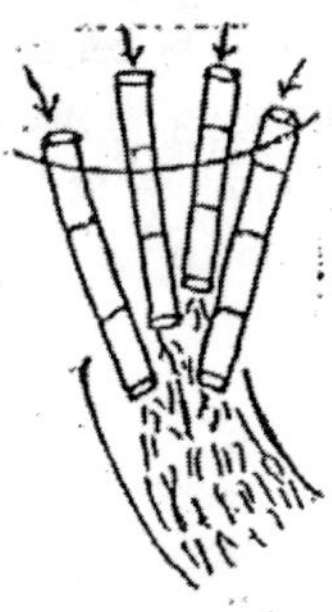

In RYFO Handout No. 57 9 a method is shown to draw sea water in bamboo pipes through a brick and mortar chamber. The purpose of the chamber is to provide a land mark, (2) level fixing, (3) double "flow block" (4) a steady out-flow and (5) a sense of community ownership and participation in getting water into their village.

However, sea water can also be drawn straightaway without building a brick chamber as shown in this page. In fact, this method is cheaper, quicker, easier and absolutely safe but can serve only one or two villages because of limited flow.

All that is required are bamboo poles preferably of uniform size and thickness and the number of poles required depends on the extent of sand to be covered on the shore by length. The bamboos may be cleared and cleaned inside, for free flow of water (RYFO Handout 323) and laid as shown in this picture.... The bamboo pipes may be joined as a carpenter joins two pieces of wood and glued (joinery). If necessary, joints may be wrapped with plastic or rubber sheets (cut car tube) or thick cloth or jute cloth and tied well to reduce leakage as much as possible.

Where the sea water enters the bamboo a strong cloth should be tied to act as a filter. The bamboo must be above 15 centimetres mean sea level - and it may have to be laid about a metre on the shore. It may be buried in and be laid slightly slanting position as shown and at the "outsflow" end the community may dig a channel to lead the water to the nearest dry lake, or pond or catchment area and from there take the water by channels to other places.

To stop the flow one has merely to go into the sea knee-deep and insert a bottle into the inlet end. Alternatively, a bottle or stopper may be inserted at the outflow end. Thus, all the materials required are some bamboos, discarded car tubes or plastic sheets, old cloth, fibre ropes, country-made glue and empty bottles.

It is better to lay four or five pipes all leading to the same channel to get a good flow into the village. Instead of bamboo, cement or earthern ware pipes used in sewerages may be used if available. Metal pipes will corrode in sea water and should not be used.

Chapter - 61

Ryfo Concept of a Town in the 21st Century

Whether one likes it or not ecocites towns and villages VIII have to be the order of the 21st century. There eco be no eco-development without water and the only source of infinite water in the sea. The sea has to be drawn into the land by man-made canals and creaks, as done in some Gulf countries and some island countries.

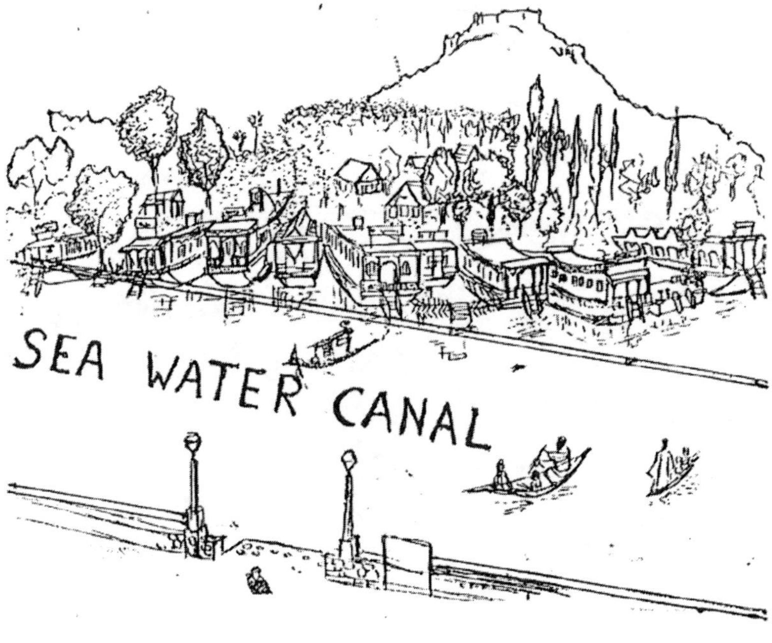

Within the towns and cities transport systems will have to be in eco-friendly row boats and pedal boats.

Sea food and sea water plants, weeds and reeds and seeds will become the staple food of the civilisation to come.

By and large, production for consumption coined Prosumption by Alwin Toffler, will be the rule within a decade from now.

Natural Sea Water Creeks Have Been Deepened and Widened Not Only in Dubai But also in Abu Dhabi, Fujairah and Sharjah and in Many Other Countries of the World.

Chapter - 62

Forestry Problem

(Only forest islands explained in the next chapter 87 can solve forestry problem to a large extent).

An expert committee of the Government of India constituted in 1986 said that watch and ward officials of the forest department are either overwhelmed by the situation and the forces behind exploitation or fall an easy prey to the temptation of making hay while the sun shines.

The committee, which had three forest specialists and the secretary of the union environment department as members with the former chief secretary of the Madhya Pradesh government as chairman, said that senior officers involved in forest management often became helpless and even when they acted they were hampered by political pressure and ultimately many of them resigned themselves to become mere watchers of the destruction".

Dwindling forests will disappear faster than is realised "unless this picture is radically altered" T.L. Shankar, former member secretary of the Planning Commission working group regarding energy policy records in the State of India's environment 1984-85 published by the Centre for Science and Environment, New Delhi, "I have seen many programmes launched in the name of the poor but soon distorted to benefit the upper classes but no programme has been diverted further away from its objective than the social forestry.

The Tamil Nadu government introduced the Social Forestry Plan (SFP) in 1981, targeted for a period of 10 years and with considerable financial commitment from the Swedish government. Today, in 2014, it is slowly limping to its half-way mark, considerably behind schedule, its effectiveness greatly hampered by corruption and lackadaisical implementation.

Involvement of non-government organisations and social service agencies discouraged by the forest department officials, says Fatimson, project director of the Social Forestry Information Project (SFIP), a voluntary agency founded by the Swedish International Development Authority (STDA) to oversee and co-ordinate social forestry activities in Tamil Nadu. 'Forest officials literally, resent our involvement; complains Arulanandam, the executive director of the Integrated Rural Community Development Society (IRCDS), funded by Christian Aid, London.

If we are to go by the reported numbers of seedlings that have been planted through the government forestation programme and if all of them were to grow into trees there may be hardly any space left for humans to live. According to Government of India statistics 1990, only 14 percent of the total land area is covert by forests - nowhere near the international average of 33 percent.

Foliage collected from forests can make heaps and mountains of briquettes for cooking and industrial use and thereby change the fate and quality of life of the rural people, and leading them to sustainable and job-oriented decentralised growth.

Eleven litres of sea water fill the tank above the false bottoms upto the disc through the inlet pipe 'C' Heat is directed through the 4 inch circular

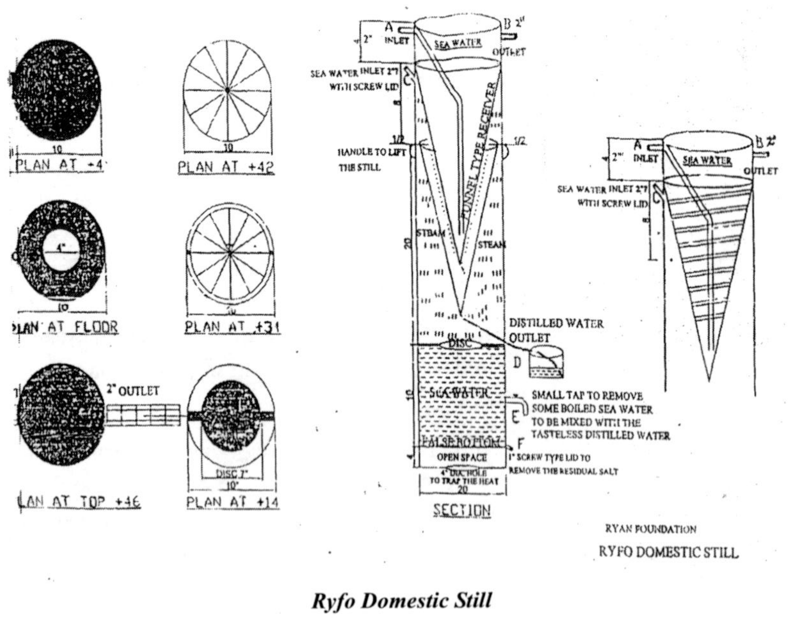

Ryfo Domestic Still

opening into the open space. The trapped heat in the open space is atleast twice as hot as the untrapped heat. The heated sea water at a temperature of 60 degrees to 80 degrees centigrade rises and hits the outer surface of the cold inverted one and condenses as water into water beeds and drips into the lower funner and collects in a container which has some water which submerges the pipe.

Sea water continuously fills the top inverted cone and flows out to waste or to irrigate plants which grow in sea water. The sea water circulation helps to condense the steam into distilled water.

Chapter - 63

Forest Islands

One of the worst incidents of terrorism happened in Terai region of Uttar Pradesh, on 1st August, 1992 when 29 innocent villagers entered the forest to collect a wild edible fruit and were ruthlessly murdered in the Puranpur forests of Pilibhit district. It was reported that the terrorists hiding in the forest there with arms and ammunitions were suspicious of anyone entering the forests and discovering their hiding places and therefore murdered the intruders.

Terrorists hiding in the dense forests and waging a guerilla warfare in Malaysia, Korea, Mayanmar (Burma) and many other countries is common knowledge. The LTTE militants of Sri Lanka were also trained in their hiding places in the dense forests of India. Undoubtedly, terrorism and guerilla warfare and coup there are on the increase oil over the world because of forests, besides other reasons. Most of the drug producing plants are cultivated in dense forests and illicit (poisonous) alcohol is brewed behind bushes.

Nevertheless we need more and more forests and trees and better ecology and environment and rain forests to save our planet. We also need more forest bushes to give habitat for birds and animals.

The solution to the problem is better and clever forest management free of guerillas, terrorists, drugs and poachers and the following module for developing and managing forests may be considered.

The conventional stone-marking of forest boundaries or barbed wire methods have failed to produce results and in fact, in many countries, they have been counter productive. After spending billions annually third world countries will be poorer by a few hundred thousand trees cut to make the uprights for barbed wire fencing.

Strictly speaking, under the order of the day the social forestry projects of most part of the globe have no social challenge, no social action, no social change, the resources are not being utilised for the welfare of the people, not for the community but only to fill pockets of poachers and politicians. They have no element of permanent resource creation.

Nevertheless we need more and more forests and trees and better ecology and environment and rain forests to save our planet. We also need more forest bushes to give habitat for birds and animals.

In most of the social forestry projects in India the rich grow eucalyptus paying a pittance to the forest labourers. They sell the timber to paper and board factories and get richer. They even shoot the animals and sell the skin.

Forestry needs a radical change and therefore some unconventional and rational methods have to be introduced. The Ryan Foundation module explained below is based on (or linked to) the "Survival of Sea Water" movement of the Foundation.

- To make a beginning, dense forests within a distance of 100 kms from the coastline in countries like Guiena, Somalia, India, Kampuchea, Malaysia or Mayanmar may be declared as "Forest Islands".

- 100 mts broad and 3 mts deep man-made canals may be dug round the forests making the forests like islands with lock (sluice)

arrangement to regulate the flow of sea water into the land, like the canal cutting across Dubai.

- There shall be only two jettys on the circular canal, one to enter into the Island forest and the other to exit.
- Authorised country boats may ply in the canal to transport forest products, fish and sea water plants and plant and tree products.
- Two, three or more speed boats, depending on the circumference of the canal may be on guard to prevent unauthorised people entering in or out.
- Watch towers may be built on the banks of the canal and also in the islands and search lights provided.
- These forest islands may be made permanent headquarters of army companies or batallions and the army may be given the responsibility of keeping law and order in the forests as their peace time operations.
- Hundreds of trees have been identified to grow well in sea water by the Ryan Foundation in the book "Plants and Trees that Generate Employment" and several of them identifed, explained and published in the book entitled "Saline Agriculture: Salt Tolerant Plants for Developing Countries" published by the National Academy Press, Washington DC in 1990. These trees may be planted on the banks of the ring canal. Several of them yield edible fruits, oil, fibre, dye, medicines, gum, nuts, fodder, fertilisers, chemical, etc. and generate employment.
- The commercial planting and exploitation and replanting may be regulated by the army as part of their peace time operation and the army men and women will love to engage themselves in such useful and purposeful work during peace time.
- On the outside of the ring canal the army may build their quarters and settlements and habitation may spread from the periphery of the forests outward. This will develop new habitations, homes, villages and towns and prevent over congestion in already congested cities and towns and retard, the growth of slums and squatter settlements in urban centres. Barren lands will become water sheds.
- Sea water fish such as sardine and the milk fish (chanos chanos) may be introduced into the canals and fish based industries such as chicken feed, cattle feed, organic fertilisers, fish oil, fish gum

etc. may be promoted based on appropriate rural technology as advocated by Ryan Foundation.

- The sand from the sea that comes into the canal may be manually lifted and shifted to fill low lying areas and this will provide employment to hundreds of women in each of the surrounding villages.
- Small rails with open wagons may be provided upto a distance and hand drawn carts, donkey drawn trucks and pack horses employed from where the rails end.
- With sand thus lifted pits and pot holes in the villages and towns may be filled, low lying areas raised and sand mountains raised along the coast lines in countries like Guiena or Maldives for sea defence. Raising such sand mountains will cost much less than erection of tetrapods as now done in Maldives, Bombay and many coastal towns.
- Administration of these forest islands may be entrusted also to the navy and air-force for peace time operations but should not be given to the forest department or civil administration.
- For this matter, the projects may first be tried in two or three party democracies and not in countries under one party rulers.
- If the forests enclosed by the canal is too vast, diagonal canals may also be dug across the forests for better transportation, communication and administration.
- When work is in progress the animal, bird and plant life of the area shall not be disturbed in any way but regulated and protected.
- Wild animals may be confined inside pockets in the island surrounded by water so that forest workers can move about without fear of being attacked. By this arrangement forest islands will become natural zoos in natural settings.
- Nurseries and saw mills may be provided outside the island in which the locals may be employed.
- Fish canning factories and foliage briquette factories may also be established.
- Sea tortoise and crabs and sea weeds may be introduced in the canals and weed based industries to make products such as caragrenan may be promoted.

- Medicinal herbs may be grown, gathered, processed and exported where possible.
- The nuts may be collected and crushed for oil and that oil may be used for making soaps.
- Fruits may be collected, dehydrated and sold in local and foreign markets.

There are indeed thousands of employment possibilities based on plants and trees that generate employment and thousands of employment possibilities to generate power, fuel and employment using sea water. For further information, see the book "Plants and Trees that Generate Employment" published by the Ryan Foundation in USA and the booklet "Survival By Sea Water" published by the Foundation in India which was declared part of UN Secretariat document for the UNCED. Earth Summit at Brazil, also see the 3 vols on "Better Life Technologies for the Poor" by the Foundation. Three more vols of the book will soon be in the market.

Richer the nature's capital, richer the human society will be

Forests: past and present

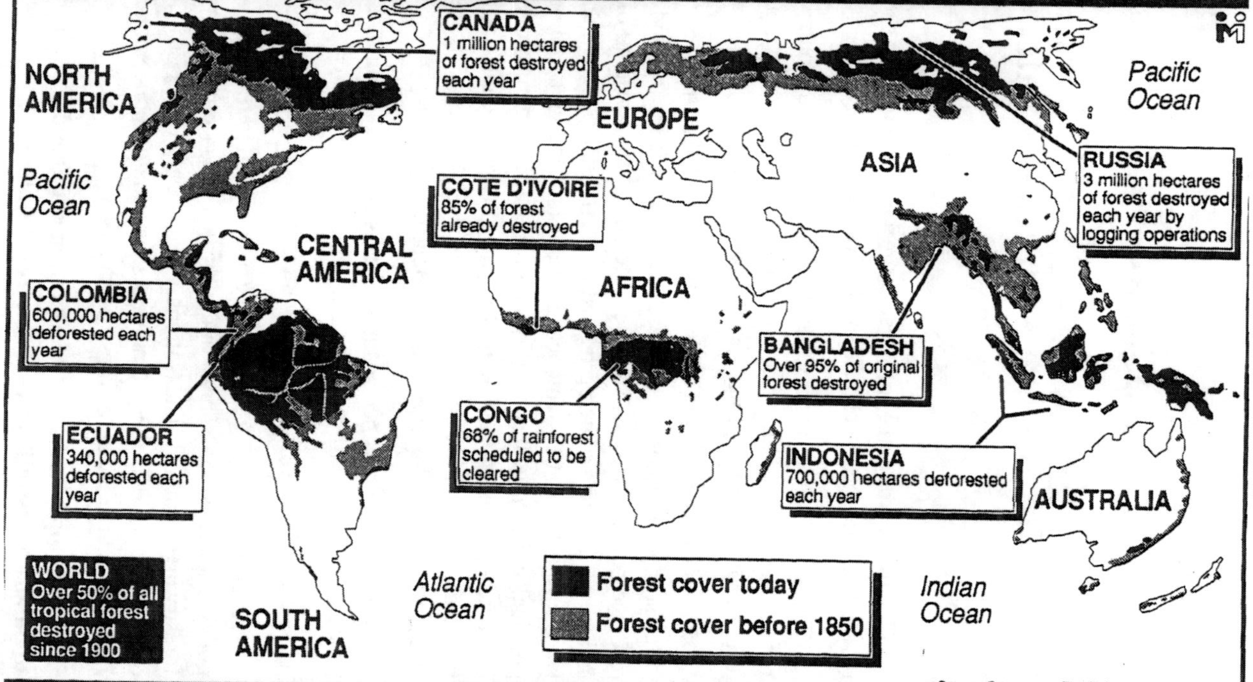

Source: Greenpeace, The Fight for the Forests, 1994

Chapter - 64

Forests are Sources of Water

The forest are sources of water as the women of Chipko showed in the 1970s. They are the storehouse of biodiversity.

Today the rule of money and greed dominates our society, economy and politics. The culture of conquest is invading into our tribal lands and forest through mining of iron-ore, bauxite and coal.

Every forest area has become a war zone. Every tribal is defined as a "Maoist" by a militarised corporate state appropriating the land and natural resources of the tribals. And every defender of the rights of the forest and forest dwellers is being treated as a criminal. This is the context of Dr Binayak Sen's life sentence. If India is to survive ecologically and politically, if India has to stay democratic, if Indian citizen is to be guaranteed, we need to give up the road of conquest and destruction and take the road of union and conservation, we need to cultivate peace and compassion instead of power and violence.

We need to turn, once again, to the forest as our perennial teachers of peace and freedom, of diversity and democracy.

Dr Vandana Shiva is the executive director of Navdariya Trust

After becoming the first country in the world to frame an agroforestry policy, India has gone ahead and allocated ₹ 444 crore to promote agro forestry in the country. The National Agroforestry Policy 2014, announced on February 10, has the potential to substantially reduce poverty in rural India and revive the agro-forestry industry.

More opportunities for corruption and fraud. Soon India is likely to import many million cubic metres of timber.

India has been meeting its growing demand for agroforestry products such as plywood, timber, pulp and paper through imports, while the domestic industries have been under-performing. According to the National Research Centre for Agroforestry (NCRA), the country imported 6 million cubic metres of timber and round logs and spent ₹ 18,000 crore last year.

India does not even grow halophytes (sea water plants and trees) and drought trees which require little or no care. India cuts more trees than what nature can replenish. Money allocated for trees encourage corruption and fraud and contributes to inflation.

> "Much of the money allegedly spent to help the Third World is actually used to finance the purchase of industrial goods from Western corporations or to finance the luxurious life-style of the international aid bureaucrats themselves," Graham Hancock -Author of Lords of Poverty."

Mangrove forests of Sundarbans (Photo: P.K. Pandit)

In several countries and wherever feasible, along the interface between land and sea, mangrove forests must be promoted to support genetically diverse groups of aquatics as well as terrestrial organisms. Mangroves can exist and flourish under a wide range of salinity conditions, tidal behavior, winds, temperature variations and muddy and anaerobic conditions.

Some important mangrove families found in the Sundarbans are Rhizophoraceae.

Ayiceenniaceae

Meliaceae

Sonneratiaceae

Sterculiaceae

Myrtiniaceae etc.

These must be tried in mangrove forests all over the world.

Chapter - 65

Some Mangrove Species

Table 1: Some mangrove species and their IUCN status

Sl. No.	*Family*	*Species*	*IUCN status of species*	*Occurrence*
1.	Rhizophoraceae	*Rhizophora apiculata*	LR	C
2.	Rhizophoraceae	*Rhizophora mucronata*	LR	C
3.	Rhizophoraceae	*Bruguiera parviflora*	VU	C
4.	Rhizophoraceae	*Bruguiera cylindrica*	LR	C
5.	Rhizophoraceae	*Bruguiera gymnorrhiza*	LR	C
6.	Rhizophoraceae	*Bruguiera sexangula*	VU	F
7.	Rhizophoraceae	*Ceriops tagal*	VU	R
8.	Rhizophoraceae	*Ceriops decandra*	VU	R
9.	Rhizophoraceae	*kandelia candel*	LR	F
10.	Meliaceae	*Xylocarpus mekongensis*	VU	C
11.	Meliaceae	*Xylocarpus granatum*	VU	F
12.	Sterculiaceae	*Heritiera fomes*	VU	R
13.	Arecaceae	*Nypa fruticans*	EN	R
14.	Acanthaceae	*Acanthus ilicifolius*	LR	C
15.	Sonneratiaceae	*Sonneratia alba*	VU	-
16.	Sonneratiaceae	*Sonneratia apetala*	VU	F
17.	Sonneratiaceae	*Sonneratia caseolaris*	VU	F
18.	Avicenniaceae	*Avicennia marina*	LR	C
19.	Avicenniaceae	*Avicennia alba*	LR	C
20.	Avicenniaceae	*Avicennia officinalis*	LR	C
21.	Mytiniaceae	*Aegiceras corniculatum*	LR	A

EN = Endangered, VU = Vulnerable, LR = Lower Risk, - = Not recorded, A = Abundance, C = Common, R = Rare, F = Frequent

Source: Kathiresan (2000) and Kathiresan and Qasim (2005)

Chapter - 66

Forests for Sustainable Development, Towards the 21st Century

Without Creating Organic Abundance we Cannot Liberate Rural People out of Poverty.

Proceedings of the XI World Forestry Congress 13-22 October 1997, Forests, biological diversity and the maintenance of the natural heritage, Protective and environmental functions of forests.

Forest Islands

We need more and more forests and trees and better ecology and environment and rain forests to save our planet. We also need more forest bushes to give habitat to birds and animals.

The solution to the problem is better and clever forest management free of guerrillas, terrorists, drugs and poachers and a module is presented in the paper for developing and managing forests.

The conventional stone-marking of forest boundaries or barbed wire methods have failed to produce results and in fact in many countries they have been counter productive. After spending thousands of millions annually, Third World countries will be poorer by a few hundred thousand trees cut to make the uprights for barbed wire fencing.

Strictly speaking under the order of the day the social forestry projects of most part of the globe have no social changes, no social action and the resources are not being utilised for the welfare of the people, not for the community, but only to fill the pockets of poachers and politicians who support them. They have no element of permanent resource creation.

Forestry needs a radical change and therefore some unconventional and rational methods have to be introduced. The Ryan Foundation module, which is explained in the paper, is based on (or linked to) the "Survival by Sea Water" movement of the Foundation.

President cum Managing Trustee, Ryan Foundation Facsimile: (91, 44) 4910746.

> A village is after all an integrated ecosystem of trees and tanks, ponds and pools, farms and fields, grasslands and wastelands, animals and birds and so the survival of rural people depend on all these and if these resources degrade lives of our village people will also degrade as it is happening today -Author.
>
> 'No person can live without the means of living, that is, the means of livelihood. Deprive a person of his right to livelihood and you shall have deprived him of his life'. The quote from Chief Justice Chandrachaud in the Olga Tellis case (Bombay pavement dwellers case).

Chapter - 67

Sea Water Pipe and Chamber Inlet

Where governments are not coming forward to build sluice gates or locks to let sea water flow into nearby villages, people had better take the initiative and lay bamboo pipes or earthenware pipes One or Two metres below ground (about 15 centimetres below mean sea level) and take the sea water to fill the dry and, by now, defunct lakes, ponds and catchment

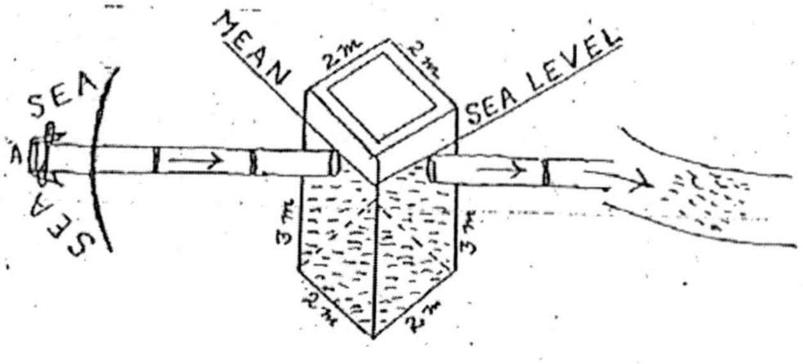

Design by the Author

areas. The pipe connection is to be laid only to a distance of about 50 metres in all (about 50 bamboo poles) and connected to a brick-and-mortar chamber as sketched above. About metres of the pipe will be submerged in the water and about 40 metres buried in the ground. The chamber will have a bamboo outlet pipe, which will drop the water into narrow open channels and the water can be led and taken wherever it is required.

This is the easiest, safest and cheapest way of letting or tapping sea water with a control system.The dimensions of the brick-and-mortar chamber may be as shown above and its depth will depend on the height of the ground above mean sea level. There will be a continuous flow of water and to stop the flow when necessary, one has to only walk knee-deep into the sea and insert a wooden stopper or plug or stuff cotton rags inside the bamboo at point "A". The chamber may be built without cement using RYFO Mortar Mix, explained in this book and the bamboo pipes may be prepared as explained in this book. The chamber may either be raised a metre above ground level or kept closed with a wooden plank or fenced to prevent children or cattle from falling into it.

Point 'A' where the sea water enters must have a cloth or mesh filter tied around with a fibre rope or wire as shown. This is to prevent any foreign matter or sand from entering. Instead of bamboo, flexible garden hose pipes may also be used.

As there is absolutely no danger of opening out the sea and marooning the village by this system, no government permission is required to build it. As the chamber can be constructed with village-made bricks and with free community labour, no external, technical or monetary aid is required and the project can be entirely a community initiative devoid of government interference and the red tape. The water may be led into open chanels, dry ponds, lakes and catchment areas for rearing fish, cultivating plants, trees and weeds that grow in sea and saline water and for desalinating the water for human consumption in one of the RYFO stills shown in the book.

Chapter - 68

Lifting Water into Overhead Tanks

Not gear wheels but two discarded car tyres of even wound light size are used. Each tyre is with a half inch fibre rope, like a coil spring. The

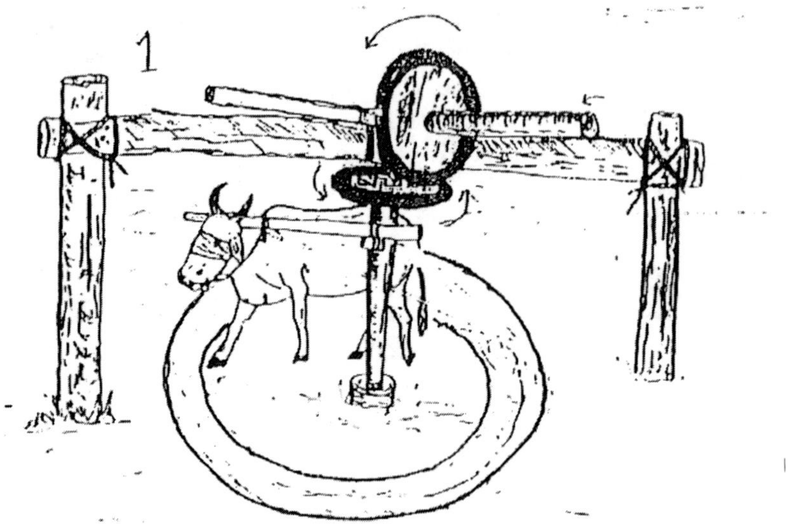

Design by the Author

animal turns the axle to which the horizontal tyre is fixed. When the horizontal tyre turns the vertical tyre also turns along with the axle to which it is fixed.

The ropes carrying the buckets of water goes over the projection from the overhead tank, drops the water at that point and returns to the tank to fetch more water. All the pulleys or rollers are made of wood, as shown.

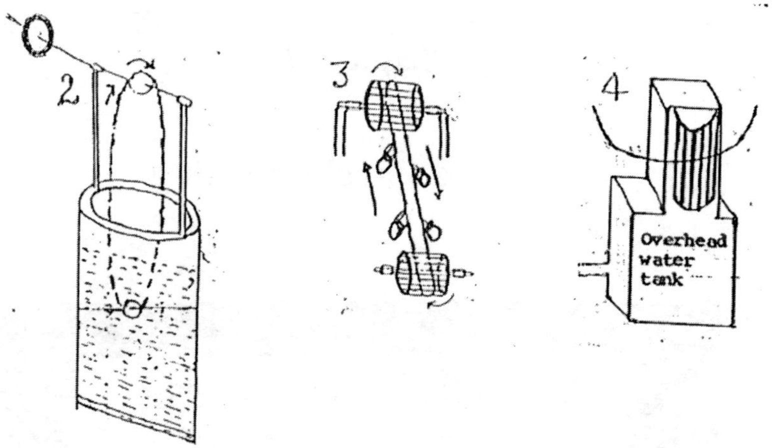

A stainless steel pipe connects the overhead tank to one of the stainless steel tanks.

A pulley rotates on the horizontal axle fixed wall to wall in the middle of the ground-level tank. The rope or ropes to which canvas or car tube buckets are fixed passes around another pulley fixed on top of the overhead tank as shown in diagram 3 & 4.

Chapter - 69

Restore Parched Lands

Bring in. The Sea For Saline Agriculture. Sea Water is Better Than No Water. Halophytes are Better Than Chemical Agriculture. The Sea Holds the Key.

Chapter - 70

Cultivation by Sea Water is feasible and Inexpensive

₹ 26,133 crores (One crore is million) was spent on major and medium irrigation projects in India till the 7th Five Year Plan. ₹ 14,470 crores was spent on minor irrigation projects, including institutional finance. So much money brought only 22 million hectares under major and minor irrigation till the 7th Five Year Plan and 34 million hectares under minor irrigation. On an average ₹ 11,976 was spent to bring just one hectare under major or medium irrigation and ₹ 4,293 under minor irrigation during the seven successive Five Year Plans (1951 to 1990).

The working group on major and minor irrigation projects set up for the 8th Five Year Plan (1990-95) has stated that the expenditure to bring only hectare under irrigation has, now increased several fold. The group has stated that about ₹ 1530 was spent to bring hectare under irrigation during the 1st Five Year Plan and ₹ 36,300 during the Seventh Five Year Plan. Certainly, pumping more and more money into water projects results in diminishing returns. In fact, the money goes down the drain and not into the field for cultivation.

In global discussions availability of water does not receive due consideration. Hydrologists talk about conserving rain water and arresting water runs off into the sea. Seldom do they apply their mind to cheap and feasible methods of desalinating the plentiful sea water for human consumption and irrigation. Discussions on the water problem have been inadequate on sustainable uses of natural water resources. The technology of today cannot and will not expand or create more of the basic resource water.

As the population grows the total amount of freshwater available per person declines. Water experts agree that when a certain ratio of human numbers to renewable fresh water supplies is exceeded, water stress and outright scarcity are all but inevitable.

Desalinating sea water is the solution to the problem, as is already being done in several countries. Ryan Foundations innovation, the Four-Pot Still, can be made in stainless steel and in giant size and powerful magnifying glasses may be used to focus the meet of the sun on the bottom pot of sea water to boil, evaporate and condense it. In Saudi Arabia the magnifying glasses enhance the heat of the sun by 40 times. In Israel also magnifying glasses are used. The 4 pots still of the Ryan Foundation will be simpler, easier to erect and cheaper than the Saudi or Israel method which work almost on the same principle.

In seaside aquifers the brackish water is less salty.

Chapter - 71

Barker's Turbine

(Reproduced from Mini Technology Vol. 11 with courtesy to the authors A. Bachmann and B.R. Saubolle.)

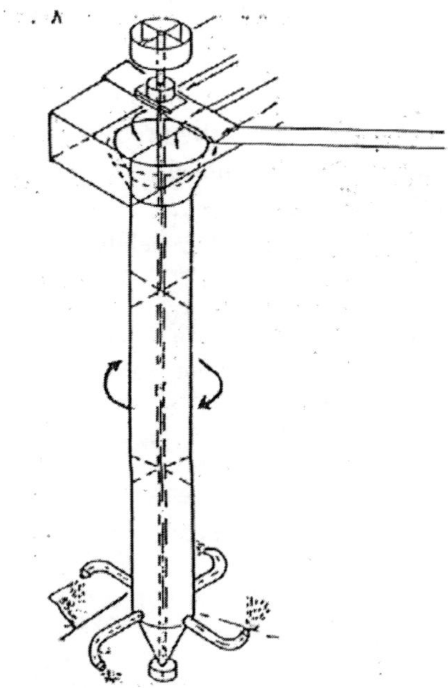

This Barker turbine was made by Balaju Yantra Shala. P.O. Box 20, Kathmandu, Nepal. Test runs were made by mid 1982.

In coastal towns and cities if sea water is carried through PVC Pipes and into sewerage junctions (after inspection/cleaning openings with a manhole cover), a Barker- Turbine like this or the sewerage generators shown in RYAN foundation Handout 213 may be installed.

By this arrangement not only the sewerage lines (gutters) will produce electricity in mini generators to light some street lights, but they will also flush out the gutters and keep them clean and less infected with cockroaches, mosquitoes and evil bacteria because of salty water. Fibreglass should be used to avoid corrosion.

In the city of Chennai concrete pipes may take the water from below the spider-like pipes at the bottom and drop the water into the sea water Cooum canal to flow away back into the sea. The outlet pipe should be double the size of the inlet pipes. The depth of the Cooum canal where the water is let out should be at least 2 feet below the level of the spider-like pipes.

Since the Cooum canal is already available in Chennai it is easy and inexpensive to set up this device. As the sea water from the Baker' turbine flows out a mini turbine may be set there to power street lights. The sea water that flows out after turning the turbine will flush the canal and keep it clean and free of mosquitos.

Wherewer there are sea water canals, it is possible to set up Barker's turbines.

Chapter - 72

Ryfo Ram

The Ryfo Ram is nothing but the famous Latin American machine with the provision of a tap an extra fitting as shown, to collect oil from nuts, starch from cassava or other juices pressed in it. It is also used for compressing foliage briquettes mixed with Cassava starch, tamarind seed or fish glue. Thus, unlike the Cinva Ram the Ryfo Ram is multipurpose.

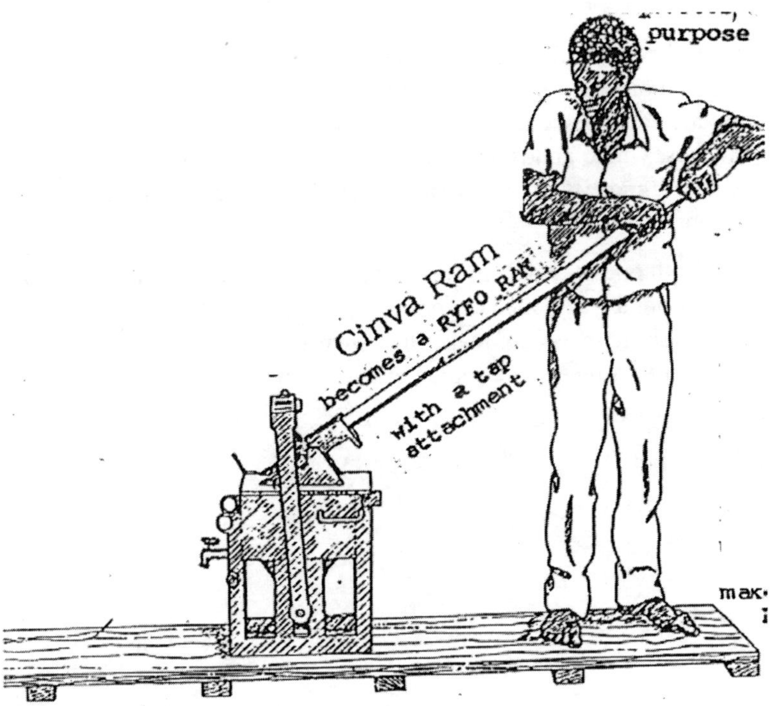

The V S Cinva Ram developed by Thomas Kuby of Gate, is certainly an improvement on the traditional Cinva Ram. A tap may be attached to a V S Cinva Ram as well to make a Ryfo Ram.

By the attachment of a tap to extract oil, starch and juices and to make briquettes the RYFO Ram does not lose its original and primary purpose of compressing mud bricks, blocks or tiles.

The process of making (extracting) starch, glue or oil and compressing foliage briquettes in a RYFO Ram is explained in this booklet. The tap hole must be blocked with a thin metal plate when blocks or briquettes are pressed.

There are many types of mud block presses in common use today, including those that work in hydraulic pressure. Any one of these may be converted into a RYFO Ram by attaching a tap and made multipurpose. The RYFO Ram may have two handles make pressing casier to make compression easier and more productive.

When pressed for oil or starch, the lever must be held in the hard-pressed position for about minutes during every operation or to enable for as much oil or starch as possible to come out. Even then the recovery will be only about percent but still it is a good device and method for backward villages, where goods and services from outside do not reach easily and regularly and even if they do, they reach for a price which the poor villagers cannot afford their price.

With one or two Ryfo Rans a village community can press its own oil from groundnuts, boiled castor nuts, and other oil-bearing nuts which yield oil merely by pressing, make their own starch and briquettes and be self sufficient for renewable source of energy made from waste.

Chapter - 73

IIT Madras - Cam Block Press

Operating the Cam Block Press

The base plate of the mould box is lowered by moving the lever down on the side opposite to the open lid. The mould walls are greased and the soil filled in with a measuring scoop. The lid is closed and locked by a simple turn of the eccentric roller type arrangement.

Two workers then lift up the lever until it reaches an indicator on the side of the machine. This completes the compression phase. With the lever held in this position, the lid is unlocked and opened.

The lever is then pushed further over the top of the mould to the other side, until it is automatically blocked by the catch of the cam. By this lime the compressed block is completely ejected from the mould and can be removed for curing.

The IIT Madras Cam Block Press is the first manually operated machine to employ a cam mechanism (instead of a toggle lever) to push a plunger plate and central piston upwards to compress loose soil into blocks. The Press was development in 1986 by Professor T.P. Ganesan and R. Chandra Sekhar (M.Sc. Thesis), as part of an Indo-German project on "Appropriate Technology for Rural Housing".

The aim was to make the manual operations as simple and fluent as possible, and allow for variations in the compression ratio. The all steel machine, with interchangeable moulds for different block sizes, is assembled with bolts and nuts, so that it can be dismantled for transportation.

Alternative models of the Cam Block Press are under development to increase the efficiency, for instance, by providing a ratchet to the lever, so that the compression and ejection operations can both be done from the same side of the machine.

Technical Details

Size of machine (length x width x height)	200 x 45 x 90 cm (79 x 18 x 36 in)
Weight of machine (without mould)	142 kg
Weight of standard sized block mould	- .16 kg
Size of crate for shipment	150 x 60 x 60 cm (59 x 24 x 24 in)
Weight of packed machine	200 kg
Standard block size (single mould)	30 x 20 x 10 cm (12 x8 x 4 in)
Maximum nominal compaction force	15 tonnes
Nominal compaction pressure	2.5 N/mm^2 (365 p.s.i.)
Compression ratio (can be varied).	1.5:1
Energy input/transmission	.manual/mechanical
No. of blocks per cycle/output rate	1 /54 blocks per hour
Labour force required (incl. excavation and mixing	5 men

Note: The technical details were provided by the producer. GATE is not in a position to verify these data and therefore cannot accept the responsibility for any inaccuracies. As the prices and exchange rates are subject to change they are only meant to serve as guidelines.

Due to increase in population and climatic changes in many parts of the world, the water table has fallen below 150 feet and energized bore pumps are reducing the level further. In most places in the city of Chennai bore well water contains excessive selenium or arsenic and fluoride, making water unfit for human consumption as per WHO standards.

Centralised water storing and distribution in pipelines does not make available adequate water to the unreached. Scanty rains, deforestation, internal conflicts, inter-state politics, terrorism and sabotage are some of the reasons for this state of affairs.Therefore, every country, region, town, village and home must make its own arrangements, Domestic and community-or family-scale distillation is the answer.

Government should provide a network of man-made sea canals and streams for people to distill sea water using mud pots. It has been demonstrated that steam from a rice pot, for instance, is easily tapped to convert three bottles of seawater into two bottles of potable water. This principle can be used on a commercial scale to cultivate fields and to supply hotels, hostels, farms and factories.

Wind, wave and solar energy must be combined to distill sea water as done in Saudi Arabia. Transporation by lorries and pipelines must be avoided.

Chapter - 74

The Bitter Cassava - Binds Briquettes

Different types of cassava satisfy a range of needs. There are early maturing varieties and late maturing varieties. There are those that yield little starch. There are giant size cassava and pygmy type. There are also sweet cassava and bitter ones. Cassava leaves are used as spinach and all unwanted parts and root of Cassava are used as pig meal, as fuel and also for brewing beer.

Some varieties are grown for the extreme "bitterness" to prevent damage by wild pigs and rats. Usually the bitter ones are grown all around the sweet varieties. In the Lake Victoria-Zone of North Tanzania people grow several varieties of Cassava, including the very bitter ones. The bitter ones yield plenty of starch, which can be used to make RYFO foliage briquettes. In the take Victoria-Zone most families two or three bitter varieties around four or five sweet varieties.

The Tuber Researh Centre of the Government of India has evolved giant and juicy cassavas each of about 60 centimetres long and 30 centimetres broad weighing 10 to 15 kg each. This variety can be promoted on a large scale for providing fuel and briquettes for the poor. But the bitter ones are the best for briquetting, they are wanted neither by people nor by the beasts and they grow without cultivation and care.

Cassavas can be left to remain underground for a long period of time, even for 3 or 4 years, and taken out for use when required. Nature stores Cassava for the poor man to save him during famine and Cassava allowed to rot in a drum is used to produce bio-gas. In fact this is now being done successfully in Salem District of Tamil Nadu, India.

Drawing inspiration from indigenous knowledge and customs, Mahatma Gandhi (1869-1948) advocated the concept of Trusteeship with regards to both money and knowledge. He wanted the wealthy to hold their surplus wealth in trust, as it were, and use it for the public good. Similarly, he wanted those with specialised knowledge and skills to share them with the illiterate and the less fortunate.

The rural poor are caught between a silent evolution of nature and a violent revolution of science.

Chapter - 75

Cassava Binds Briquettes, to Steam Distill Sea Water

Cassava starch is good for binding foliage briquettes. To extract the starch, mature tubers are washed and, on a cottage scale, the skin is peeled by hand. Then the roots are chopped and crushed and the milky starch is drained out and separated from the fibrous tissues. Water is added to the slurry while crushing to bring out much of the starch. The starchy water and slurry are all mixed well to make a semiliquid -paste. On the following

BALL TYPE FLY PRESS

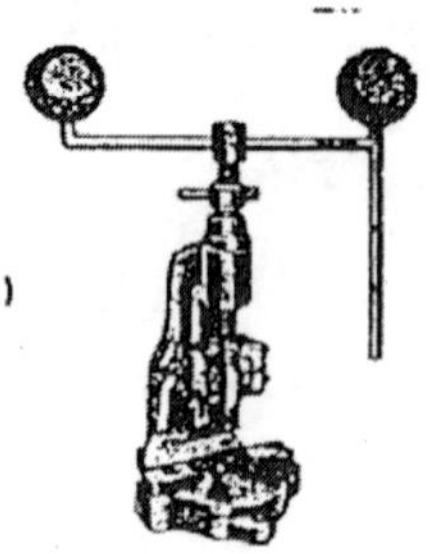

day the surface layer will look yellowish green with an unpleasant smell. Water is added to the mass and stirred vigorously and left to settle to bring the liquid glue to a proper consistency (like wall paint ready to use).

Before the starch solidifies it is mixed with the pulverised or otherwise granuled garbage (solid waste) and the matter is filled into the female die and compressed by the male die to harden as briquettes. This may be done either by hand or in a Cinva Ram, fly press, or ball press (see diagram).

The die must be oiled with waste oil beforehand. The formed briquettes are taken out of the die and dried in the sun. To make the middle portion dry quickly the briquettes are broken into three or four pieces. After drying they are ready to be used to cook food or boil sea water for steam distillation in the innovation designed by Ryan Foundation.

> Bittern, (the waste produced in the production of salt) must be put in a impoundment (a confined body of water) as big as a large swimming pool to be washed by rain water, dried and mixed with molten tar for laying or topping public roads. More and more sand and solids from the sea should come into the land to fill pits. Potholes and low-lying areas..."I will make a road in the desert and rivers in the dry desert sand"-Isaiah 43.19.

Chapter - 76

Bitter Cassava

As cassava is a commercial crop eaten both by people and animals, it fetches a price for the rural family income. So the bitter cassava, which is not a commercial commodity and not eaten by people or animals, is recommended for making briquettes. It may be cultivated in a sector of land in every village and made available for the poor. It grows and multiplies without water or care.

There are so many ways of extracting the starch and compressing the briquettes but those explained in this page are ideally suited a village, cottage industry. The machines recommended are inexpensive and easy to operate by women and children. The residue, including the skin, makes a good fertiliser.

Tamarind Seed

Glue from tamarind seed also yields good starch for making briquettes and the starch is extracted in a similar way after crushing and boiling the seeds. There is no need to boil cassava for extracting starch.

> After years of the world becoming "modern" as many as 2.67 billion people - over 40% of the world's people - still burn biomass in their insufficient and dirty cook stoves, writes Sunita Narain in her editorial of DOWN TO EARTH, April 2015.
>
> In the same issue of the journal, Alex Thomas writes, "In Bangaluru, for instance, the going rate for 1 kilolitre of potable water (obtained from private borewells and sold to households through tractor drawn tankers) is ₹ 100. This price is increased to ₹ 150 on summer. Bangaluru is facing the challenge of rapidly depleting ground water due to encroachment of lakes.

Chapter - 77

Ryfo Two Drum Still

Instruction, (l) Any thick, foldable plastic sheet, preferably transparent, may be used. (2) It should be tied to the drum air-tight so that steam does not leak out. (3) The inverted plastic sheet cone should be filled with cold water and should not be allowed to get warm. With a cup or mug keep

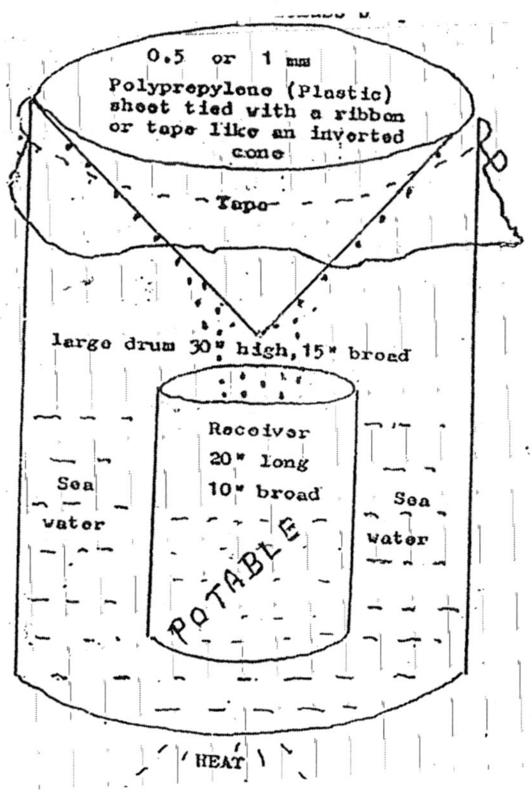

removing the warm water that gets in it and replace it with cold sea water. You may have a siphon arrangement. (4) Heat the big drum with cheap fuel, such as foliage briquettes, juliflora, twigs, wood shavings, waste oil including non-edible fish oil and biomass briquettes made out of garbage (city sweepings will be ideal. They can be compressed into lumps by an hand-operated Sinva ram or RYPO ram. (5) You will get anything between 5 and 10 litres of pure drinking water depending on the heating and cooling system. (6) Remove the plastic sheet and the safe water from the small container and reset. Each cycle will take about one hour. (7) for every litre of condensed water add half a litre of heated and filtered sea water to restore the lost ingra dients and taste. Alternatively, add half teaspoon of sugar and salt. For improving the smell add juice of tulasi (basil), curry leaf (karuvail coriander ginger or (8) Paint the big drum black inside and outside.

Chapter - 78

Ryfo Clay Still

This is not as complicated as it looks in the diagram. Even an aborigine can see a model functioning and build one for herself or himself without any help. Every poor hut all over the world can build one in the front or backyard to convert sea or saline water into drinking water.

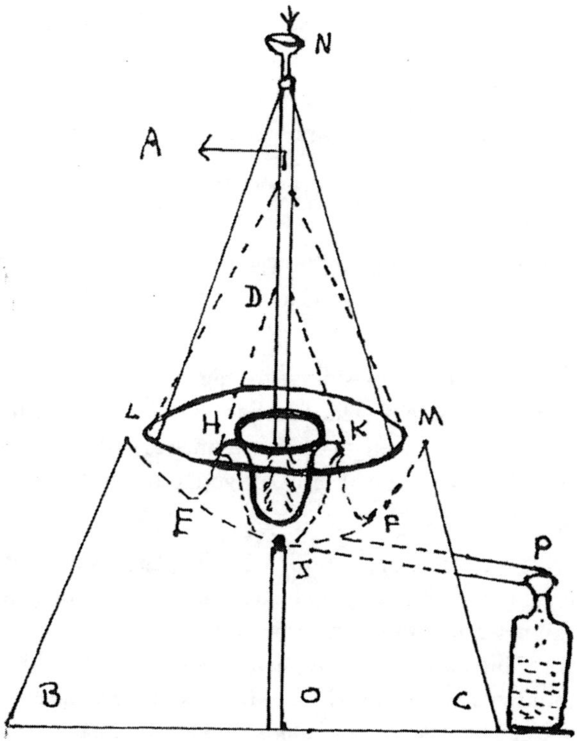

High Ground

Materials Required

(1) About 100 kilograms clay. (If clay is not available, mix earth with other materials and make clay as explained in RYFO Handout 110. Alternatively, prepare mortar mixture as explained in Handout 270.

(2) One and only-metre long bamboo of 6 to 10 centimetres. It must be like a pipe without blocks in the middle.

(3) Another bamboo pipe almost of the same length but 3 to 6 centimetres in dameter or (smaller diameter.)

(4) A large empty glass bottle, preferably colourless.

(5) From 4 to 5 metres of thin colourless transparent polythene (plastic) sheet.(Used cement, fertiliser or milk powder bags may be opened out and heat welded into sheets and used for lining the clay basins. The bags used must be laminated bags and they usually are). Instead of lining the clay basins with such plastic sheets, they may be painted with RYFO glaze paint to prevent seepage.

How to Construct

(1) Build a clay or mortar-mix cone. (ABC in diagram)

(2) Before it dries, cut half the cone with a big knife at LM and discard the portion ALM.

(3) Make (scoop) a cavity (bowl) like LEJFM as deep as point J and make the diameter of the bowl as long as possible without weakening the side wall of the bowl.

(4) Before the bowl dries, drive the big bamboo pipe ON in the middle up to point O as shown. The bamboo should have a-one centimetre perforations above J to a length of 10 centimetres and should be completely blocked at J. Before the bowl dries, drive the bamboo JP also in position should be.

(5) Line the inside of the bowl with the plastic sheet cut to size. Tree gum, fish gum, tamarind seed or cluster been gum may be used for pasting, (for fish gum see RYFO Handout 501). Build the small cone EDF in the middle of the big bowl, cut it into half and make the bowl HJK as shown. The height of this narrow bowl should be about the same as that of the big ball and the gap between the circular side walls of the two basins should be only 10 to 15c.

(6) Cut transparent plastic sheet into a round piece of 1.5 centimetres diameter. For joining bits if necessary, heat seal. If it is hard to find thin plastic sheets in villages, opaque or transparent, to line the basins as already explained, paint the basins with village-made "RYFO paint for glaze effect," For details regarding glaze paint, see page 3 in the Ryan foundation booklet "Do-It-Yourself Rural Toilets. Add some charcoal powdered to a line mesh to the paint to make it pitch black. Black container quickens evaporation.

(7) Fold and stitch or heat seal the boarder end of the transparent plastic sheet like the ends of a handkerchief but the folding should be 5 to 6 centimetre broad to be loaded (inserted) with weights and closed. The weights may be pebbles or stones.

(8) Place the central point of the circular transparent sheet at point N and tuck the hanging circumference around LM into the bigger bowl.

(9) When in position, the sheet going into the big bowl should not touch the small bowl. This is very important.

(10) To prevent air from getting in, pack clay between the big bowl and the sheet all around or pack with cotton or rags.

(11) Make a small hole in the plastic sheet at point N and insert a small funnel as shown and always keep the hole in this funnel closed with a piece of cloth or cotton.

(12) Pour sea or saline water into the funnel and fill the small bowl HJK and close the funnel.

(13) Hot sun falling on the conical plastic sheet during the day will evaporate the water which will rise, hit the plastic sheet, get condensed and roll down the sheet along NL and NM, drop into the big bowl and find it's way out through the bamboo pipe FP and fill the bottle.

The African proverb that, "Filthy Water Cannot Be Washed", is not true. It can be distilled to get pure and safe drinking water.

(14) A small rubber tube, long balloon, cycle tube bit or tube made of canvas or umbrella cloth may be used to connect the pipe to the bottle at point P.

(15) When the bottle gets filled another bottle may replace it.

(16) As the water in the small bowl evaporates, more water may be poured into it.

(17) If saline or unclean water is poured, after it evaporates a dirty residue will be seen in the small bowl. Remove the plastic sheet, clean the bowl and set it again every fortnight.

(18) If sea water is used, salt deposit will be found, which can be recovered and used for cooking. If the salt is dirty, dissolve it in clean water, filter and evaporate again to get clean salt (see Handout 64 of Ryan Foundation).

(19) The base BC may be one metre in length and the still should be built in an elevated place or on a platform.

(20) It is very important to enclose the still with in a thorny fence to prevent children and stray cattle pulling it down as it happened in Somalia when a model unit was constructed at Jalalaqsi by this author.

(21) The water collected in the bottle need not be boiled and for taste a bit of salt or sugar may be added to it.

(22) If a little flavour is desired, cummin seeds, basil leaves, curry leaf or any aromatic leaf or root including lemon grass or Khas-Khas may be added.

(23) Though each family can make it's own still, in every village one or two women or unemployed and underemployed youth may be asked to specialise in constructing these clay stills and they may be paid ₹ 10 for constructing a still if materials are provided and ₹ 50 if materials are not provided.

(24) If big mirrors or large shiny metal sheets are available, they may be set high on a table or wall to reflect the rays of the sun on the water in the middle basin to expedite evaporation.

(25) NGOs may organise the supply of plastic sheets and bamboo pipes for the village community and dealers may be named in each village.

(26) NGOs may construct one model-cum-demonstration unit in every water scarce village for the village community to copy.

Mr. Eilas Diaz Pena of Paraguay said "We can build a World where water is secured for all or a World where water is a cause of War". Now many sensible people repeat this statement. The next major war will be for water Only and its warning is writ large and being sounded loud and clear presently in the Gulf Countries.

Chapter - 79

Ryfo Pit Still (sos Still)

The Ryfo pit still is perhaps the best, cheapest, easiest, safest and sanest method of converting sea and saline water into drinking water. God willing this should solve the problem of drinking water anywhere in the world where the sun shines and sea and saline water is available. It is expensive to transport potable water from one point to another. So the poor must be taught to make water potable at their door step.

This idea is similar to what I have written under the heading "Drinking Water from Nowhere" On page 25, where water from the desert air is evoporated and condensed inside a pit covered with a thin transparent plastic sheet, as illustrated in the diagram.

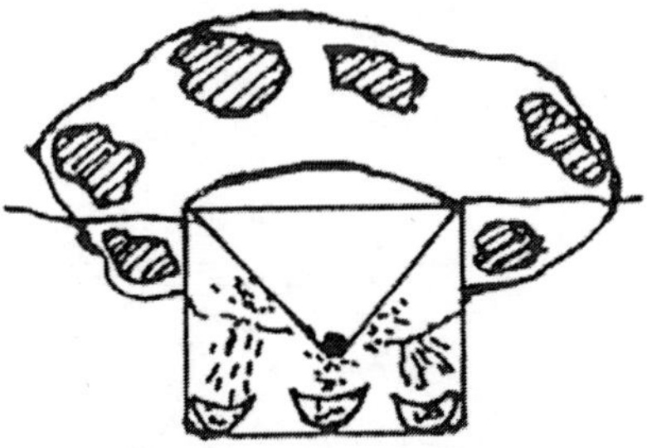

The pit still required to place several vessels of sea water inside the pit along with a clean empty vessel in the middle (right under the cone). The water in the surrounding vessels inside the pit will evoporate when the sun is hot, condense and trickle down the cone and fill the vessel in the middle and this will be clean pure water and need not be boiled. If a family has 3 or 4 pits each of one metre diameter and half a metre deep, there will be enough water for the family to drink and cook.

The Rays of the sun may be focused on the water inside the pit from all sides by arranging mirrors, metal sheets or aluminium foil stuck on cardboard to augment the beat. It is simple science that the trapped heat inside "the pit will be more than the heat side". (green house effect). Vessels containing sea water may be made of any material but those made of metal and painted black will augment heat. Tin sheets (cans opened out and straightened) may be used to line the bottom and side wall of the pit also to augment heat. The solar box (solar cooker) works on the same principle.

Every morning before the sun comes up the pit still may be set and the next morning - the condensed water taken out with the vessel and the still set again. The vessel may be removed easily by lifting the plastic sheet on one side without disturbing the setting.

To achieve this, all that the poor remote villager requires is 2 metres of plastic sheet and sea water close by and some tin cans the anthas appeals to NGOs working in warm the author countries to promote pit stills and provide water to the people.

It is estimated that about 30 to 40 percent of water systems, in developing countries, are broken at any one time, and that a third or more of piped water in towns and cities, is lost before it ever reaches the tap.

Chapter - 80

Bore Pumps-a Tool for Corruption

(Excerpts from a front page article (report) which appeared in the popular Indian daily, The Hindu, dated March 11, 1991).

"The state Government has taken in sinking of over 5000 deep bore wells in chronically shortage zones, particularly in the Central and Western parts of the state.

Over the years the Tamil Nadu Water Supply and Drainage Board (TWAD) alone has sunk nearly 1.15 lakh borewells in the state in addition to several thousands put up by different agencies, such as Department of Social Forestry, Slum Clearance Board, municipalities and panchayats.

"Maintenance of assets (borewells) already created now vests with the local bodies left much to be desired, leading to repeated expenditure in the same localities.

"A recent study in Coimbatore district reveals that owing to over exploitation of ground water the water table is going down by 25.3 mm every year.

"Although the financial allocation provides for deepening of the existing open wells, experience in recent years has indicated an all round clamour for sinking of borewells and so much of political pressure used to be brought on the officials that often the basic rules and norms were given the go-by.

For detailed and revealing information on borepumps read the booklet 'Pumps without Water' by Dr. Felix Ryan which is now in circulation in most English speaking countries and in the hands of agencies which give aid for hand pumps, and borewells without getting the facts (and the truth) from the field.

While planning for water we have to balance what is technically sound, environmentally friendly, economically viable, socially acceptable and politically possible.

The Water Racket

In some Third World countries, India in particular, drinking water drawn from underground is being commercialised and blackmarketed by middlemen and politicians to the detriment of the poor and lower middle-classes and the environment. (See "Creating Water Markets: Solution or threat for farmers?" by John Colmey in the November-December issue of Development Forum).

We cannot say that drawing groundwater unscrupulously is illegal as there is no legislation in these countries preventing people from boring the ground and drawing water. But the point is that both the groundwater and the poor are being over exploited. In several areas, like Cuddalore and Saurashtra in India, for instance, available potable water has been sucked out and machines on borewells are now pouring out saline and sea water.

Local authorities take a commission from the water supply contractors for every tanker load of water drawn from these aquifers or government land leased out to them. Exploiters are making capital out of this scarce drinking water situation. It is time the United Nations, World Bank or UNICEF brings out a draft legislation on the use of ground water and insists that the member countries introduce and enforce the legislation to qualify for any United Nations aid for water-related development.

Better to the editor by Dr Felix Ryan

Reproduced from the Development Forum - March-April 1991. (Journal of the UN Secretariat published by the United Nations Department of Public Information for the Joint United Nations Information Committee).

> In the United States, Although Its Per Capita Water Resources are More Than Six Times Larger Than India's, Water Disputes Have Gradually Spread From The West to the East.

Country-to-Country Water Sale Has Started

(UNI reports from Singapore - 1st July 1991)

SINGAPORE:- Indonesia is to sell water to Singapore under a long term water agreement that will last well into the 21st Century. Under a landmark umbrella accord signed in Jakarta on Friday, Singapore will be able to draw upto 1000 million gallons of water a day from the Riau Province."

Water-rich countries may or may not build up their economic power like the oil-rich countries and their success depends on the longevity of their potable water supply.

Water scarce countries must exploit the sea and not their neighbouring countries. "Dam the neighbour" is not a good national policy for international co-existence and water for all by the year 2000. One must live at all cost but not at other's cost.

In the United States, Although Its Per Capita Water Resources are More Than Six Times Larger Than India's, Water Disputes Have Gradually Spread From the West to the East.

Chapter - 81

Solar Ponds to Desalinate Sea Water

(the principle and a concept)

A solar pond is a shallow pond with water about one metre deep for collecting the heat of the sun. All ponds with water convert solar radiation into heat but the heat so converted is lost due to convection and evaporation.

A solar pond is designed to reduce convectivity and evaporation and to collect heat and store it till late in the evening.

Solar ponds are classified as convecting and non-convecting types.

Conventing ponds are quite shallow - only about a metre deep. A cannon method followed is to arrange in the water rows of plastic bags filled with water and tied. The top of the bag is transparent but prevents evaporation losses. The bottom of the bag is painted black to absorb sunlight. Insulation beneath the plastic bags minimise heat losses to the ground. The temperature of the pond by this method rises to nearly 90 degrees Celsius in warm climates.

In the Non-Conventing ponds salt is disolved in high concentration at the bottom with reduced concentration towards the surface. This salt density gradient (top to bottom) suppresses convection and the temperature in the pond reaches to about 90 degrees Celsius when the sun shines hot and bright.

Presently, the largest and most successful solar pond of the world is in Israel near the Dead Sea. The pond of 250,000 square metre operates a 5 megawatt station. Such solar ponds are the cheapest solar heat collectors and should be promoted in all warm countries.

A convecting pond and a non-convecting pond may be designed side by side and the combined heat may be drawn and used to heat sea water in giant size RYFO four pots stills to desalinate sea water for both human consumption and minor irrigation. All warm countries should copy the Israel model, which has been working very successfully for the past several years.

The cost of a solar pond is said to be ₹ 200 per square metre or about US$ 7 (1995 prices).

Heating the pond, or enhancing the heat of the sun falling on the pond by mounting magnifying glasses around the pond, is possible and should be experimented. Twenty two miles from Jadaan, and in Saudi Arabia water in giant size boilers is heated with magnifying glass.

Thus, solar ponds offer excellent possibilities to generate electricity, to desalinate water and for producing cooking energy with very low investment. Such ponds can be commissioned in remote villages to help the rural poor. A trained maintenance man may be required to be in charge until the village community takes over.

Chapter - 82

Cultivation by Sea and Sewage Water Mix

Experiments conducted by the University of Arizona, the Ben Gurion University of Israel, the National Academy of Sciences, USA, Ryan Foundation International, the Marine and Salt Research Institute of the Government of India at Jamnagar and several other agencies in China, Philippines, Japan, Jakarta and other countries have proved that hundreds and thousands of plants, trees, shrubs, weeds and algae can be grown using sea water or in sea water canals, ponds and lagoons and many other kinds of vegetation can be cultivated by mixing sea and sewage water.

If parched lands such as the one shown in chapter 15 is actually the cracked mosaic of an Ethiopian salt lake, several saline and semi-saline soil plants can be introduced there and cultivated for human, cattle, ecology and environment welfare.

The common seashore plants and trees of the tropics are:-

(1) Casuarinas (Casuarinas equisetifolia), (2) Cashew nut (Anarcardium occidentale), (3) Palm (Borassus flabellifer), (4) Palm grass (Setaria glauca), (5) Coconut, (6) Salicornia, (7) A triplex nummularia, (8) Sesuvium portula castrum (9) Wild tomatoes of Ecuador's Galapagos Islands, (10) Mesquite, (11) Calatropis gigantia, (12) Autumn melon, (13) Brocoty sorghum, (14) Jujube (pronounced as Ho-ho-ba), (15) Jackfruit, (16) Pongamia Pinnata, (17) Barley, (18) Accacia holosericea, (native of Australia), (19) Red seaweed, (20) Ablat of Benin, (21) Yock plants and cactus, (22) Colophospermum Mopane, (23) Atriplex salt bush, (24) Callophyllum Inophyllum, (25) Moringa peregrina from the deserts of Oman priced for oil, (26) Sweet Melons, (27) Cupparis decidua, (28) Isticlis, (29) Salicornia, decidua (30) Sporbolus, (31) Tree cotton, (32) Sea cotton, (33) Vicoa Indica which prevents

conception in women, as established by the Indian Council of Medical Education and Research, Government of India, New Delhi, (34) Vinca osca, used to treat leukemia, (35) Datura used for asthma relief, (36) Custard apple, the seeds of which contain oil, (37) Ailanthus excelsia used for making packing cases, (38) Castor plants priced for oil, (39) Ipomoea canneb Jacq used for reinforcing mud walls, (40) Peltophorum, (41) Delomix, (42) Rain tree, (43) Tabubia,

> Africa has the largest reserve of uncultivated land without clear land ownership and access to water power and transport.

(44) Spathodea, (45) Sporobolus marginatus, (46) Suaeda fruitiosa, (47) Haloxylon Salicornicum, (48) Haloxylon brachita, (49) Mangroves, (50) Juncus sp. (51) Saliomia brachita, (52) Simmondssa chinensis and so on. Zoztera Marina grows fully submerged in sea water and the grains of this sea plant, which float on the shore, are edible. High quality bread is made out of the grains. It is a native of the South American coast but perhaps can be introduced and propagated in other parts of the world.

The book Haloph by J. Arson published by the office of the Arid Land Studies of the university of Arisona in 1981 lists 1,560 halophytes (salt loving trees) which grow well on the sea shore or in saline soil.

Then, there are seaweeds, which belong to a group of algae such as Gracilaria and Caulerpa. There are also seaweeds of a larger variety, such as Cystoseira, Sargassam, Turbinaria and red weed. All these are valuable sources of food in China and Japan, where such foods are called Nori, Jellies, confectionary items, medical products and photographic film, are made out of some types of Seaweed. The well-known food base cum industrial chemical agar-agar, alginic A acid, and Carrageenan (food colouring agent mass-produced now in the Philippines with UN Aid) are all extracted from types of seaweeds. They are important reservoirs of mineral nutrients such as nitrogen and phosphorus. Dried seaweeds produce high-quality methane gas. Port Okha in western India and the Gulf of Mannar in the south have plenty of these weeds to make agar-agar. Geildiella Acerosa is also a variety seen in these areas, which can be propagated and exploited.

There are other group seaweeds such as Ulva luctuca reticulata Geldicella, acerosa and Gracilaria edulis. The weed called Sargassum myriocystum is a common algae which grows like a small tree in great depths. The uses for this have not still not been established.

Not that all these seaweeds can be introduced and promoted in all sea coasts as they all require different climatic, ecological and ocean conditions. But certainly many of the all weather ones can be introduced and promoted and many perhaps will also take to sea water.

> Unimaginative blind policies of several governments, are the major cause for holding back sustainable development, failire to and carry failize to forward projects that support it.

Failire to canals, ponds and parched lands (filled with sea water) such as the one in Ethiopia shown in the picture. This ought to be a new field of research for botanists and environmentalists.

Regarding those that are not really sea plants and trees, which are mentioned earlier in this paper, experiments have sufficiently proven, that many of them grow by sea water cultivation and some of them grow well or even better when saline and potable water are mixed with a small quantity of potable water in the initial stages, such as in the case of pomegranate. The best results, however, are obtained when sea and sewerage water are mixed in the proportion of three parts of sea water to one part of sewerage water, according to some informal experiments and formal experiments done in Israel and by the Ryan Foundation in Africa.

A good proposition is to flush city or town sewerage water by the sea water canals into parched fields such as the one shown in the picture on page 24.

Six species of salt-resistant acacias which grow in sea water have been identified and described by the CSIRO (Commonwealth Scientific and Industrial Research Organisation) of Canberra, Australia. It is very necessary to popularise these six species in the coastline of countries to build up sea defense and along sea water canals and around sea water ponds and lagoons to provide habitation and a home for birds, animals, crabs, turtles and mangroves and most important of all, to provide employment and food for the people. Among the Prosopis species, five are very good for saline soils and seashores. They are Prosopis Juliflora, P. Chilensis, P. Tamarugo, P. Cineraria and P. Pallida. Among the hundreds of varieties of Bamboo species, Bombusa Vulgaris and Bombusa Arundinacea may be tried on seashores and saline soils and these are identified and singled out by the Technology Mission of the Government of India.

> Whether our senseless attention is focused on the destruction of tropical rain forests, the depletion of the ozone layer or wasting of water, the basic cause of

the environmental crisis is the same. It is extreme scarcity of potable water. People are consuming thoughtlessly more water than what the planet can give or replenish. Our consumption and waste in every direction, starting from water, is rapidly destroying the ecosystem and will further hopelessly destroy living conditions for the coming generations.

Chapter - 83

Plant Potentialities

Marine algae has great potential as livestock feed, fertiliser and raw material for several industries. They are a rich source of numerous minerals and life saving medicines. The liquid sea water fertiliser recently researched and developed is becoming popular in several countries as reported in The Hindu dated 5th June 1990. Some fresh algae, including Spirollina, Chlorella and Scenedesmus, rich in edible protein, are widely sold as health food in Western countries and Japan.

1. Jojoba for Coastlines

The Central Salt and Marine Chemical Research Institute of Bhavnagar has been experimenting with the cultivation of Jojoba - an oil bearing plant in coastal sand dunes under semi-arid conditions in the west coast of Gujarat and in sub-humid regions of Ganjam district of Odisha. Experiments have established that Jojoba could be grown successfully in coastal sand dunes and the oil obtained from the plant could be used as a lubricant and for other purposes.

2. Red Seaweed

In 1990 the Philippines ranked fourth among the World's producers of Red Sea-weed, exporting about US $ 48 million worth of dried sea-weed and CARAGRENAN - a sea weed derivative used as a binding agent in food products. If dry and drought-ridden ground is moistened with sea water, more of this weed can be grown, not only in the Philippines but also in several parts of the world.

3. Ablat of Benin

In Benin people cultivate a bean called Ablat, which requires very little water to grow.

From time immemorial a native tribe in Karnataka, India, cultivates salt tolerant rice drawing water from the creeks on the sea shore and the cultivation produces good rice with little care. This variety has to be investigated, researched and introduced in other parts of the world, especially in Africa, which is dying for want of food and water. The Director of Agriculture in Karnataka can help in propagating this variety.

Chapter - 84

The Neglected Castor Oil

(good for steam distillation)

Castor is a drought-tolerant plant which is common in tropical countries and requires a warm climate. It can be seen growing wild on either side of roads in villages, towns and cities. In Chennai on the Grand Trunk Road from the city to the airport the plants are seen on both sides of the road growing uncared for. The pods ripen, dry on the plant and fall to the ground. When they are collected and pounded or crushed, they yield oil and the oil obtained is nearly half the weight of the dry pod.

The high-density castor oil can be used to burn stoves and burners to steam distil sea water. The burning oil is said to be good for the eyes. Since the plant is thorny and tickly, children and animals don't go near the plant. Villagers in the tropics should promote this plant and extract the oil, for use as an alternative source of energy.

There are other uses for castor. It is mainly used as a lubricant and for making soaps and paints. It is also used in the making of typewriter ink for the ribbon, for making varnish, phenyle and hair oil.

Powdered leaves are used as insecticides and the seed cake as a fertiliser. Paper is made out of the pulp of the wood.

The possibility of mixing castor oil with diesel oil is to be investigated and researched. Andhra, Maharashtra, Odisha and Karnataka are exploiting the tree but it is sadly neglected in other parts of India and in Africa.

The prevailing price of castor in the year 2014 is ₹ 750 per 10 kg. Gujarat and the Saurashtra region are the largest importers of castor seeds from all over India.

About 50 castor plants can give enough oil for a family to steam distill sea water for 2 or 3 days. Instead of encouraging villagers to produce castor for oil, Third World governments make them depend on kerosine oil, and bottled gas. This simply damages the rural economy and make the poor starve for water and energy.

(For more information on oil-yielding nuts and seeds refer to author's book, "Plants and Trees That Generate Employment). Gyan Publishing House.

> A senior advisor to the United Nations used to say, If you want to know what will grow in the soil of a particular village don't ask the principal of the engineering college but ask the farmer's wife who lives on that soil and lives by cultivating that soil.

Chapter - 85

Jojoba for Shore Cultivation

Jojoba, a native shrub of Arizona, USA pronounced as "Ho-Ho-ba", is Botanic ally known as Simmondsia Chinemes. R&D done for about 4 years by the Central Arid Zone Research Institute (CAZARI) India, confirmed that Jojoba is highly drought tolerant and salt resistant and can convert desert and sea shore sands into an oasis of prosperity. Several years of research done by the Central Salt and Marine Chemicals Research Institute, Bhavnagar, in the waste and coastal lands of northern Gujarat also confirms that Jojoba is highly drought and salt tolerant.

Agricultural economists consider this shrub a gold mine and the FAO of the United Nations and the World Bank recommend the introduction and propagation of this shrub in wastelands and saline soils.

In the international market a tonne of Jojoba seeds fetches about US$ 8000. The money is paid for the oil which the seeds contain. The oil is very much like the whale oil in physical and chemical properties and better than almost all commercial substitutes for whale oil, which is getting scarce duet to the restrictions laid on whaling.

Israel is cultivating the shrub already on a large scale. Australia, Costa Rica, Brazil, S. Africa, and Mexico, are among the forerunners in the promotion of Jojoba. India, China, the Philipines and some East African countries are also promoting the shrub now.

Jojoba oil, golden in colour, is totally free from alkaloids and resins. The cosmetic industry in the West consumes a large quantity of it. The oil is also used in the preparation of high-quality candles, polish, varnish, lubricants, antifoam agents, fire-retardants and in soaps. The oil is edible and has low calorie. But the most important and indispensible use for the oil is as a high pressure lubricant by the engineering industry, to replace sperm whale oil which is scarce and costly.

The Ben Gurion University of Israel reports that the oil relieves pain and cures skin diseases.

The Jojoba shrub thrives in temperatures ranging from 5 to 50 degrees Celsius and lives for 50 to 200 years. The shrubs are propagated by direct sowing, stem planting, or tissue culture. A plant yields about 125 grams of seeds in the first 5 years and thereafter about two kilograms. The plant tolerates a salinity level of 10,000 parts per million, making it absolutely suitable for desert and coastal areas. CAZRI estimates a return of 1.8 lakh rupees per year from one hectare of Jojoba.

The Poor are not Responsible For the Destruction of Natural Resources. It is Their Poverty Caused By Their Governments.

Chapter - 86

Perhaps Carrageenan can be used also to Bind Foliage Briquettes

Seaweeds on Fibre Ropes

Thousands of fishermen in the Philippines coast have been shifting from fishing to seaweed cultivation during the past decade and claim that seaweed farming is more profitable than fish farming. The farming is based on appropriate technology. They do not go into the sea neck-deep in search of weeds but have them grow on coconut fibre ropes tied to casuarina poles planted 5 metres away from the shore and drawn to the shore and tied to a line of poles on the shore.

Small lengths of fresh seaweeds are tied to the ropes and in about six weeks under water the weeds are harvested. Some weeds are left on the ropes to grow again and the cycle continues.

In 1990, these weeds were taken from the Philippines and introduced in the coastal areas of Tanzania in Zanzibar Island. Reports say that the harvest is much better in the African region and two business companies are promoting the weed and selling them to a factory which extracts from the weeds Carrageenan, which is a popular binding agent in food products. Appropriate Technologists now see possibilities of promoting this weed in several regions and also promoting other quick growing weeds and China grass on fibre ropes. Much research and investigation has to be done, but seaweed farming has already thrown open several alternative employment possibilities for poor fishing communities. What we need now is collecting, processing and exporting centres.

The United Nations Children's Fund (UNICEF) reports that one-third of the funds spent at the end of the 1980s, were destined to keeping existing schemes going in the Gorks the district of Nepal. Save the Children reports that 80 percent of taps in a rural piped water system became dry. Two years after installation, because of broken pipelines and sediments in the tanks.

They were a gift from aid agencies, and the realities are true to the dictum of the Survival by Sea Water Movement, launched by the Ryan Foundation International, that: "Aid does more harm than AIDS".

Chapter - 87

Callophyllum Inophyllum

Botanically known Callophyllum Inophyllum is a common tree in most tropical countries. In Seychelles the tree is called Takamaka and in South India as, "Pinai". In North of India people call it "Dilo". Different African countries and the Island countries of the Pacific have their own names. (Callophyllum Parisorum is slightly different within the same family). The wood which is considered good timber is used for furniture and for roof and wall panelling. There is a dense growth of this tree in most African countries, Madagascar, Mauritius, Seychelles, India and Sri Lanka and it is also found in South America and in the Caribbean countries. Being a sturdy tropical tree it can be introduced (by seeds) in countries having a warm dry climate with plenty of sunlight and moderate rainfall. The tree is of a large size and under normal conditions grows beyond a 100 feet hight and spreads its heavy branches of deep green large leaves in all directions.

The perfectly round nuts of this tree are smaller than a lime fruit (the fully developed nut has a diameter of about One inch). It is green on the tree and brownish and smaller when fallen and dried. The shell weighs about one-third of the seed weight.

The kernel and shell together contain 40 to 70 percent oil. The moisture content is about 25635 percent. The oil contains about 10% resinous substances which give a strong olive green colour and bitter taste thereby making the fat inedible. The oil is used in India for lighting lamps (especially in Hindu temples) for making soaps and in native medicines for skin disorders, including the treatment of leprosy.

Refining of such high f.f.a (free faty acid) oils by conventional alkali refining is possible but uneconomical. Refining methods such as miscella refining may be used but have not been tried so far. To make it edible the active toxic principles are to be determined first and its sub-acute toxicity will have to be studied and established. The active principles are.

(a) Calophyllolide

(b) Inophyllolide, and

(c) Calopyllic acid. Calophylloide is found to possess some active anti-inflamatory properties.

Since the crude oil contains considerable amount of free fatty acids, it would perhaps be better to split the oil and fractionate the fatty acids instead of trying to make it edible. Typical physical and chemical characteristics of the crude oil as well as refined oil are as under:

Crude Oil

Density	20	0.9377	-	0.9527
R.1	20/n	1.4779	-	1.4857
Acid number	30			
Saponification number		191 -195		
Iodine number		85 - 97		
Unsaponifiable matter		0.6% 0.8%		

Neutral Oil

Density	20			
R.1	d20/D	0.9196	-	0.9208
		1.4706	-	1.4714
Acid No.		0.5 - 1.0		
Sap. No.		196 - 200	(Saponification)	
Iodine No.		79 - 88		
Unsaponifiable matter		0.67- 0.8%		

Fatty acid composition of neutral oil

Palmitic acid	-	17%. - 19%
Stearic acid	-	6%- 11%
Oleic Acid	-	48%- 52%
Linoleic	-	21%- 24%

(Analysis was done by the National Botanical Research Institute, Lucknow - 226001, India, and studies were carried out by the A/C College of Chemical Technology, Guindy, Chennai, S. India).

A hand operated screw press of the type explained in the Appropriate Technology Quarterly published by the ITDG, England, Vol. 5 No. 4 Feb. 79 issue will be ideally suited for extracting the oil. Alternatively, a ready-made screw press (size No. 6 is adequate) may be purchased from any machinery supplier. In India the present cost of this machine is about ₹ 2,000. The soap industry may be encouraged to make better use of this oil till more important and profitable uses are established.

The oil serves also as an alternative cooking energy and it does not have a bad odour. In its crudest form the oil can be poured in an old basin or a flat empty tin and if a piece of rag twisted and tied to stay like a rope is soaked in it and used as a wick it burns for hours together giving light and heat. In the interior villages of LDCs this can bring light and warmth into the homes and cook meals for the poor.

The IPT (Institute de Pesquisas Technologies. Box 7141 Sao Paulo, Brazil, has the know-how for using vegetable oils as a substitute for diesel oil. The UNIDO in its publication ID/WG-282/65 on technologies from developing countries focuses attention on such a substitution, Callophyllum Inophyllum is a very thick oil and since it is available in large quantities it will perhaps be worthwhile to experimenting this oil on the diesel engine.

Water Hyacinth is A Microphyte and can be used as A Renewable Source of Energy.

Thick and Big Leaves of this Tree can be Dried and Mixed With Water Hyacinth for Callophyllum Inophyllum Making Good-quality Briquettes.

Chapter - 88

Most of the Plants Mentioned Here Tolerate Saline Soil and Sea Water

Fight Mosquitoes

Water and mosquitoes cause most of the diseases in the world, especially in the tropical countries. Several ways and means are being tried to control the mosquito menace, especially nowadays when, Dengue, Chikungunya, Malaria, lephantiasis and other diseases are taking an increasing toll. Water-borne diseases are likely to increase further during monsoons. Surveys conducted annually reveal that an alarming number of water samples are unfit for drinking, particularly during monsoons.

Introduction of mosquito larvicidal oil, a brainchild of Dhanaraj, Chief Vector Control Officer of the Corporation of Chennai, is in the right direction. But the larvicidal oil is available only in restricted quantity and is costly. Moreover, the method is labour-intensive. And how much oil can be dripped into the flowing water?

The cheapest and easiest way of keeping mosquitoes away is to burn green or dried leaves that have insecticidal active ingredients, especially to drive away or kill mosquitoes while, at the same time, proving good for human inhalation. The shrubs and leaves are given in the following table.

Species Name	Tamil Name
1. Adhatoda vasica Nees	(Adhatoda)
2. Trigonelle foenum-graeeum	(Vendhiam)
3. Aziodivachta indica Adrjuss	(Neem, Vepam)
4. Moringa oleifera Adams	(Murungai)
5. Ocimum	(Tulasi)
6. Carica papaya	(Pappali)
7. Catharanthus roseus	(Nithya Kalyani)

(Contd...)

8.	Peganum har malall	(Simai yaravandi)
9.	Eucalyptus	(Ncelagiri thalai)
10.	Vitex negundo L,	(Vellai nochi)
11.	Acacia holosericia,	(Olosia)
12.	Datura	(Voomatham)

A handful of a mixture of at least four of these leaves should be burnt in a tray every night before bedtime in any closed room and, it necessary, once again around midnight.

All, excepting one or two, are shrubs and grow wild and uncared for. If they are planted intensively all along the bunds and banks of in-land waterways and open gutters they will prevent mosquitoes multiplying.

Acacia holosericea, which is leguminous and drought-resistant and has a high calorific value, grows in different types of soil and climatic conditions. It is a medium-size tree, ornamental and fast-growing and belongs to Australia. The tree has a natural, self-regenerating property, attracts honey bees with and its small yellow flowers and repels moquitoes. Auroville has thousands of these trees in the township, multiplied by the seeds and saplings brought from Australia. This tree and the neem tree must be planted in rows in the middle of our inland waterways and on either side of their banks and bunds. They can be uprooted easily when not wanted.

Many of the shrubs mentioned have medicinal properties also and some of them are exported from India. **The leaves can be Sold.**

Felix Ryan

The Twelve Species Listed Here are Shrubs, Herbs or Drought tolerant Trees and Grow Uncared For in The Tropics. They Tolerate Saline Soil And Bad Weather Conditions. Villages Must Propagate These Species.

Chapter - 89

Sea Water Kitchen Gardens

Plants and trees that cannot stand saline soil or brackish water can be raised by this method innovated by the Ryan Foundation. Experiments conducted in some African countries have proved very successful.

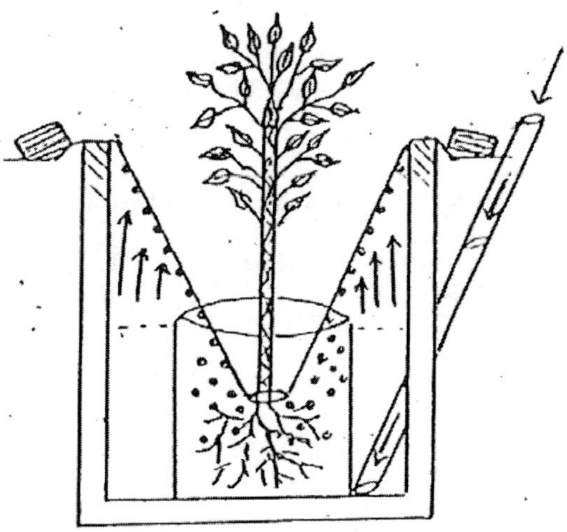

Materials required are 3 sq. metre of thin transparent plastic sheet, 200 hand-compressed and sun-dried bricks, a 20 litre garden or kitchen mud pot, RYFO mortar (no cement) and a 6 cm. dia hallow bamboo about 120 cm. long.

First, a one metre by one metre pit is dug and the floor and sides are lined with bricks and plastered with mortar. While building a side wall the bamboo is placed slanting, as shown in the diagram. The mud pot is placed in the middle of the pit as shown and filled with good earth mixed with organic manure. (RYFO manure is recommanded).

In the twiddle of the plastic sheet a circular hole of 10 centimetres diameter is cut. The pit is covered with the sheet and the central portion of the sheet taken down into the mud in the middle pot as shown. Weights are placed on the colour of the sheet around the pit as shown and sand heaped around the pit on the plastic to prevent air from entering the pit.

The required vegetable, say the quick growing Okra (lady's finger) is sowed or planted in the middle of the central pot as shown.

Sea water is poured through the top end of the bamboo and the mouth of the bamboo is plugged.

During the day the sea water evaporates, hits the sides of the plastic sheet, gets condensed and rolls down the sheet to the roots of the plant. Thus, the plant gets something like pure rain water and flourishes. The method can be adapted; also to growing trees in the middle of deserts, provided man-made sea water canals are taken into the deserts. The sea and the tree hold the key to human progress.

Fused-out tubes of tubelights the metal ends removed, may be used in the place of bamboos. Table and kitchen waste may be mashed and put around the plant as manure.

> The more easily accessible supplies of water have already been exploited and depleted, and tapping new sources and transporting water, costs too much, and the treatment of water to acceptable standards and delivering it in pipelines to the vast majority of masses will cost several billion dollars.

Chapter - 90

Dryland Shrub Reduces Village Population

Leaf That Controls Birth

Botanically the leaf is called Vicoa indica and in North India as Bhanjauri. Dr. P.L. Sharma, Dean and Head of the Department of Pharmacology of the Postgraduate.

Institute of Medical Education, and Research New Delhi, of the Govt of India who has been researching on this leaf for the past several years, declares that the leaf can not only stall pregnancies but create permanent sterility in women without disturbing the normal menstruation or producing side effects. The plant has no toxic effect and has been declared absolutely safe for human consumption.

It is reported that bhanjauri has been used by the Adivasis (tribals) of the area called Sokhodeora in the Nawada district of Bihar for creating temporary or permanent sterility in women and to prolong the interval between births.

The entire plant-the roots, stem, leaves and flowers is dried and powdered, mixed with some water, sugar or honey and a pinch of black pepper and administered in three doses during three successive menstrual cycles. Only 15 gms of the powder is given in every dose. Dr. Sharma has announced that this has proved quite successful in controlling pregnancy for at least 6 months. When taken immediately after child birth for 3 months it causes almost permanent sterility.

Several experiments on fast multiplying rats have been absolutely successful and new experiments and observations are being done on women in the Institute. Over a hundred women who have taken this powder from the institute are absolutely normal, physically and mentally and no side effects have been reported so far.

The Department of Pharmacology of the Graduate Institute of Medical Education and Research is optimistic about this discovery. Dr Sharma, believes that this discovery will reduce the population in India and the cultivation of the plant will give rural employment to tribals and Adivasis. Perhaps the powder may be made into capsule or tablets and exported to other countries. Some say that even if eaten as spinich the leaves reduce fertility. However, no authentic information has been released so far about this.

Let there be a new approach to development that will bring about a better civilisation, better life style and better sense of values that take us closer to Nature and God and away from destruction and the devil.

There is no excuse for famine or poverty. They are not just "bad luck" but are totally preventable by bringing sea water into land by man-made canals for navigation, saline agriculture and desalination on a commercial and domestic scale. Saline agriculture is a sensible science and can generate employment and income for millions of families in hundreds of countries Felix Ryan. (Chief Monitor, "Survival By Sea Water" global movement).

If Resources Like Water are Degraded, Environmental Refugees Will Follow. Sana City in Yemen Risks Becoming The First Capital City To Run out of Water.

Chapter - 91

Leucaena Leucocephala

(Subabul) or (Subabool)

This is a drought resistant tropical tree with several uses. It grows well in different soils and varying temperature upto an altitude of 1000 metres under various weather conditions. The tree gives not only excellent fuel wood and timber for furniture but also fodder from its leaves, honey from

its flowers and dye (colours) from its bark. In some parts of Africa the tree is known as EPIL-EPIL. This fast-growing tree reaches a height of five metres within a year and 20 metres in about 3 to 4 years.

The pods contain hard seeds and African women make necklaces, handbags and fancy items out of them. Big bunches of white flowers on the tree look beautiful. The wood makes good pulp for the paper industry and for the production of rayon of cellophane.

Leucaena, which belongs to South America, has been introduced in a Indonesia, The Philppines, Australia, India and several other countries, and the tree lends itself to various types of inter-cropping.

A very important feature of this tree is that it fixes nitrogen from the air through fungi adhering to its roots and this enriches the soil in which it grows and benefits plants and trees planted around it. It can be used for flooring, fencing, packing and also as beams and poles.

The leaves of the tree fall throughout the year and farmers use them as mulch (green manure) to increase soil fertility and also as fodder for animals and feed for chicken. But only one-fourth of the total quantity of feed given to livestock should be Leucaena leaves. If more is given, live-stock suffer ill health.

RYFO recommends the promotion of this multi-purpose tree on a large scale in rural areas to develop the rural economy.

The survival rate of trees planted by people is many times more than those planted by government agencies and when ownership of the trees is given to the people the results are even better.

Put the environment first for development to last.

Hundreds of millions of men and women, whose daily survival on this planet is a daily challenge, have little time to wait. Tree planting can help them fight their battle.

In Benin people cultivate a bean called ABLAT, which requires very little water to grow. They just put some compost around the foot of the plant and it produces bean all round the year. Can this bean be introduced in other countries?

Chapter - 92

Mopane -drought Tree Which Fights Poverty

Colophospermum Mopane is the botanical name of an exotic tree taken to many countries from Zimbabwe. It is a fast growing, drought-tolerant, multipurpose tree and improves soil fertility. From a distance the tree looks like an Acacia. It is planted as a windbreaker, shade provider and soil binder, especially where the soil is sandy. The tree grows 10 to 15 metres high and makes good fire wood.

The flat, thin kidney-shaped pods, which are about one centimetre long, contain single seeds which are highly resinous. Seeds get propagated by nature and germinate in any moist soil but not in cold, wet weather.

In Africa, low-quality furniture and building materials are made out of the trunk of the tree which is not attacked by white ants. Carpenters powder the resinous seeds, mix the powder with country brewed liquor and use the semi-liquid to give colour and polish to furniture. They also dry and powder the bark of the tree for the same purpose.

The leaves make good fodder. More than half the leaves of the tree may be lopped at a time and frequent lopping does not stunt its growth. Botanists have researched and reported that Mopane leaves have a remarkable quality of retaining nutritional value for soil and fodder long after they fall and dry on the ground.

> Plants and trees shed leaves which dry and become foliage. Foliage with other burning matter is crushed and made into briquettes. Briquettes are used to boil and steam-distil sea water into safe drinking water. Residual impure salt is dissolved in distilled water and evaporated to get clean white salt as a by-product.

The tree can be planted as avenue trees in village roads. Heavy winds do not uproot it even during the rainy season, when the soil is loose. The roots let out carbon dioxide which helps lower soil alkalinity. Indeed, mopane is ideal to fight poverty, provide fuel wood, fodder and wood for furniture. It should be introduced and propagated in waste lands in poor and warm countries.

> The poor are not responsible for the destruction of natural resources: it is their poverty caused by their governments.
>
> As grass root corruption has defeated bare-foot revolution, we now need a self-help revolution to defeat grass-root corruption.
>
> Thailand, the leading supplier of tropical mushrooms to the world markets produces about 80,000 tonnes of mushrooms per annum in small farms and rural homes.
>
> The luna plant or spineless cactus. (Opuntia ficusindica) is found in Mexico and the African coast. The fruit is processed into liquor jams, sweet meal and so on cattle relish the funa plant. Can this be introduced in other countries?

Chapter - 93

Save Firewood

"Millions all over the world live by the survival economics of the poor, especially by cutting trees." Photo: Hermann.

Heavy Burden: Garo tribal women carrying firewood in Meghalaya. Inefficient cookstoves mean that women spend 5 to 8 hours per day on cooking activities, 20 percent of that time collecting fuel-time that could be spent on educational or other activities.

They cut trees to live

Firewood should not be used to steam distil sea water until our trees and forests are doubled and more of halophytes and drought-tolerant trees are grown in all villages and along road-sides as avenue trees all over the world. Trees can live without man but man cannot live without trees.

Carrying firewood is a heavy burden on village women. But there is no use telling village women not to cut tree branches without giving them alternative cooking fuels. Foliage briquettes, fish oil, and garbage gas are the best alternatives and they are do it yourself alternatives. These alternative sources of energy are explained in this book for the purpose of steam distilling sea water, progress is in finding sensible alternatives. The author received United Nations Global award for innovating several alternative sources of energy, including the use of solar energy with reflectors and magnifying glasses to enhance the heat.

Chapter - 94

Grows Well on Coastal Belts

Indian Almond Tree

(Otherwise called Bengal almond)

Botanical name Terminalia catappa. Native of India and Malaysia but now found in almost: all countries with a tropical climate. The islands of Seychelles and the semi-deserts of Somalia have plenty of it.

Uses and Employment

1. The leaves are broad and big and are commonly used for serving meals. The African variety has much bigger leaves than the Indian and Malaysian varieties.
2. The fully ripe fruits axe Turkish-red in colour and taste like plums and children relish the fruits.
3. Jams and pickles are made out of the fruits and this is a village industry for urban markets.
4. Good-quality wine is also made out of the fully ripe fruits by the usual process of making grape wine and wine making can give employment for some urban communities.
5. The kernel inside the nut is very tasty and children break the nut and eat the kernel.
6. When the kernel is collected in lots, fried in ghee, Vanaspati or any edible oil, peppered and salted, it gains in commercial value. It can be packed like cashew kernels and sold as evening eats. Village women can process the kernels for sale in city markets and even for exports (Frenchmen used to export them from Puducherry to Paris).
7. A valuable oil, almond oil, is extracted from the dried fruit and the kernel by simple steam distillation by a village process and

this oil has medicinal and commercial value and can also be used as a cooking medium.

8. Butter paste or almond butter can be made out of the kernel similar to peanut butter, and it tastes like peanut butter. (See RYFO Handout 23.)
9. The leaves can be used as fodder and animals like them.
10. Plates and tumblers are made out of the leaves by a simple hand-operated machine invented by the Central Food Technology Research Institute of the government of India. Mysore, and many families are engaged in this industry now in India.
11. The softwood of the tree is excellent for making wooden toys and Montessori toys.
12. All parts of the tree can be chopped and used as manure.
13. When the fully ripe fruits are boiled, they give a beetroot coloured liquid, which is used to dye fibre-made articles like bags and rugs.
14. The twigs and leaves when dried burn well.

Provided politicians don't exploit the very poor in the poor countries and give them false promises, innovation, waste utilisation and self reliance will raise them from their penury. Aid does more harm than AIDS and damages the metabolism of communities and countries at large.

The Peredovik variety of Sunflower now imported into India from Russia is very high yielding and drought resistant. It must be propagated in all countries.

Chapter - 95

Silk Cotton Tree

Botanical name		Bombex ceiba Linn
Family	-	Bombacaceae
Common		
Indian name	-	Simul
English	-	Silk Cotton Tree

Description: Grows to a height of about 20 metres Deciduous, with wide spreading branches. The wood is soft and the bark light in colour - almost grayish. The large leaves are bright green in colour and divided into five to seven the leaflets spread out like a fan. The leaves drop in January, just before the tree begins to flower in February.

Strangely, the colour of the flowers of this tree varies. In some trees, and usually, they are bright red but there are trees that give yellow flowers and white flowers also. Trees giving yellow flowers are credited with divine powers. Flowers appear when the tree is leafless and then they appear in large numbers. Petals are thick, more than 12 centimetres and 2 to 3 centimetres. The fruit is more or less egg shaped, 10 to 12 cm long and 2 to 4 centimetre thick. Each fruit consists of stout woody valves, which at maturity open out and fall away from the rest of the fruit. When this happens the black seeds scatter round the tree.

Uses: The soft timber is used for making toys and match sticks. It is also used to make paper. The cotton fibre from the seeds (pods), too short for spinning, is used for stuffing pillows and matresses. White ants attack the wood. The gum that oozes from wounds made on the tree is known as mocharas and it is used in Ayurvedic medicines. The calyx of the young flower is cooked and eaten as a vegetable. Leaves are eaten by cows and goats.

This tree comes up almost in any soil and it is propagated either through cuttings or through seeds. It comes up very easily and if planted during the rainy seasons needs no water as long as it lives. Nature takes care of it. Israel has demonstrated that on the sea shore and in very saline soil the cotton tree yields 20 percent more cotton.

The rural poor can come out of poverty if taught to exploit and replenish around their settlements organic matter such as plants trees. Birds, livestock, fish and so on for their raw materials and livelihood without depending on outside raw-materials and if NGOs (not governments) assure limited procurement of their produce/products to meet common consumer and industrial needs in nearby urban centres, (Government buy-back agreements are almost never kept in the rural sector).

The Central Salt and Marine Chemicals Research Institute, Bhavnagar, Gujarat, India, has successfully cultivated the oil bearing plant Jojoba in the coastal sand dunes under semiarid conditions in the west coast of Gujarat and in the humid regions of Ganjam District of Odisha. Several experiments have established that Jojoba could be grown successfully in coastal sand dunes and the oil obtained from the plant could be used in soap making, as a lubricant and so on.

Some trees fix-nitrogen in the soil; a water-harvesting pond and a biogas plant in big farms will help to improve enormously the productivity and profitability of dryland farming. In addition, this will help to mitigate climate changes.

Chapter - 96

Praying for Rain

Almighty God, if you don't want to give us rains at least give us access to sea water that we may distill the salty water and grow saline agriculture and eat sea food.

The United Nations expert RYAN who was sent to Gujarat sees no other possibility. He has recommended to the government to let in sea water into wells in the coastal belt and fill the dry wells, lakes, ponds and rivers to make sea water easily accessible to people and to takes sea water across the country by open-cut man-made canals to enable people to fetch the water and distill it in their homes and huts using three mud or metal pots, which are usually available in their kitchen, as he demonstrated in Gujarat.

Chapter - 97

Flourishes Where There is Saline Moisture

Coral Leaf (as rabbit meal)

Botanical name	-	Erythrina indica Lam
Family	-	Papilionaceae
English	-	The Indian scoral tree
Tamil	-	Kaliana murungai or maruka
Hindi	-	Pangli
Marathi	-	Pangara
Malayalam	-	Mandaram

Native of India grows well near the sea in saline soil. Very bright and large beautiful red flowers bloom in March-April and May. Grows to a height of 15 to 20 metres. Young branches have sharp spikes. Leaves are large in long stalks, each stalk holding three leaflets. Flowering takes place when the tree is completely bare of leaves and each flower has five petals. Pods are stout and each pod holds about 10 seeds. The fruits remain on the tree till the leaves fall. Fallen seeds grow by themselves if soil conditions are favourable.

Cultivation: Very easy to grow by seeds or cuttings. Usually, sticks of about a metre long are planted as soon as the monsoon sets in. Trees grown out of seeds are healthier and live longer than those grown out of cuttings. Ryan Foundation recommends both the methods as seed growing takes a longer time to grow and mature.

In many villages Erythrina is grown for shade and as a support tree for pumpkins, betel leaves, grapes or beans. The tree is leguminous and enriches the soil for kitchen gardens and for tea plantation.

Uses: The tree has at least five excellent uses. (1) Rabbits love to eat the leaves and the leaves have many basic nutrients required by rabbits.

But if rabbits are fed only with these leaves, their growth is small and slow. Therefore, other leaves mentioned in the Ryan Foundation booklet on rabbit raising must also be given. (2) The leaves, with strong nerves. lend themselves ideally to be made into plates and cups. The small inexpensive machine invented by the Central Food Technology Research Institute (CFTRI) of the government of India, Mysore, makes excellent plates out of these leaves, which are now being exported. (3) The dried leaves make excellent green manure and fodder. (4) The dried twigs make good fuel wood and (5) The flowers yield good honey.

Development of water resources is the surest and shortest path towards poverty alleviation and improvement of the quality of life of the people - sea water must be brought into land and salt-tolerant plants and trees must be promoted and exploited.

Change is possible and it takes place all the time. It is a spontaneous response if the benefits of a change are sufficiently attractive.

In parts of Odisha (India) an alcoholic drink made from the MAHUA flower is very popular and it is consumed by men and women in large quantities. If the people use the contraption that they use to brew alcohol (illicit liquor) to boil, steam and condense sea water they will have clean and healthy water to drink.

Chapter - 98

Waste Land Tree in Warm Countries

Flame of the Forest (Or Butea-Gum Tree)

Botanical name Butea frondosa. Family -Leguminosea. Hindi-dhake or palasha. Tamil at-tumurukku, alarasu, Bengali-palas. Sanskrit-palasha. Telugu-Papssamu. Butea Superba Roxh. (Hindi-palsa lata) is a different species of the same family).

Description: Many people wrongly call the flamboyant tree "Flame of the Forest", but the two belong to different species, even though both have large bright red or deep orange flowers; which look alike from a distance. Both bloom in midsummer and have plenty of flowers densely clustered and both are highly drought resistant and thrive in different climates and soils. The stalk of the Butea flower is dark brown and the Calyx jet black and velvety. The immediate striking difference between the two trees is that the Flame of the Forest has very big leaves while the flamboyant has very small leaves. Flame of the forest has twisted branches and is found in jungles. It is medium in size and deciduous. The trunk has thick brownish bark on which lac insects thrive. The wood is not durable unless kept in water. Each leaf of the tree contains three leaflets and withers in winter. The pods are flat with a single seed called monosperma. Mature pods have white silvery hair and dried ones are carried by wind for natural propagation. Saplings transplanted along with the root nodules thrive better than what is grown by sowing seeds.

Distribution: Dry forests and wastelands throughout India and countries with a warm climate and minimum rainfall. They are dense in forests not far away from the sea and many tropical countries have this tree.

Now is the time to take up community driven plans to use village assets like ponds, trees, kitchen gardens, animals and birds to regenerate rural ecology and environment.

Uses: The wood is excellent for fuel and the roots yield strong fibre. Both wood and fibre are used in building low-cost houses. The dark brown seed gives an inedible fatty oil (18 to 20 percent), which is good for making soaps. Flowers boiled in water yield a dye which is not a fast colour. The gum, called Bengal Kino or Butea Kino is used in native medicines and in the tanning industry and thin gum is taken from the stem of the tree. The value of the tree is really in the lac that thrives on it as lac, which has very many uses, is in demand all over the world. French polish, sealing wax and several other fastmoving consumer and industrial items are made out of lac. The flowers yield good honey. Thus, Flame of the Forest generates employment by yielding lac, honey, oil, fibre, fuel wood, medicines, leaves for making plates and several products in which these are used as raw materials. From the inner bark a strong fibre is extracted for making cordage and paper. The root fibre is so rough that tribals weave and make slippers out of it. Seeds are insecticidal and cattle eat the leaves.

Medicinal: Seeds are used in the treatment of roundworms, tapeworms, and ringworms. A durable blue dye is made out of the gum, which contains tannin and gallic acid, used in the treatment of diarrhoea. Research done on this oil by the Oil Technology Research Institute, Ananthapur, reveals that the oil of Palas seeds stops balding and strengthens hair roots. Five full grown trees in a village can give a living wage for one family all through the year.

Development must be conservation-based and people centered.

Gandhiji said that India lives in her villages but in reality, India dies in her villages.

Chapter - 99

Some Prosopis Species

The genus prosopis includes more than 40 species out of which five are prominent in tree planting projects and social forestry.

1. The most popular, Prosopis fuliflora (also known as mesquite in America and "Mad Tree" in several countries because it grows in deserts, in between rocks, on the sea shore, with water and without water and with stands severe changes in climatic conditions.
2. P. chilensis.
3. P. lamarugo
4. P. Cineraria
5. P. Fallida

All these five trees (species) thrive well in light and sandy soils, adapt well to arid conditions, root well in hard and rocky soils and most important of all, tolerate alkaline or saline soils and grow well along the seashore. They withstand drought for nearly. If months in a year in desert temperatures. Most of these trees have thorns, and fruits appear as pods containing several seeds (about 10 on an average). The pods contain 13 percent crude protein and 4 percent carbohydrate. Ripe pods ground into flour can even be baked as food for humans.

Most of the Prosopis listed above yield good honey. A coffee-like drink is made out of the dried seeds. The bark of the tree is crushed and used in leather tanning. Some of them have medicinal properties.

The species mentioned above reproduce easily from seeds or cuttings. The pods gathered from the tree or collected from the ground must be dried in the sun for at least ten days and then used in a nursery.

Before sowing, the seeds must be doused in boiling water and allowed to soak for a day or two until the natural selling that results bursts the endocarp from within.

> Principle 23 of the Rio Declaration says, "The environment and natural resources of people under oppression, domination and occupation shall be protected." Let the affluent usher an austerity revolution to save our planet. Poverty degrades not only those who suffer it but also those who tolerate it.
>
> People need work and water, not weapons and wars.
>
> The yield of soya as an intercrop with banana, groundnut, sugarcane, tapioca, coconut, cotton, cambu, turmeric, cholam, castor or maize is very high. On an average a quarter tone of soya can be raised in 1 acre of land along with a main crop without any extra watering. Soya roots fix nitrogen in the soil and put up the yield of the main crop.

> Thailand Produces a large Quantity of soaps using the ash of water hyacinth which contains potassium chloride.

Bamboo Pipes to carry water

Bamboo, of the grass family, grown fast and high under hazardous conditions of soil and climate. It is a highly drought resistant forest shrub-tree offering several livelihood possibilities and yet neglected. Bomboos make excellent sewerage pipes in remote rural areas where they can be grown and processed.

Ryan foundation recommends standardisation of bamboo pipes to lengths of 3 mts and 8 cms.internal diameter mature bamboo may be cut at both ends uniformly with a saw and projections outside may be chopped and broadened. To do this. A long iron rod long sharpened at one end like a chisel may be used. The bamboo may be placed on two 'Y' stands as shown in the diagram and chiseled from either end. Only in this position can the see through easily and knock out internal projections and smoothen the ruggedness whether used above ground or under all bamboos have to be treated. The best rural technology for treating is to boil castor oil with neem, vasaka and datura leaves (3 parta) and turmeric, chilli powder and wood ash (one part) and paint the bomboos with the solution inside and outside. On a commercial scale the solution may be prepared in a long concrete pib-tub and the bamboos put into it and soaked for about a week. This treatment prevents bamboo from cracking or being attacked by white ants. If bitumn (tar) is available it may be melted and painted outside (not inside) the inside must be as smooth as possible if far is not available, used cement or fertilisar bags (Plastic or fibre) or locally woven crude fibre cloth may be wrapped around the bamboo and tied.

Bamboo bits 30 centimetres long and 10 centimetres in inner diameter may be similarly prepared and used as sleeves to join the pipes where required. Portar mixture, explained in RF Hanoout 294, may be used as a binder to plaster the joints.

These bamboo pipes may also be used to take water to the fields and for domestic supply of drinking water.

In the Tanzania region of Arusha, Rukwa, Morogoro and Dar-ee-Salam several bamboo processing and installing units have been established financed by the Dutch government (US $ 140,000) Norad (USS 196,000), SIDA (US $ 280,000), and the EEC. The processing method is different in Tanzania from what is recommended by the RYAN foundation. For RF processing only local materials available in the village need be used and nothing need be imported.

Raw materials being free, each 3 metre long pipe may be sold for ₹ 10 or Uses 1 (one) and the industry can provide full employment for atleast a dozen families of the village in the cultivation, processing, laying, repair, replacement and maintenance and it is a ideal industry for rural women.

Delivering safe and healthy water at a reasonable and affordable price, and providing sanitation for everyone is technically and economically feasible. But do we have the universal political will to do so?

Chapter - 100

Dryland Tree Needs No Watering

Glynicidia Maculata

Family name - Leguminosae (papilionaceae)

Tamil - Kona maram Telugu - Madra or madura

This is a fast-growing dryland tree and needs no watering after the roots catch the soil. It is tolerant to all sorts of soils and weather conditions. Being leguminous, it enhances soil fertility.

Usually, it is propagated through seedlings. When the sapling is about 50 centimetres it is transplanted during rainy season into a 30 cm hole made in the soil. The tree takes a year to grow 3 metres high and should be protected from cattle. If propagated by cuttings, watering for about months is necessary if the soil is not wet. It can be raised in problem soils which are saline, alkaline, or calcareous.

Maculata sheds its leaves during winter and its flowers during February-March. Its whitish pink flowers arc in clusters and contain honey. During April, pods with seeds appear. During summer, the tree develops several branches and plenty of leaves, which are rich in nitrogen and used as green manure.

Dry branches of this tree burn well and give high calorie of heat. The weak branches must be cut and the tree pruned every year before the monsoon. Pollarding (cutting the tree to the height of a man) should not be done even though the tree survives if cut that way. It does not flower if pollarded.

In rural areas the tree can be planted in rows with a gap of 3 metres between, and the leaves of this tree make excellent fodder and increase the yield of milk. Goats love the leaves. The leaves contain about 27 percent crude protein, 4 percent nitrogen, 2 percent calcium and vitamin A Phosphorus, potassium and amino acids.

The great advantage with Maculata is that the weak and irregular branches can be cut regularly every year throughout one's lifetime and the tree keeps on growing and flowering and providing fuel and fodder. Moreover, the roots of this tree go deep and do not spread out and so can be kept near homes in small open grounds.

Maculta leaves make better foliage than sub-abul or eucalyptus. Unlike suba and euca, Maculata develops horizontal branches and provides good shade and flowers.

This is a tree to be propagated in rural areas, in urban homes that have open space, and also as avenue trees to line city roads. It can also be planted as avenue trees inside residential colonies and new settlements.

The whole earth clamours for an environmental revolution.

> Near Nakhon Ratchasima in Indonesia, when forests were being wiped out and social forestry and agriculture were posing problems the Indonesian government and FAO of the UN found that the only way to encourage a reforestation settlement was to give the settlers the right to use wastelands for 5 years on the condition that the right would be extended and even passed on to sons and daughters if the lands were cultivated satisfactorily. This assurance led about 1,500 families move and settle down in the area and Nakhon Ratchasima is now a flourishing settlement. Inspired by this result the government of Indonesia has declared it as a national policy to promote agriculture and social forestry in the country under the same system. Why do not other countries follow this system?

Chapter - 101

Needs A Water Bore Atmosphere

Ailanthus For Saline Soil

Botanical - Ailanthus excelsca

Family - Simarubaceae; Hindi - Ardu

Tamil - Peru maram or Pimaram

Native of South India, this highly drought-resistant large tree grows well in hot and dry climates, in saline or calcareous (containing calcium carbonate) soil, on the seashore and river beds. The tree is now popular all over India and can be easily introduced into other tropical countries. The average height of the tree in about 25 years is 20 metres with a girth of about a metre. It grows excellently well in farmyards and tank bunds and does not take away the soil nutrient for other plants. Its softwood is commonly used for match sticks and the leaves are used as fodder. The tree can be lopped once in two years and over 1000 kilograms of green leaves can be gathered from each tree and sold in village markets as fodder. Cattle eat the dried leaves also and the protein content of the leaf is about 19 percent. Honeybees are active where these trees are in plenty and silk worms can be grown on these trees. The worms eat the leaves as they eat the leaves of mulberry plants. So where mulberry does not thrive these trees may be introduced or both the trees may be cultivated side by side. The leaves have about 2 percent calcium. (Silkworms eat castor leaves also).

The tree is propagated through seeds. Raised in nurseries, the seedlings are transplanted in about months, at intervals of 10 metres and given plenty of farmyard manure. Chemical fertilisers are not necessary. Stump planting is done in some areas but nursery saplings are the best.

Tribals claim that the bark of the tree is medicinal but nothing so far has been scientifically tested and established. The leaves mixed with paddy straw and cow dung make good humus. Even though Ailanthus does not spread out to give shade it is considered a good avenue tree. Seeds of ripe fruits are collected during May-June, dried and sowed immediately. They should not be stored. Germination takes places within 2 weeks.

The trunk of the tree is long and straight like that of a casuarina or eucalyptus and therefore it is used in low-cost building industries and in thatched houses. Only low-quality furniture can be made out of it but the wood is ideally suited for making wooden toys and packing cases. Softwood packing cases a scarce item in several industrial centres and therefore this tree, if promoted in the villages all around industrial cities, can provide employment for hundreds of village carpenters in making packing cases. Planks made of this wood can be cut to uniform size, say, 2 feet by 1 foot, and used in the place of roofing tiles (shingles) in low-cost houses.

Appropriate technology is technology that is suitable to provide basic needs for people of all strata of society of the present and future generations.

Chapter - 102

Evergreen Drought Tree

Uses of Pongamia Glabra (Pinnate)

Pongamia glabra is an evergreen and fast growing tropical tree. It belongs to the botanical family Leguminosae and known in Tamil as PUNGA and as KARANJ in Hindi. In Karnataka it is called and in Andhra Pradesh as ANUGACHETTU. In Kerala, it is called Unne. The tree is

quite common in India and in many tropical countries. It is usually planted as avenue trees and in parks for shade, it is a hardy tree, drought resistant, and can be propagated by seeds and cuttings during the rainy season. People use different parts of this tree for different purposes and some common uses are given below:

1. Fresh leaves are ground to a paste and applied for skin diseases.
2. Dried leaves kept in wardrobes and drawers keep away white ants and small insects.

3. Green twigs from the tree are used as a toothbrush by villagers.
4. Honey bees collect good honey from the flowers of this tree.
5. The leaves make excellent green manure, especially for flower gardens.
6. Oil is extracted from the nut of the tree which is used for preparing soaps and skin ointments. (Many doctors in India advise their patients to use soaps that contain this oil to keep their skin healthy). Hamaam soap is made out of this oil.
7. The oil cake serves as a rich manure.
8. Trunks of well-grown trees are used as timber.
9. Small branches are used as fuel wood.
10. Cows and sheep don't eat them.

Ryfo recommends the propagation of this tree in rural settlements and refugee camps for creating self-employment opportunities for the poor people. It is a popular avenue tree on highways in peninsular India.

> The simple truth is that Third World governments have neither the will nor the ability and resources to tackle the devastating degradation of land, water and forest resources.

> If one nation or community has the knowledge and experience to get things done quicker, easier and with less inputs in terms of land, labour, capital, organisation, water and fuel, why don't the other countries or communities acquire that knowledge and experience instead of trying to reinvent the wheel and why don't government and church agencies and NGOs spread such knowledge and experience for the welfare of the poor and for the progress of humanity? Dr. Ryan.

> Man's Salvation is to be Found in Group Activities and Economic Co-operation.

Index

C

I

J

P

Q

R

T

Y

Z